W9-BXY-734

LIAR
TEMPTRESS
SOLDIER
SPY

LIAR
TEMPTRESS
SOLDIER
SPY

FOUR WOMEN UNDERCOVER IN THE CIVIL WAR

KAREN ABBOTT

HARPER

An Imprint of HarperCollins*Publishers*

HarperCollins books may be purchased for educational, business, or sales promotional use. For information, please e-mail the Special Markets Department at SPsales@harpercollins.com.

All photographs are courtesy of the Library of Congress unless otherwise noted on page 435.

FIRST EDITION

Designed by Michael P. Correy

Library of Congress Cataloging-in-Publication Data

Abbott, Karen.
Liar, temptress, soldier, spy : four women undercover in the Civil War / Karen Abbott.—First edition.
pages   cm
ISBN 978-0-06-209289-2
1. United States—History—Civil War, 1861–1865—Secret service. 2. Women spies—United States—Biography. 3. Women spies—Confederate States of America—Biography. 4. United States—History—Civil War, 1861–1865—Participation, Female. 5. United States—History—Civil War, 1861–1865—Women. 6. United States—History—Civil War, 1861–1865—Biography.
7. Boyd, Belle, 1844–1900. 8. Greenhow, Rose O'Neal, 1814–1864. 9. Van Lew, Elizabeth L., 1818–1900. 10. Bowser, Mary Elizabeth, approximately 1840– .
I. Title.
E608.A22   2014
973.7'85—dc23   2014013602

14 15 16 17 18   OV/RRD   10 9 8 7 6 5 4 3 2 1

FOR CHUCK, FROM HIS UNEQUAL HALF

Contents

PART THREE: 1863

PART FOUR: 1864

PART FIVE: 1865

Splendid and Silent Suns
(A Note)

For a period of thirty-three hours, from just before dawn on April 12, 1861, to mid-afternoon the following day, sleep was hard to come by, in both North and South. In Manhattan, Walt Whitman left the Academy of Music and strolled down Broadway, where he heard the hoarse cries of the newsboys: "Extry—a *Herald*! Got the bombardment of Fort Sumter!"

Passersby broke into small groups under the brightly blazing lamps, each huddled around a paper, unable to wait until they got home to read. So it was true: the Confederates had opened fire on Fort Sumter, a federal fort in South Carolina, the first shots of the first battle of the American Civil War.

In Charleston, so close to the awful roar in the harbor, ladies solaced themselves with tea and a firm faith that God "hates the Yankees" and was clearly on their side. In Washington, DC, President Lincoln, in office barely six weeks, prepared to call 75,000 volunteers to quell this "domestic insurrection." One hundred miles away, across the rolling Virginia countryside, the citizens of Richmond celebrated and cried, "Down with the Old Flag!" Within the week they got their wish: Virginia became the eighth state to join the Confederacy, with vessels in the James River flying not the Stars and Stripes but the Stars and Bars. By early June the South had added three more: Arkansas, North Carolina, Tennessee.

The new enemy countries settled into a war that many predicted would be over in ninety days. The twenty-three Northern states had 22.3 million people to the South's 9.1, nearly four million of them slaves whom their masters dared not arm. Jefferson Davis, former U.S. senator from Mississippi and new leader of the South, moved his pregnant wife and three children to the Confederate capital of Richmond. He was more prescient than most, expecting "many a bitter experience" before all was said and done.

Troops poured into the two rival capitals and began making themselves into armies. Morning brought the reveille of the drum; night, the mournful notes of taps. Nothing was seen, nor spoken of, nor thought of but the war. There was work for everyone to do, even the women—especially the women. They had to adjust quickly to the sudden absence of fathers and husbands and sons, to the idea that things would never be as they had been. They had no vote, no straightforward access to political discourse, no influence in how the battles were waged. Instead they took control of homes, businesses, plantations. They managed their slaves in the fields, sometimes backing up orders with violence. They formed aid societies, gathering to darn socks and underwear for the soldiers. To raise money for supplies they hosted raffles and bazaars, despite widespread resistance from the very men they aimed to help (protested one general, "It merely looks unbecoming for a lady to stand behind a table to sell things"). They even served as informal recruiting officers, urging men to enlist and humiliating those who demurred, sending a skirt and crinoline with a note attached: "Wear these, or volunteer."

Some—privately or publicly, with shrewd caution or gleeful abandon—chafed at the limitations society set for them and determined to change the course of the war. In the pages that follow I tell the stories of four such women: a rebellious teenager with a dangerous temper; a Canadian expat on the run from her past; a widow and mother with nothing left to lose; and a wealthy society matron who endured death threats for years, and lost as much as she won. Each, in her own way, was a liar, a temptress, a soldier, and a spy, often all at once.

This is a work of nonfiction, with no invented dialogue. Anything that appears between quotation marks comes from a book, diary, letter, archival note, or transcript, or, in the case of Elizabeth Van Lew, from stories passed down by her descendants—details about her incredible operation that have never before appeared in print. Characters' thoughts are gleaned or extrapolated from these same sources. In any instance where the women may have engaged in the time-honored Civil War tradition of self-mythology, rendering the events too fantastic, I make note of it in the endnotes or in the narrative itself.

Beneath the gore of battle and the daring escapades on and off the fields, this book is about the war's unsung heroes—the people whose "determin'd voice," as Whitman wrote, "launch'd forth again and again," until at last they were heard.

Karen Abbott
New York City

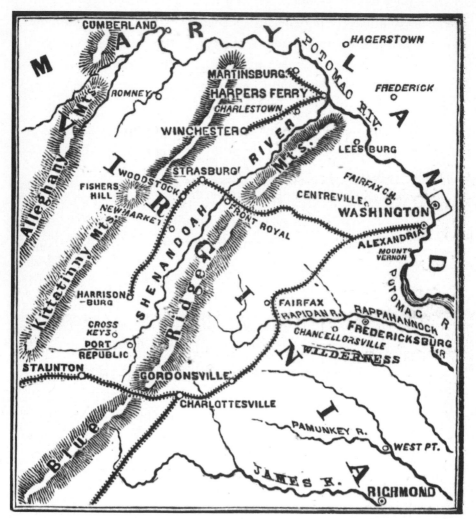

Shenandoah Valley, 1861.

[PART ONE]

1861

THE FASTEST GIRL IN VIRGINIA
(OR ANYWHERE ELSE FOR THAT MATTER)

Belle Boyd, circa 1861.

THE SHENANDOAH VALLEY, VIRGINIA

In the town of Martinsburg on the lower tip of the Valley, a seventeen-year-old rebel named Belle Boyd sat by the windows of her wood-frame home, waiting for the war to come to her. It was July 4 and the war was still new, only two and a half months old, but Belle—known by one young rival as "the fastest girl in Virginia or anywhere else for that matter"—had long been accustomed to things operating on her schedule, and at her whim.

She tracked the progress of Union forces as they stormed down from the North, all those boys sweating and filthy under blue wool coats, lean as the rifles slung at their sides—nearly fifteen thousand of them, a few as young as thirteen, away from their mothers for the very first time. She felt they had no respect at all, waving American flags with the stars of thirty-four states when eleven no longer belonged. Two days prior, on July 2, about thirty-five hundred of them crossed the Potomac, slipped through a gap in the Blue Ridge mountains, and trampled across the lush sprawl of the Shenandoah Valley to face the Southern army at Falling Waters—a "romantic spot," in Belle's opinion, eight miles from her home. There Confederate colonel Thomas Jackson was waiting with four cannon and 380 boys of his own. When the rebels retreated, they left the field scattered over with blankets and canteens and, most regrettably to Belle, only twenty-one Yankee wounded and three Yankee dead.

She took the loss at Falling Waters personally. She had family in this war, uncles and cousins and even her forty-five-year-old father, a wealthy shopkeeper and tobacco farmer who depended on a team of slaves to grow and harvest his crop. Despite his age and social prominence he'd enlisted as a private in Company D, 2nd Virginia Infantry, part of Colonel Jackson's brigade. The mood in her home shifted overnight, with Belle noticing a general sadness and depression in her mother and younger siblings, all of them too consumed by worry even to sleep. The entire town seemed unsettled. Berkeley County (of which Martinsburg was the county seat) had voted three to one against secession, the only locale in the Shenandoah Valley to do so. Seven companies of soldiers were recruited from the county, five for the Confederacy and two for the Union, and now neighbor fought against neighbor, friend against friend. No one dared trust anyone else. Citizens formed a volunteer Home Guard, sitting up all night and arresting anyone prowling about, an enterprise that lasted until one member was fatally shot by a stranger passing through town.

The women of the Valley got to work supporting the war effort, gathering to sew clothing and raise money for supplies. At first Belle joined them, wielding her needle and laundering sheets, but she soon found such activities "too tame and monotonous." Instead she scandalized the ladies of Martinsburg by openly waving to soldiers on the street, and organized trips to the Confederate camp at nearby Harpers Ferry, where all of them temporarily escaped the gloomy atmosphere of their homes. They danced the Virginia reel and sang "Dixie" and forgot about the prospect of impending battle. Belle herself exchanged "fond vows" with several young soldiers, even as she wondered how many of them would soon be dead. "War will exact its victims of both sexes," she mused, "and claims the hearts of women no less than the bodies of men."

Occasionally she wandered around camp, handing out religious tracts denouncing everything from profanity to gambling to procrastination (soldiers, one cautioned, must avoid the "sin of being surprised" by either the enemy or the devil), not because she objected to such vices but because she longed to be useful. Any unfamiliar man might be a Yankee spy, and she believed it was her duty to entrap him.

"Be very careful what you say," she warned one trespasser dressed as a photographer. "I was born at the North, but have lived among these people seven years. My sympathies are all with the Northern people. I am trying now to get a pass from General Beauregard that I may visit my sister in New York, who is a teacher in one of the public schools. I will gladly take any message you may want to send to your friends."

The stranger declined her offer, but she would have other opportunities to dupe Yankee men.

This respite at camp was interrupted by reports that the enemy was marching down the Shenandoah Valley; the men went to fight at Falling Waters on July 2 and the women went home. After the Confederates retreated, the Union continued on south toward Martinsburg, scheduled to arrive in time for a victory parade on the Fourth of July. Belle recognized that this day now belonged only to the Yankees—the eighty-fifth birthday of a nation that had amputated a third of itself, split into uneven halves.

Staring out her window onto South Queen Street, she heard the soldiers before she saw them. They announced their presence with laughter and song, hollering about that damned Yankee Doodle riding on his pony, booted feet stomping to the burst of bugle and the grumble of drums. The beat thrummed in the air, keeping time with the tap of her heart against her ribs. It was late afternoon, the sun shedding its heat layer by layer, hunkering down toward the baked dirt roads. The soldiers' song grew louder, their laughter more brazen. They slashed bayonets at the pale Virginia sky, marching closer and closer still.

House "servants," a common euphemism for slaves, rounded up children in the public square and hustled them to safety. John O'Neal locked the doors of his saddle and harness shop. The church bells sat untolled, the hour unmarked. The Baltimore & Ohio railroad depot stood deserted; rebel troops had destroyed forty-eight locomotives and three hundred cars, wrapping one of the engines in an American flag before setting it afire, all to prevent Union supplies from arriving by train. Field hands hid in their quarters instead of harvesting wheat or quarrying native limestone. Clusters of homes sat darkened and deserted, the owners having packed up their silverware and their help and fled farther south. A few bold spectators arrived on horseback from neighboring towns, waiting for whatever came next.

General Robert Patterson's Yankees were everywhere, winding through the cemetery and around the jail, pausing to shatter the windows of a church, pillage the offices of the local newspaper, claim the county courthouse as Union headquarters, and raid the distillery of a Confederate captain to guzzle his whiskey. There were thousands and thousands of them, an endlessly advancing blue line, a menacing horizon almost upon her.

To Belle's side, within reach, lay a Colt 1849 pocket pistol.

Since the abolitionist John Brown's attempt to start an armed slave rebellion, Belle had been terrified of "an uprising of the negroes," and believed that "Northerners were coming down to murder us." She told herself she would not hesitate to use the pistol; she had never hesitated at anything. All her life she had been blissfully unburdened by doubt or introspection. She believed her plain face was striking, her defiance charming, her wit precocious, her every thought clever and significant. "I am tall," she once boasted to her cousin, lobbying him to find her a husband. "I weigh 106½ pounds. My form is beautiful. My eyes are of a dark blue and so expressive. My hair of a rich brown and I think I tie it up nicely. My neck and arms are beautiful & my foot is perfect. Only wear [size] two and a half shoes. My teeth the same pearly whiteness, I think perhaps a little whiter. Nose quite as large as ever, neither Grecian nor Roman but beautifully shaped and indeed I am decidedly the most beautiful of all your cousins."

She had the quickest answers in class at Mount Washington Female College (where, using a diamond ring, she carved her name in a window of the Octagonal Room); the most graceful curtsy at her debutante ball in Washington, DC; and a distinguished lineage comprising politicians and Revolutionary War heroes. When Belle was eleven, her parents declared she was too young to attend their dinner party, given for a group of Virginia officials. Instead of pleading or protesting, she went to the stable, saddled up her horse, Fleeter, and rode him into the dining room, interrupting the second course. Fleeter whinnied and sidestepped. A startled servant dropped a tray. Sweetbreads skittered across the floor, and pigeon soup splattered across the walls.

Belle looked down on everyone, watching her mother's mouth gape, her hand rising to cover it. She yanked at the reins and cleared her throat.

"Well," she said, "my horse is old enough, isn't he?"

In a dry, tight voice her mother ordered her to return the horse to the stable and head directly to her room. But a guest intervened.

"Surely so high a spirit should not be thoughtlessly quelled by severe punishment!" he exclaimed, and turned to Mrs. Boyd. "Mary, won't you tell me more about your little rebel?"

And for the rest of the evening Belle seized the spotlight, redirecting its focus anytime she sensed it veering away. She scarcely knew herself without it, neither then nor now.

Her Negro maid, whom she called "Mauma Eliza," now stood poised at the bottom of the parlor stairs, holding Belle's Confederate flag in her arms, properly and respectfully folded. Belle would love Eliza even if she didn't own her; at night, in secret, she defied the law and taught her to read and write. "Slavery, like all other imperfect forms of society, will have its day," Belle believed, "but the time for its final extinction in the Confederate States of America has not yet arrived." Eliza was thirty-three and had raised Belle from birth, protecting her and soothing her and tolerating her nonsense. Without being asked, she hurried up to Belle's room and hid the flag under her bed before returning to her mistress's side. In an adjacent chamber five other slaves huddled with Belle's three younger siblings; Belle had urged them to lock the doors. From the corner of her eye she spotted her mother sitting tense and alert on a velvet settee, and Belle could trace the course of her thoughts: four of her eight children had died within the span of five years, from 1846 to 1851, and she was terrified of losing another. She always told Belle she was too "saucy" for her own good.

The air hung thick and unstirred. The wooden floors were warped from the heat. Belle wore nine items of clothing, all assembled by Eliza every morning—chemise, pantalettes, corset, corset cover, crinoline, petticoat, a two-piece dress, silk stockings, and side-button boots—and drops of sweat crept down her back, soaking through the layers. She tried to hold her body still. She heard the clatter of gun carriages, the fervent thud of drums. Fine china quivered behind the doors of a rococo hutch. And here they came, a massive serpent of blue and steel. There were gunshots and splintering glass, doors being hacked off hinges. Chairs and tables soared into the street. The warbled refrain of "John Brown's Body" mingled with the sound of children's screams. They were just one door away.

Belle caught a swatch of blue blurring past the window. There was a thundering of fists. The front door gave way and there was no divide now. She saw tracks in their dirty faces carved by falling sweat. Mary Boyd jumped from the settee. Eliza stayed put by the stairs, gripping the banister.

One of the soldiers, "a great big Dutchman"—a common term for a German—focused his gaze on Belle. She could tell he'd been drinking.

"Are you one of those damned rebels?" he asked.

The word "rebel" was not yet one Southerners used with pride. They lived in sovereign states, and in their view this war was not about "rebellion" but about defending their homeland against coercive foreigners. Coming from a Yankee, the word was a mockery Belle would not abide.

She drilled her fists into her hips and said, "I am a secessionist."

He demanded to know if there were any rebel flags on the premises. Belle didn't respond. Another soldier pointed out that the town was Federal property now, and they would hoist a Union flag up over the house.

At this, Belle's mother stepped forward.

"Men," she said, "every member of my household will die before that flag shall be raised over us."

The circle of men contracted, fencing her in. Eliza peeked through a latticework of fingers. Belle noticed the Dutchman at the head of the pack. His arm coiled around her mother's body and yanked her close. He aimed his slack mouth at hers. Belle considered her mother a "very handsome woman," and she knew the Yankees would stop at nothing. There were reports throughout the South of "Yankee outrages," as the papers called them, soldiers invading homes and destroying property and assaulting women. In Maryland, a border state with a large secessionist population, one woman claimed a Union soldier thrust his hands against her bosom, under the pretense of looking for concealed arms. Another Yankee, in broad daylight and on a public street, pinned a girl's arms behind her back and asked, "Is it true that you're the prettiest girl in Baltimore?" In one farmer's home they found a Confederate uniform coat and, in retaliation, took the man's two young daughters as hostages, treating them "in a manner too inhuman and revolting to dwell upon." Communities beseeched Confederate president Jefferson Davis to send in troops to protect their "defenseless women."

Belle did not consider herself one of them.

"Let go my mother!" she screamed.

The Dutchman looked up at her and grinned.

Belle could stand it no longer. Her indignation was "roused beyond control"; her blood "literally boiling" in her veins. The room seemed to skid to a stop, and Belle became the only moving thing inside it. Her hand grasped her pistol, finger curling around the trigger. She found a clearing amid the tangle of limbs, her target offering himself up. Letting instinct dictate aim, she bucked from the force of her shot. The circle split, bodies retreating, and there was nothing to break the soldier's fall.

Belle heard the terse crack of bone against wood. She saw the blood pulse from his neck. She looked at her pistol in her hand, smoke still wisping from the barrel, and realized what she had done. She let it slide from her fingers, landing by the toe of her button-up boots. She heard screams, her mother's and Eliza's, sounding miles and miles away. All of her seventeen years seemed crammed into those seconds. Her heart scrabbled in her chest. Several soldiers shifted in her direction, threatening to kill her.

She returned to herself, then, the moment sliding into focus. She remembered who these men were, why they were there, what they had almost done.

She heard herself speak before she had a chance to contemplate her words: "Only those who are cowards shoot women," she said, and spread open her arms. "Now shoot!"

OUR WOMAN

Soldiers in front of the Capitol, Washington, DC, 1861.

WASHINGTON, DC

By July 4, Franklin Thompson had been a Union soldier for six weeks, undergoing basic training in the Federal capital and waiting for orders to march on Virginia. He'd never expected to join an army or fight in a war—although a sharpshooter, he had yet to aim his gun at a man—but when President Lincoln called for volunteers, he posed the question to God, who made the decision for him.

Frank, as he preferred to be called, always believed that God was with him, protecting him even during—perhaps especially during—his transgressions. At night, after taps at 9:00 p.m., when the last light was extinguished and the last voice silenced, his mind sometimes conjured his most recent and serious trespass: the afternoon, six weeks earlier, when he took his place in line at Fort Wayne in Detroit, waiting for his turn with the medical examiner.

Official protocol of the US War Department dictated that all recruits strip and undergo a thorough physical examination, but doctors across the country flouted these rules. They had quotas to fill and needed bodies, quickly. It didn't matter if a recruit was prone to convulsions or deaf in one ear or suffering from diphtheria. He merely required a trigger finger, the strength to carry a gun, and enough teeth with which to tear open powder cartridges. One recruit recalled the doctor pinching his collarbones and asking, "You have pretty good health, don't you?" before passing him. Another was welcomed into the army after receiving "two or three little sort of 'love taps'" on the chest.

Frank stepped forward. He was five foot six, two inches shorter than the average Union army recruit, solid but thin. He told the doctor he was nineteen years old, twenty come December. The doctor's eyes skimmed his shoulders and back, torso and legs. He coiled his fingers around Frank's wrist and lifted up his hand. He turned it over as if it were a tarot card, studying its nuances, noting the absence of calluses, the smooth palm, the slim and tapered fingers. For the first time he looked Frank directly in the eye.

"Well," the officer said, "what sort of living has this hand earned?"

Frank was suddenly and strangely conscious of his voice. He willed his words to flow smoothly, to sound convinced of their own authenticity and tone. They would get him to the other side.

"Up to the present," he replied, "that hand has been chiefly engaged in getting an education."

Without further questioning the doctor marked Frank Thompson fit to serve as a private for Company F, 2nd Michigan Infantry.

Frank took the oath of allegiance to the United States, solemnly swearing to Almighty God to support the Constitution and maintain it with his life. He assured himself that this was a calling, that he had to do what he could "for the defense of the right," and that if he was careful no one would discover his secret: Frank Thompson was really Emma Edmondson, and had been posing as a man for two years.

Emma Edmondson as Private Frank Thompson.

Emma was one of fifty thousand Union soldiers currently in the nation's capital, and she had never seen so many men in one place. Trainloads of fresh recruits clattered into Union Station each day; some came with pieces of rope tied to their musket barrels to use as nooses for Southern prisoners. Hotels overflowed with businessmen angling for government contracts and pursuing the city's most eligible women. Soldiers lounged on the cushioned seats of the Capitol and reclined in easy chairs inside the White House. They spilled from the doorways of saloons and brothels and drove convoys of army wagons along the dusty streets. After dark, they accosted passersby in alleys for spare change and fired gunshots into the sky. Thousands of white army tents dotted the hills surrounding the city. Runaway slaves from Virginia and Maryland began slipping into the capital, and District police jailed those who lacked sufficient documentation of their freedom. The air reeked of garbage and manure and the contents of overburdened sewers. On July 4, she and the other recruits paraded for President Lincoln, who faced pressure to follow up the recent success in the Shenandoah Valley with a decisive attack that would end the rebellion.

In preparation they began each day at 5:00 a.m.: reveille, breakfast, and drills, endless drills: getting accustomed to orders and marching in columns; practicing how to "dress the line" by turning heads right or to the center; learning the drum and bugle calls that signaled whether to charge or retreat. "The first thing in the morning is drill," one private complained. "Then drill, then drill again. Then drill, drill, a little more drill. Then drill, and lastly drill." During the exercises the soldiers' boots collected as much as fifteen pounds of mud and clay each. Emma kept pace with all of them; she was lithe and hard-muscled from a childhood spent working on the family farm. The city boys did not even know how to load their cartridges ball foremost, and she took furtive pleasure in teaching them. Members of the Seventh New York Infantry, known as the "silk stocking regiment," had arrived in Washington with sandwiches from Delmonico's and a thousand velvet-covered campstools on which to eat them. Many of the immigrant regiments—the Italian Legion, the German Rifles, the Steuben Volunteers—couldn't understand orders in English. Some commanders conducted drills without live ammunition for fear that the neophyte troops would injure themselves before they even saw battle.

Over time Emma had *become* Frank Thompson—it was impossible to measure where one ended and the other began—but she now regarded her creation from a strange and subtle distance. Frank had never been tested this way, living in such tight quarters and under continuous scrutiny, and she grew keenly aware of each honed mannerism, the practiced and precise tenor of his voice. She took careful note of her comrades' first impressions: they all knew that Frank hailed from Flint, Michigan, by way of Canada, and that despite his slight stature and oddly smooth face he had enjoyed a reputation as a ladies' man before the outbreak of the war, squiring them around town in the finest horse and buggy, complete with a "silver mounted harness and all the paraphernalia of a nice turnout"—the fruits of a brief but successful career as an itinerant Bible salesman. They nicknamed him "our woman" and occasionally joked about his falsetto voice and small feet—so small he couldn't wear standard-issue boots and had to bring his own—but they seemed to believe Frank was one of them, and considered him their equal. Emma liked to think they had more in common than not, the greatest distinction being that she had already died once and was willing to die again, this time for a cause much greater than her own.

If Frank faltered, Emma could be arrested and jailed. Even worse, in her mind, she would disappoint God and lose the chance to act in "this great drama," as she liked to call the war. She was grateful for a few small mercies. To save time to prepare for roll call in the morning, most of the officers remained partially clothed for bed. Some even turned in wearing coats and boots, so no one would consider her odd if she didn't strip down to her linens. She was lucky to count one tent mate as an old friend: Damon Stewart, a twenty-six-year-old shopkeeper from Flint, Michigan, whom she'd known before the war. They'd even double-dated on occasion, a history that now provided Emma some comfort; during downtime, gathered around the fire, her stories of sweethearts back home had a witness. Men were accustomed to seeing women in crinolines and bonnets and had no concept of what one would look like wearing trousers and a kepi cap. They could go for months without bathing or changing. Most of them avoided the long, filthy trenches that served as latrines and instead took care of "the necessaries" privately, in the woods. The stress and physical demands of her new role would almost certainly stop her menstrual cycle, and if she did bleed, her soiled rags could be passed off as the used bindings of a minor injury—hers or someone else's.

Her smooth face and high voice were attributed to her youth, as was her disdain for swearing, drinking, and smoking. She had to be careful not to appear too adept at domestic skills like cooking and washing. Like the others, she sponged her plate with a piece of soft bread dampened with a few drops of coffee, or scoured it with dirt and rinsed it with her spit. Occasionally an enterprising officer would offer to clean or sew for others, charging a steep price, but Emma couldn't afford to take the risk. It was the crucial details that might give her away.

She didn't personally know any other female recruits, but she was not alone. As many as four hundred women, in both North and South, were posing and fighting as men. Some joined the army with a brother, father, sweetheart, or husband; one couple even enlisted together on their honeymoon. Some, like a twelve-year-old girl who joined as a drummer boy, were fleeing an abusive home situation. For poor, working-class, and farm women, the bounties and pay ($13 per month for Union soldiers, $11 for Confederates) served as an incentive. A small number of women had been living as men prior to the war and felt the same pressure as men to enlist. One Northern woman was a staunch abolitionist who fought because "slavery was an awful thing." A Southern counterpart sought adventure, yearning to "shoulder my pistol and shoot some Yankees."

Not so bloodthirsty, Emma had no intention of aiming her musket at any rebel, male or female. Every morning after roll call the medical staff had sick call, a task for which she eagerly volunteered. Believing there was a "magnetic power" in her hands to "soothe the delirium," she tended to the soldiers suffering from typhoid fever and dysentery and the resulting chronic diarrhea, illnesses that would ultimately kill twice as many Union troops as would Confederate weapons. ("Bowels are of more consequence than brains" was a common jest.) She examined tongues and pulses, dispensed quinine and "blue mass"—a common compound of mercury—and offered "a little eau de vie to wash down the bitter drugs." Once they received marching orders, she planned to work alongside army surgeons, hauling the wounded off battlefields, assisting with amputations, serving as the lone witness to whispered last words.

The last time Emma aimed her gun at any living being was two years ago, in 1859, before she'd ever imagined fighting in a war. She had just reinvented herself as Frank Thompson, selling Bibles door-to-door, and longed to see her mother, Betsy, who had no idea she was now living as a man. On one brisk October day Emma returned to the family farm in Magaguadavic, New Brunswick, dressed, ironically, in an army uniform, a cap tilted atop her tight brown curls. She approached the front door of her childhood home, introduced herself as Frank Thompson, and asked for something to cat. Betsy invited her inside and called Emma's sister, Frances, to help prepare a meal for "the stranger." Watching her sister, Emma glimpsed what her life would be had she stayed behind and bent to her father's will. She would be married by now to her neighbor, the elderly widower who'd always watched too intently as she tended to her chores on the farm.

Emma's brother, Thomas, sat down to join them. Her father, Isaac— "the stern master of ceremonies of that demoralized household"—was not around. Years earlier he had gone about selecting a wife who would be a good breeder in much the same way he chose a female animal for a stud, hoping to raise a large family of sons to help grow potatoes, the most important crop for New Brunswick farmers. Instead his wife gave birth to four daughters in quick succession and a son who suffered from epilepsy, an affliction Isaac viewed with disgust. He had always been prone to violent rages, a tendency that increased with the birth of each child, and when Betsy became pregnant in 1841 for the sixth and final time, she prayed every day for a boy. Instead, on a cold December day came Sarah Emma Evelyn, Emma for short. Betsy wept as the midwife cleared away the evidence of her betrayal. "My infant soul was impressed with a sense of my mother's wrongs," Emma later said, "although I managed to outgrow it immeasurably." She learned to hunt and fish and break wild colts, trying her best to be the boy her father wanted, but never heard one word of praise.

Emma was relieved by his absence, and after an hour or so of idle chat, Betsy turned to Frances and without warning began to cry. "Fanny," she said, "don't you think this young man looks like your poor sister?" At this, Emma rushed to kneel beside Betsy and asked, "Mother, dear, don't you know me?"

Betsy stared at her. She removed Emma's cap and ran her fingers through her hair. One fingertip trailed down her daughter's face. "No," she said, "you are not my child. My daughter had a mole on her left cheek, but there is none here."

"Mother," Emma replied, "get your glasses and you will see the scar. I had the mole removed for fear I might be detected by it." She jumped up to retrieve the glasses herself, reveling in her mother's reaction: "She cried and laughed both at once, and I caught her up in my strong arms as if she were a baby, and carried her around the room, and held her and kissed her until she forgave me."

Emma stayed as long as she dared, glancing frequently at the door, fearing that her father might walk in at any moment. Her family assured her they would say nothing of the visit, lest Isaac Edmondson set off to hunt his daughter down. Thomas accompanied his sister to the train depot, and along the way they spotted six partridges. He had never learned how to handle a gun. One by one Emma shot the birds herself, cleanly in the head, and gave the quarry to her brother to take home.

She would lose her aversion to shooting rebels sooner than she could have imagined.

Despite their inexperience Emma and her fellow recruits were eager for action; they worried that the war would be over before they even had a chance to meet the enemy. Lincoln agreed that it was time. Every day Horace Greeley, the outspoken and influential editor of the *New York Tribune*, egged him on with the same bold headline: "FORWARD TO RICHMOND! The Rebel Congress must not be allowed to meet there on the 20 of July! By that date the place must be held by the National Army!" The fighting in the Shenandoah Valley had made headlines but resolved little, and the Union army had a tentative plan: Major General George McClellan, the commander of the Department of the Ohio, would bring his troops eastward to support General Patterson in Martinsburg, Virginia. Together they would prevent Confederate forces from leaving the Valley to reinforce Brigadier General P. G. T. Beauregard at Manassas, twenty-five miles west of Washington. Brigadier General Irvin McDowell, commander of the Union troops around the capital, would dispose of Beauregard and push on to Richmond. A Union victory at Manassas and the capture of the Confederate capital would effectively crush the rebellion.

But McDowell was not yet ready to fight. He had an undersized staff and didn't even possess a map of Virginia that showed anything beyond the main roads. Most of his recruits were ninety-day volunteers and still too ill trained to go to war.

"This is not an army," he warned President Lincoln. "It will take a long time to make an army."

"You are green, it is true," Lincoln replied. "But they are green, also; you are green alike."

It was decided: thirty-seven thousand Union recruits, Emma included, would soon march onward to Virginia. "No gloomy forebodings," she wrote, "seemed to damp the spirits of the men." They looked, to her, as though they had an overabundance of life, and all that life was darting about inside them, making every word louder and movement quicker. How "many, very many" of them, she wondered, would never go home? She felt suddenly and strangely out of harmony, but Frank Thompson buoyed her along.

A Shaft in Her Quiver

Rose O'Neal Greenhow, circa 1861.

WASHINGTON, DC

As soon as Emma received her marching orders, Rose O'Neal Greenhow summoned a courier to her home at Sixteenth and K Streets near Lafayette Square—"within easy rifle range of the White House," as her friend, Confederate general Beauregard, liked to joke. She had to warn him that the enemy was coming, that swarms of Yankee soldiers would soon be filing from the city and heading to Manassas, Virginia, ready to fight the first major land battle of the war. The courier, Bettie Duvall, was just sixteen years old, but Rose trusted her to make this delivery. Bettie was a Washington socialite, the daughter of an old family friend, but for this mission she would play the part of a simple farm girl returning from the market, a disguise unlikely to arouse suspicion among Union guards. The Confederacy was counting on Rose to provide intelligence about the enemy's plans, and she needed to secure a victory for the South. The new nation would not survive if the rebel army lost.

Rose led Bettie to her vanity and pulled the girl's hair as if it were a bridle, cinching the strands around her wrist. Her other hand held a purse stitched of the finest black silk. After her husband's death seven years before, she'd used the same silk to make a mourning dress—a dress she now wore again in memory of her twenty-three-year-old daughter, Gertrude, who'd succumbed to typhoid fever two weeks after Lincoln's inauguration. Gertrude was her fifth child to die—she'd lost four within three years—and she had three daughters left but only one living at home. Little Rose, the third daughter she'd named after herself, was eight years old, and would become an important part of her plans.

She buried the purse in Bettie's hair, coiling it into a tight cocoon and clasping it with a comb. The hair would not come undone until Bettie arrived at Beauregard's headquarters and shook it loose.

Rose gave the girl meticulous instructions: Wear this tattered calico frock and drive a milk cart along a dirt road on the Washington side of the Potomac River. Pass by the endless stretch of Union camps—the 1st Massachusetts, the 2nd Wisconsin, the 2nd Michigan—and, after the last, veer left onto the Chain Bridge. Union artillerymen will be perched atop their fortification, guarding the area, but she should just proceed calmly, passing into Virginia as if she has nothing to hide. Once she reaches the countryside, she must be wary of Yankee scouts and pickets; it would probably be best to stop somewhere for the night.

In the morning, she'll continue on to Beauregard's headquarters at Manassas Junction. Tell the Confederate pickets she has important information for the general. Once inside, unwrap her chignon and deliver the contents of the silk purse: a slip of paper with a coded message: "McDowell has certainly been ordered to advance on the sixteenth. ROG." The Confederates would understand that the Union forces under Irvin McDowell planned to leave Washington for Manassas one week later. Rose had gotten the information from a "reliable source," possibly a high-ranking Northern official, who had a copy of the order to McDowell. Another source provided her with a map, used by the Senate Military Affairs Committee, showing the Union army's route to the battlefield.

Rose had been the head of the city's Confederate spy ring since April, when Captain Thomas Jordan, a West Pointer, distinguished veteran of the Mexican-American War and quartermaster in the US Army, came to call on her. Jordan was forty-one, six years younger than she, with close-set eyes and a tangled beard. She invited him into her back parlor, pulling a curtain of red gauze behind her, and offered him tea. She sensed he did not wholly trust her; she had earned her nickname, "Wild Rose."

Jordan, like the rest of Washington, had heard gossip about her late-night gentlemen callers: abolitionists, secessionists, senators, representatives, diplomats, and even their lowly aides. Perhaps he had heard similar gossip about himself, rumors that he enjoyed "the same kind of intimacy" with her. If Jordan stayed late, she might replace the tea with brandy. She might unwrap her own chignon, letting her black hair skim the small of her back.

Jordan told her he had decided to switch sides to fight for his native Virginia and needed to create an espionage ring in the Federal capital—a ring he wanted Rose to organize and oversee. He and President Jefferson Davis considered her the ideal candidate for the job, both despite and because of her occasional indiscretions, and Rose accepted on the spot. No woman in Washington knew more men of power and influence, of both the Northern and the Southern persuasion.

In 1835, at age twenty-two, she'd married Dr. Robert Green-how, a physician and scholar who served as a high-ranking official in the State Department. His tenure spanned twenty years and seven presidential administrations, and over time Rose became a greedy prospector of the powerful. She dined with Martin Van Buren and counted the late former vice president John C. Calhoun as a mentor. An avid proponent of slavery, Calhoun would sip old port and muse about the South's "peculiar institution," insisting it was "indispens-able to the peace and happiness" of the entire country and a "posi-tive good." Rose worshipped him, calling him "the best and wisest man of this century," and let his politics shape her own. She served as a close confidante to President James Buchanan and attended masquerade balls with then Mississippi senator Jefferson Davis. The *New York Times*, in reviewing one such event, declared her "glorious as a diamond richly set."

As Washington became increasingly polarized over the issue of slavery, the Greenhows considered settling in the American West to prospect in land and gold. In February 1854, during a trip to San Francisco, Robert slipped off a section of planked road and fell six feet to the ground, badly injuring his left leg. An infection set in and he died six weeks later. Rose learned of the news in a letter from the head of the California Land Commission: "Robert Greenhow Esq," it read, "is no more." Little Rose was only ten months old.

After Robert's death Rose strove to maintain her social position and relevance by any means necessary, disregarding neighbors' catty talk about her "confidential relations." She heard whispers that her "widowed tears were soon dried up" by the $10,000 settle-ment she received from the City of San Francisco (money she lost speculating in stocks). Although there was no relation, some com-pared Rose to the notorious Peggy O'Neale Eaton, who reportedly cheated on her husband and wed too soon after his death; this second marriage, to a Tennessee senator and adviser to Andrew

Jackson, led to the dissolution of that president's cabinet. But like Peggy, Rose retained numerous admirers, men who appreciated her wit and savvy as much as her figure, who knew she needed no one to come to her defense. Confederate naval secretary Stephen Mallory marveled at the way she "hunted man with that resistless zeal and unfailing instinct . . . she had a shaft in her quiver for every defense which game might attempt." Union colonel Erasmus D. Keyes called her "one of the most persuasive women that was ever known in Washington"—persuasive enough to wheedle classified information from him in between their alleged trysts. Next came Senator Joseph Lane, Democrat from Oregon, who came to call on her as often as his health would allow. "Believe me, my dear, I am not able to move as a young man should," he wrote. "Please answer."

Rose's latest conquest was Henry D. Wilson, an abolitionist Republican senator, Lincoln's chairman of the Committee on Military Affairs, and future vice president. Wilson was a particular challenge, said to be devoted to his wife and reticent with his opinions. Rose lapped at him with her questions, increasing their force and frequency, smoothing his resistance and wearing down his will, a process that felt more like erosion than seduction. Afterward he allegedly sent torrid letters, some written on congressional stationery, all signed only "H."

"You know that I do love you," read one. "I am suffering this morning. In fact I am sick physically and mentally and know nothing that would soothe me so much as an hour with you. And tonight, at whatever cost, I will see you."

If "H" was indeed Wilson, he was fully aware of Rose's secessionist proclivities and apparently didn't care. The letters contained no breach of national security, but neither he nor Rose ever divulged what was said behind closed doors.

Jordan taught Rose a rudimentary cipher of the type used in Edgar Allan Poe's "The Gold-Bug," in which mysterious-looking symbols each conceal a different letter, number, or word. She was to address all correspondence to his alias, Thomas J. Rayford. Certain combinations became familiar: "Infantry" was two parallel lines adorned with small circles; "Pennsylvania Avenue," a series of squiggles framed by dots; "President Lincoln" (whom Rose alternately called "Beanpole" and "Satan") looked like an inverted triangle bisected by a single slash. Jordan pointed out that the upper windows on one side of her home were easily visible through a telescope on the Virginia side of the Potomac. He gave her a copy of the Morse code and demonstrated how to manipulate the window shades to indicate dots and dashes. Each day at 10:00 a.m. and 4:00 p.m., he would order a scout or picket to watch the windows for any important news. On the street she could achieve the same effect with precise flutterings of her fan.

Rose Greenhow's cipher.

Rose immersed herself in her new role. Days that had once been spent shopping and making social calls to other women—the same women who gossiped about her late-night visitors—were now devoted to practicing her cipher and plotting her next move. The work overtook her mind, muting the dull, relentless drumbeat of thoughts about what her life had become. She was a widow with no steady income, forced to take seamstress jobs and put up furniture as collateral for rent money. With the end of the Buchanan reign and nearly thirty years of Democratic rule she'd lost her place as the doyenne of Washington society. Her daughter Florence, who was married to a captain in the Union army, sent carefully worded letters pleading with her to choose a different path: "I am so much worried about the latest news from Washington. They say some ladies have been taken up as spies. . . . Dear Mamma, do keep as clear of all Secessionists as you possibly can. I so much fear everything for you all alone there." Her very way of life was in danger of extinction, with the Yankees intent on liberating those Negro "beasts of the field." This war was the only hope of restoring her past and securing her future, and she would employ "every capacity with which God has endowed me" to aid the rebel cause.

She understood that the Confederate espionage system had certain advantages, circumstances she could exploit to gather intelligence. Washington, DC, was a Southern city in both character and origin; a third of its residents had been born in the slaveholding states of Virginia or Maryland. In Maryland, the Lincoln ticket received less than 3 percent of votes cast; in Virginia, barely 1 percent. Nearly every member of the Confederate government had once been a Federal official and, as such, possessed intimate knowledge of government operations. Jefferson Davis himself had served as secretary of war under President Franklin Pierce. Even the city's mayor, James G. Berret, was said to "smack of sympathy with secession."

The office of Attorney General Edwin M. Stanton, soon to be Lincoln's secretary of war, was so riddled with Southern sympathizers that he had to walk to its entryway to speak confidentially with a Republican senator. Even the White House was vulnerable, more or less open to any citizen seeking a meeting with the president, who held court in a second-floor office behind a pigeonhole desk. Mary Todd Lincoln, a Kentuckian with a brother, three half brothers, and three brothers-in-law in the Confederate army, once discovered that a Southern guest was in the habit of eavesdropping outside the cabinet room doors. Such ill-gotten information would be delivered via the Underground Railroad—not, in this context, a secret route for escaped slaves but a complicated chain of couriers, boats, and horses connecting Baltimore and Washington with Richmond.

The Confederate government had no fund set aside to employ spies, but volunteers abounded, and the willingness to work without payment was a test of sincerity and loyalty. Rose had to choose carefully; the currently disorganized nature of Union intelligence did not preclude a savvy spy or two from gathering information to use against the rebels. At least one Union agent, who secretly worked as a telegraph operator, had ingratiated himself with Beauregard, persuading the general to introduce him at the Manassas Junction station as "a railroad man willing to undertake any work you may have for him to do." Meanwhile, he was already busy sweeping out the telegraph station, listening to every message that came in.

In early May, as Union troops began to arrive in Washington, Rose began assembling her network. She called them her "scouts" but they were technically spies, the difference being that a scout was an official member of the army and a spy was a civilian. If a scout was captured, he would be treated as an enemy combatant, while a spy would be summarily hanged. Her group included a lawyer with roots in South Carolina; a clerk in the Post Office Department; a dentist who, during frequent visits to his son in the Confederate army, delivered messages and documents; a banker who traveled throughout the South under an assumed name; a professor, a cotton broker, an opium smoker, and a Sister of Charity; a widow who agreed to deliver messages for the money, completely ignorant of the true nature of her job; and various society women who had known Rose for years and their young daughters, among them

Bettie Duvall, who, on July 9, began the journey from Washington to Beauregard's headquarters, carrying news of the enemy's plans.

Bettie rode her farm cart through Georgetown, passing by a series of Union camps, a thread of canvas tents stretched as far as she could see. Bold curls of black smoke spiraled toward the sky, carrying the scents of grease and burned sugar. She heard the jumpy plink of a banjo, the hypnotic skip of a fiddle. Confederate and Union soldiers sometimes exchanged fire across the river, but she refrained from snapping her reins to push faster; panicking would only draw attention. She turned left and rattled onto the rickety boards of Chain Bridge. If she peeked upward, past the brim of her bonnet, she would see the Union artillerymen at Battery Martin Scott, a two-tiered stone-and-turf fortification overlooking the bridge. Iron twelve-pounder guns mounted at the end of the bridge could sweep the span, and three forty-two-pounders loomed a hundred feet up the hill. No one stopped her as she rumbled across, pushing into the Virginia countryside. She had left Rose's home hours earlier and was still not even halfway to her destination.

At sundown Bettie worried about Yankee scouts and pickets; she, like Belle Boyd, had heard of "Yankee outrages" against women. For safety's sake Bettie stopped for the night at Sharon, a plantation owned by a family friend just west of the village of Langley. In the morning she shed her calico dress for a riding habit, abandoned the cart, and borrowed a saddle horse. She cantered off, passing abandoned wood houses and weathered ox fences, reaching a Confederate outpost near Vienna. The provost marshal brought her to Beauregard's top aide, Brigadier General Milledge Luke Bonham, who was surprised to see the "beautiful young lady, a brunette with sparkling black eyes, perfect features, the glow of patriotic devotion burning in her face." Bettie told Bonham she had important information for Beauregard. Would he receive it and forward it immediately? If not, might she deliver it herself?

Bonham promised that he "would have it faithfully forwarded at once."

Satisfied, the girl took out her tucking comb and let fall the "longest and most beautiful roll of hair" he had ever seen. From the back of her head she retrieved a small package, not larger than a silver dollar, sewed up in silk. Without checking its contents, Bonham rushed the packet to Beauregard, who sent it by courier to Jefferson Davis.

A piece of Rose Greenhow's silk purse and part of an encrypted message.

On July 16, six days after Bettie's delivery, Rose felt a hand press on her shoulder. She awakened to see her maid, Lizzie Fitzgerald, who whispered that a gentleman was at the door. Rose tightened her dressing gown, checked on Little Rose sleeping in her room, and hurried down the stairs. The night had thinned and the sky cracked open a dim yolk of sun, lighting a familiar face: George Donnellan, former clerk at the Department of the Interior and a close associate of Thomas Jordan's. As an extra precaution, Rose requested further identification. Without a word Donnellan dropped a crumpled scrap of paper into her hand. She closed her fingers around it as she raised it to her eyes. Just two words, written in her cipher, but the two words she needed to read:

TRUST BEARER

Rose nodded, the messenger stepped inside, and they closed the city out.

She hurried to her library, sat down at her desk, and conjured the proper symbols for her next enciphered message, confirmation of the intelligence she'd sent a week earlier:

"Order issued for McDowell to march upon Manassas tonight. The enemy is 55,000 strong."

Donnellan stuffed the message into the hollowed-out heel of his boot and scuttled out the door. Once he disappeared around the corner of Sixteenth Street, Rose pictured the rest of his route: a few miles by buggy, a few more on horseback, down the eastern shore of the Potomac to a ferry in Charles County, Maryland, where he crossed into Virginia and handed the message to a Confederate cavalry officer, along with these fervent instructions: "This must go thro' by a *lightning express* to Beauregard." Confederate troops stationed along the Potomac included fifty-eight cavalrymen ready to participate in a relay system designed for maximum efficiency. Stations were ten to twelve miles apart, with fresh horses available at each, and by 8:00 p.m. Beauregard was studying Rose's deciphered note. Within a half hour the general telegraphed the information to President Jefferson Davis. The Southern army had a few days to prepare before the enemy arrived.

The following morning, Donnellan returned to Rose's home with Jordan's reply: "Let them come. We are ready for them."

As If They Were Chased by Demons

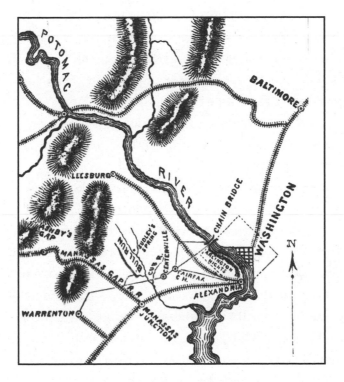

The Battle of Bull Run, 1861.

MANASSAS, VIRGINIA

And so they came. Thousands of Union soldiers crossed the Long Bridge from the capital into Virginia, regiment after regiment after regiment, looking, said one witness, like "a bristling monster lifting himself by a slow, wavy motion up the laborious ascent." They took their time, marching only twenty-five miles in two and a half days, wandering off to pick blackberries (considered the best cure for diarrhea) and wrestle beehives from stands, devouring the honey and half the bees at once. Thick woolen stockings chafed their skin raw. They wore boots that didn't distinguish right from left and had to break them into being one or the other. Infantrymen stumbled and wilted in the sun, carrying on their backs fifty pounds of equipment, including luxury items like breastplates. Several emptied cartridge boxes to lighten their load. They waved to the Washington civilians who rode out in carriages and barouches, hoping to see a real battle—hundreds of men and women and children peering through opera glasses, digging into picnic baskets, and sipping champagne, waiting to toast a Union victory.

Beauregard dispersed his troops along the south bank of Bull Run Creek, forming an eight-mile line that guarded the river crossings, perfectly positioned.

General Irvin McDowell's Federal troops began crossing the creek at 9:00 a.m. on July 21. They reached a clearing and were greeted by a volley of musketry, bullets whirring and whining above their heads. "By the left flank, *march!*" a colonel ordered. The bullets descended and began hitting their targets. Cannonballs plowed up the ground around them. Shells exploded, tossing men ten feet into the air and tearing off their limbs. Rifle-muskets launched minié balls that flattened upon contact with human flesh, scything a swath through muscle and bone. Union cavalrymen, waiting to be ordered to the front, tried to avert their eyes as the first wounded men were carried past them to a surgeon's tent. They glimpsed bones sprouting through skin, faces without features. Many vomited from their saddles.

Emma's division swung northward to try to enclose the rebels' left flank, near the bridge. She took position on the field next to the chaplain's wife, waiting to attend to the wounded. A shell burst nearby and struck three men and two horses. She ran toward the battery and knelt by a gunner who had been shot in the chest. He lay facedown in the dirt, inhaling his own blood. As she cupped and lifted his head, she recognized the boy who had led a prayer meeting the night before. He took his last breath in her arms.

Uniforms were inconsistent on both sides—some Union regiments in gray, some Confederates in blue—and dozens of men fell from friendly fire. Yet the battle seemed to be going as McDowell had planned, his boys tearing at the Confederate left flank, driving the rebels back, crying out, "We've whipped them!" "We'll hang Jeff Davis to a sour apple tree!" "They are running!" "The war is over!" Two miles from the fighting, the civilian spectators twirled hats and raised glasses, straining to see through the smoke, breathing in the scorched-metal scent of gunpowder.

But the Confederates were not yet ready to surrender. Standing guard at the center of the Southern line were Brigadier General Thomas Jackson and his men (Belle's father, Benjamin Boyd, was currently detailed as a clerk, although she would later claim he took a bullet in the shoulder). A South Carolina general, trying to rally his own troops, pointed and yelled, "Look, there is Jackson with his Virginians, standing like a stone wall!" For the next several hours the fighting oscillated, bullets crisscrossing over the hillside. One private from the 4th South Carolina Infantry was spared death when a bullet pierced the Bible tucked inside his left coat pocket, slowing its progress through his body until it exited out his right side. Between the armies stood a farmhouse, the home of an elderly widow too ill to move. Union shells tore through the wall of her bedroom and severed her foot. She died later that day, the first civilian casualty of the war.

There were rebels all around them now and reserves to back them up, arriving first on horseback and then by train, waving Union flags to add to the confusion, a furious vortex streaming in from nowhere and everywhere. And that sound, that rebel battle cry—the inhuman yelp of something unidentifiable, something nearly dead and already halfway to hell: "There is nothing like it this side of the infernal region," one Union officer said. "The particular corkscrew sensation that it sends down your backbone under these circumstances can never be told. You have to feel it, and if you say you did not feel it, and heard the yell, you have *never* been there." It closed in around them, overpowering their own mannered cry of "Hoo-ray! Hoo-ray!"

The Union troops fumbled backward and the Confederates rammed forward, a brutal and uneven dance, with soldiers felled like rotting trees. Some hadn't drunk water in nearly twenty-four hours. Their lips were cracked, tongues coated black with gunpowder. "In reckless disorder the enemy fled in frantic confusion," witnessed one Southerner, "as if they were chased by demons rather than men." The spectators abandoned their sideline posts and charged into the creek, leaving behind fringed shawls and dainty parasols and shattered champagne flutes, many of them drowning along the way.

Every angle, every viewpoint, offered a fresh horror. The rebels slashed throats from ear to ear. They sliced off heads and drop-kicked them across the field. They carved off noses and ears and testicles and kept them as souvenirs. They propped the limp bodies of wounded soldiers against trees and practiced aiming for the heart. They wrested muskets and swords from the clenched hands of corpses. They plunged bayonets deep into the backsides of the maimed and the dead. They burned the bodies, collecting "Yankee shin-bones" to whittle into drumsticks, and skulls to use as steins.

Corpses were strewn for miles, faces blackening in the sun. The braided jackets and crimson fezzes of the New York Fire Zouaves dotted the battlefield, festive even in death. A Union surgeon found shelter beneath a tree, tying a green sash around its girth to mark the space. Within fifteen minutes he cared for thirty men, wounds hideously open, flaps of flesh blooming outward. His coat was adorned with blood and stringy ligaments and jagged flecks of bone. He amputated four legs, three arms, a hand, and a foot, each procedure taking no longer than ten minutes, the pieces tangling together in the grass. When rebels discovered the space, they hacked off each patient's remaining limbs. Horses galloped about riderless or lay writhing on the ground hitched to guns, blood leaking from nostrils and ears, gnawing wildly at their flanks.

The "great skedaddle," as the Union army ruefully called it, was in full force by 6:00 p.m., with no hope of persuading the soldiers to turn around and fight; the plan to advance to Richmond and end the war had failed. In their haste they overturned wagons and left behind a cache of ammunition. The 2nd Michigan was called to guard the retreat from a potential Confederate counterattack. The Michigan men stood watch well into the night, drenched by a sudden storm, while the rest of the army fled back to Washington. Emma was stunned by the Union army's cowardice. "Many that day who turned their backs upon the enemy and sought refuge in the woods some two miles distant," she wrote, "were found torn to pieces by shell, mangled by cannon ball—a proper reward for those who, insensible to shame, duty, or patriotism, desert their cause and comrades in the trying hour of battle, and skulk away cringing under the fear of death."

Rose was not among the Washington spectators at the battle, having picked up her middle daughter, Leila, at her Maryland boarding school and escorted her to New York. From there Leila would take two steamships and a train to San Francisco to stay with her older sister Florence until it was safe to come back. When Rose returned to Washington, she found a message awaiting her from Thomas Jordan: "Our President and our General direct me to thank you. We rely upon you for further information. The Confederacy owes you a debt." She heard that her Union contact Henry Wilson had been among the spectators, carrying a picnic basket of sandwiches. He fled back to Washington with strangers after a Confederate soldier disabled his carriage with a shotgun blast.

In the end more than 4,500 men on both sides were killed, wounded, or captured, one of the last being New York congressman

Alfred Ely, whom the rebels discovered hiding behind a tree. "The Yankee Congressman came down to see the fun," said one rebel soldier, "came out for wool and got shorn." Ely was one of a thousand Union captives delivered to Richmond, where rebel officials had converted a series of tobacco warehouses into military prisons, and where a wealthy Richmond woman named Elizabeth Van Lew waited for them.

Never as Pretty as Her Portrait Shows

Elizabeth Van Lew, circa 1861.

RICHMOND

Two days after the First Battle of Bull Run (or First Manassas, as Southerners would come to call it), Elizabeth Van Lew left her Grace Street mansion and set out for Ligon's Prison, a converted tobacco warehouse four blocks away on the corner of Twenty-Fifth and Main. The city was in a "palpable state of war," she observed, the somber notes of Handel's "Dead March" accompanying an endless procession of military funerals, warhorses with empty saddles walking toward Capitol Square. Every arriving train carried the same ghastly freight: plain pine boxes filled with the dead and hundreds of the wounded, who staggered along with powder-stained faces and bandaged heads, using muskets as crutches while searching for hospitals with empty beds. She walked among them, incongruous in her finest day attire: a silk dress trimmed with ribbon and a matching scalloped parasol, armor for a battle of her own.

A curious crowd gathered around the perimeter of the prison and stared up at the grated windows. "Whenever they caught a glimpse of a Federal officer," said one witness, "[they] hooted at and insulted him . . . men, women, and even little children scarcely old enough to walk, united in heaping scurrilous abuse upon them." They called this practice "stirring up the animals." One lady shook her parasol at the prisoners and shouted, "What did you come here for? We will have you know that if you kill all of the men, the women will make more soldiers." The bystanders misinterpreted her remark and began snickering, laughing even harder when she tried to clarify: she meant the women would take up arms when all the men were killed—nothing so bawdy as what they had in mind.

Elizabeth followed a guard to the office of Confederate lieutenant David H. Todd, the prison's commander and a brother-in-law of President Lincoln. She silently rehearsed what to say, her heart hammering inside her ears. It was possible the rebel officer would recognize her; everyone in Richmond knew the Van Lew name. Richmond society had always tolerated her, partly because of her father's legacy as a prominent businessman and slave owner and partly because she was perceived as a benign oddity, an eccentric old spinster destined to die alone in her house on the hill. She was forty-three, had never married, and still lived with her mother. She had a frail, knobby frame and blue eyes set deep within the sharp angles of her face, a face that once passed as beautiful. She was invariably described as "nervous" and "birdlike," and, according to one contemporary, was "never as pretty as her portrait shows."

Elizabeth had to be careful how she presented herself at this meeting, as her neighbors—not to mention her sister-in-law, an ardent Confederate sympathizer—would surely hear of it by day's end. Although a native of Richmond and one of its wealthiest citizens, she had Yankee roots, a pedigree that prevented her from achieving the standing that came with birth into the right families. Her father, John, hailed from Jamaica, Long Island, and her mother, Eliza, from Philadelphia, where her own father, Hilary Baker, served three terms as mayor and was an early member of the Pennsylvania Abolition Society, a pioneering and influential antislavery organization. In the early 1830s they sent Elizabeth to school in that city under the care of an abolitionist governess, and by the time she returned home to Richmond, her world had irrevocably changed. "It was my sad privilege to differ in many things from the perceived opinions and principles in my locality," she wrote in her diary. "This has made my life intensely sad and earnest."

Nearly every day she walked past Odd Fellows' Hall, just a few steps away from the Capitol, where dozens of slave owners conducted business, and the nearby Lumpkin's Alley, with its auction blocks and "Negro jails." She scanned the notices in the *Richmond Dispatch* about runaways: "Left the Tredegar Iron Works about three weeks ago," read one. "A NEGRO MAN, calling himself 'CHARLES BLACKFORD,' the property of Mrs. F. G. Skinner . . . about 5 feet 6 inches high, well built, very dark skin, looks confused when spoken to." Slavery, she believed, "is arrogant—is jealous and intrusive—is cruel—is despotic—not only over the slave, but over the community, the State." She once gave a tour of the city to the renowned Swedish author Frederika Bremer, including a stop at a tobacco factory. Bremer was horrified by the conditions of the slaves, toiling ceaselessly amid the dirt and "murderous" smells, and saw that "Good Miss Van L. could not refrain from weeping."

Elizabeth came to understand the importance of appearances, and the intricate subterfuge required to maintain them. Despite their Northern connections her family had achieved the social respect that comes with prosperity, and her father aggressively vied for his place in Richmond society. As the proprietor of a hardware business whose clients included Thomas Jefferson and the University of Virginia, he amassed enough of a fortune to purchase a three-story mansion in the prestigious neighborhood of Church Hill, a property that became a showpiece of the city and a stage for elaborate parties; Chief Justice John Marshall, Swedish opera star Jenny Lind, and Edgar Allan Poe all mingled in the parlors and wandered through the exquisite gardens. Her mother expertly played the part of the gracious Southern hostess and Christian lady, bringing her three children—Elizabeth and younger siblings Anna and John—to the venerable St. John's Episcopal Church, where Patrick Henry gave his "liberty or death" speech standing in the Van Lew family pew. They understood that slaveholding was a prerequisite for wielding any political or financial influence in the South, and by 1843, the year of John's death, they had a staff of fifteen. He bequeathed them all to his wife with the stipulation that she was not to sell or free any of them, a provision she decided not to honor.

Eliza, Elizabeth, and Elizabeth's brother John (her sister Anna had moved north to Philadelphia) participated in the system of "hiring out," in which slaves could find their own employers and keep a percentage of their wages, eventually earning enough to purchase their own freedom. Elizabeth also began spending her $10,000 inheritance (about $300,000 today), buying slaves for the express purpose of freeing them. Many of them chose to stay and work for her—in the Van Lew house, "servant" did not mean "slave"—and Elizabeth's humane treatment of her staff did not go unnoticed. "From what I have seen of the management of the Negroes of the place," said one neighbor, "the family of Van Lews are, I am satisfied, genuine abolitionists."

Elizabeth never made a secret of her loyalty to the Union; after all, two thirds of the Virginia convention had voted the same way until after the Battle of Fort Sumter, and more than a third stood firm even then. But after Virginia seceded and Richmond became the Confederate capital, such opinions had potentially dangerous consequences. Richmond society's polite tolerance of Elizabeth's background and eccentric behavior degenerated into open hostility. "Loyalty was called treason, and now cursed," Elizabeth wrote. "Surely madness was upon the people."

The *Dispatch* warned of marauding Yankees intent on "butchery, rape, theft, and arson." Young girls carried daggers and pistols in their crochet purses and fired at marks in the street. Dr. George Rogers Clark Todd, another brother-in-law of Lincoln's and a devoted Confederate, was arrested for merely declaring in a bar that President Davis had treated him "damned rascally." Her moneyed neighbors visited the troops encamped at the Old Fair Ground and beseeched the men to "kill as many Yankees as you can for me." Elizabeth once accompanied them and spoke at length with the soldiers. "I longed to say to them, 'be not like dumb driven cattle'. . . . They kindly informed me that Mr. Lincoln had said he was coming down to take all our negroes and set them free, and they were going to protect us women."

The ladies gathered for afternoon tea and spoke of slaughtering all Richmond residents who had been born up North, a thinly veiled threat against Elizabeth's mother. They wondered why her brother John hadn't volunteered for the Confederate army along with other men of his stature and class; having to run the family business was hardly a legitimate excuse. They coveted "Mr. Lincoln's head or a piece of his ear." During one recent visit, when the discussion turned to secession and slavery, a neighbor named Mrs. Watt became so "offended and disgusted" by Elizabeth's "obnoxious" opinions that she left immediately, exiting in a swirl of silk and a stomp of high-button boots. Even Elizabeth's own sister-in-law, Mary, who lived with her in the Church Hill mansion, would not hesitate to report the Van Lews for any disloyal activity, either real or perceived. The women issued Elizabeth one final invitation, asking her to help sew clothing for the rebel army. She refused—an act of defiance that only confirmed suspicions among people who never quite considered her one of their own.

Elizabeth made one public concession to her new reality, removing the family's American flag from its pole on the chimney, visible from Grace Street on the right side of the mansion. It was uncommonly large for a private flag, eleven by twenty feet, and she had first raised it in 1850, more than a decade before. With the addition of each new state to the Union she sewed a representative star, taking particular pride in Oregon and Kansas, which joined as secession began sweeping the land. With the flag down and packed away, she wrote in her diary of her "calm determination and high resolve" to aid the Union, aware that her family's position would be crucial to her success. Only a member of Southern society would know how to turn its prejudices into weapons, and have the opportunity and access to defeat it from within.

She recalled the phrase she'd doodled over and over again as a child, practicing her penmanship in her notebook: "Keep your mouth shut, your eyes and ears open."

When the war broke out, Lieutenant Todd was working as the overseer of a plantation, managing and disciplining slaves with a particularly brutal hand, a tactic he carried over into his work with the prisoners. One inmate accused him of entering the prison with his sword drawn and striking men with the flat of its blade, and others alleged that he ordered seven Union prisoners to be shot just for innocently leaning against the windows. He sat behind his desk, gray wool jacket fully buttoned, lips barely visible behind a thicket of black beard. Elizabeth thought he had a "violent appearance."

"Lieutenant," she began, "I would like to be made hospital nurse for the prisoners."

It was the truth, and her inquiry wouldn't seem unorthodox; every Southern woman had opened the doors of her home to the Confederate sick and wounded, rolling cloth bandages and collecting lint for packing wounds. She needed to begin somewhere, and this was a practical and inconspicuous first step.

Todd asked Elizabeth for her name and wrote it down. He realized who she was and looked up at her, surprised. "You are the first and only lady to make any such application," he said, meaning no one had yet volunteered to minister to the Yankees. He refused to grant her request. Perhaps, he suggested, she could appeal to Mr. Christopher Memminger, secretary of the treasury for the Confederacy.

She did, that same afternoon, taking her family's carriage, an exquisite barouche drawn by four snow-white horses, fourteen blocks west. Traffic was maddeningly obstructive—the influx of Confederate volunteers had more than doubled the city's population since the outbreak of the war—and she shared the dusty roads with convoys of wagons, some piled high with the dead, their rigid feet splayed in all directions, the occasional stiffened arm raised in an eternal salute to the sky.

The War Department Building was housed in the Mechanics Hall at Ninth and Bank Streets, a venue that in peacetime hosted concerts by the Richmond Philharmonic Association and a farming contest called the "Trial of the Ploughs." Sitting across from Memminger, Elizabeth summoned skills she had last used twenty years ago, back when she was considered one of Richmond's most eligible belles and had her pick of suitors. Only one of them had been worthy, but he'd died before they could marry, succumbing to the yellow fever epidemic in 1841 when she was twenty-three; she still kept the cameo brooch he'd given her. Subsequent prospects failed to appreciate her candor and obstinacy and unorthodox convictions, her willingness to subvert social codes if she believed they were wrong.

"Please sir," she said, and smiled. "Let me see the prisoners."

Memminger shook his head. "I could not think of such a thing. Such a set and such a class—they could not be worthy of or fit for a lady to visit."

Elizabeth let her eyes dampen and clasped her hands. "Once I heard you at a convention, in peace times, speaking beautifully on the subject of religion," she said, sotto voce. "Love was the fulfilling of the law, and if we wish *our cause* to succeed, we must begin with charity to the thankless and unworthy."

At the words "our cause," his face relaxed. He wrote a note instructing her to see General John Winder, the commanding officer of the prison system. For the third time that day Elizabeth found herself in the office of a Confederate authority—in this case, one who would become an important figure in her plans.

Winder was sixty-one, eighteen years Elizabeth's senior, and had been inspector general of the Richmond area camps for a month. Every related job that did not automatically fall to another office—chasing deserters, issuing passes, enforcing curfews, housing prisoners—became his responsibility. His official photograph, which for a century would serve as the only image of him as a Confederate general, showed a menacing figure with hooded eyes and a thin, taut mouth who looked well capable of committing all of the crimes eventually laid at his feet, including the deaths of nearly thirteen thousand Union prisoners at Andersonville in Georgia. Adding to the effect was a scar along one cheek, the remnant of a wound acquired during the Mexican-American War, when a projectile struck a nearby soldier and caused his brains to spatter across Winder's face, a piece of skull slashing his skin. He had ended his forty-year career with the US Army with great regret, ultimately deciding that attempting to restore the Union by force was unconstitutional. Elizabeth knew that he, like so many Southerners, had divided loyalties within his family; Winder's oldest son, William, was around the same age as her brother and serving as a captain in the Union army.

General John Winder, provost marshal of Richmond.

She knew, too, that he had already earned the enmity of Richmonders by recruiting a force of civilian detectives—men so boorish and violent they were nicknamed the "plug-uglies" (or, occasionally, the "alien plug-uglies," a reference to members who hailed from the foreign and unsavory cities of New York and Philadelphia). In addition to raiding gambling dens and brothels and confiscating citizens' guns and swords, the plug-uglies conducted counterintelligence operations, identifying and intercepting Union spies.

The general motioned for Elizabeth to sit down. Silently she studied his hair, the white tufts swelling and peaking like miniature waves. Her own hair, once swinging golden and thick down the length of her back, was now faded and clipped. She no longer expected flattery but still recognized its worth, the power in elevating the ephemeral.

"Your hair would adorn the temple of Janus," she told him. "It looks out of place here."

Winder's lips stretched into a smile, and for a moment she couldn't tell if he felt patronized or pleased.

"I should be glad to visit the prisoners," she added, "and I'd like to send them something."

Her approach worked. Winder wrote a note and pressed it into her palm: "Miss Van Lew has permission to visit the prisoners and to send them books, luxuries, delicacies, and what she may please." Elizabeth thanked him, surprised by how greedily he accepted her praise. "I can flatter almost anything out of old Winder," she boasted to one Unionist friend. "His personal vanity is so great."

She was aware, though, that the general might be motivated not by vanity but by his own brand of deception—a desire to know exactly what she brought into the prisons, and if she began taking anything out.

Elizabeth gathered fresh fruit, cake, books, and clothing for the Union prisoners, her carriage heavy with provisions as it pulled up to the gate. Her ministrations soon caught the attention of the press. Without naming names—although the implication was clear—the *Richmond Examiner* printed the following item:

"Two ladies, a mother and a daughter, living on Church Hill, have lately attracted public notice by their assiduous attentions for the Yankee prisoners confined in this City. Whilst every true woman in this community has been busy making articles of comfort or necessity for our troops, these two women have been expending their opulent means in aiding and giving comfort to the miscreants who have invaded our sacred soil, bent on raping and murder, the desolation of our homes and sacred places, and the ruin and dishonor of our families."

The *Richmond Dispatch* ran a follow-up: "They are Yankee offshoots, who had succeeded by stinginess, double-dealing and cuteness to amass out of the credulity of Virginians a good, substantial pile of the root of all evil." If these women weren't careful, the paper concluded, they would be "exposed and dealt with as alien enemies to the country."

It was one thing for Elizabeth to feel threatened by her neighbors and sister-in-law but quite another to see such language in print, tacit permission for both government officials and private citizens to enact any manner of revenge. At this point, early in the war, being "dealt with" was a nebulous but nonetheless terrifying prospect: it could mean permanent exile from the South, away from family and friends; it could mean prison, with all of its attendant horrors; it could mean public execution at the gallows. Given her gender it could also mean nothing at all, since Jefferson Davis's newly composed Alien Enemies Act referred only to the deportation of male citizens unwilling to take an oath of allegiance to the Confederacy. Union soldiers suspected of spying were languishing in Richmond prisons, but no action had yet been taken against any woman— let alone one of Elizabeth's prominence and wealth—suspected of disloyalty. For now, at least, her social position and gender served as her most convincing disguise. No one would believe that a frail, pampered spinster was capable of plotting treasonous acts, let alone carrying them out right under the government's nose.

Elizabeth clipped the article and scrawled in the margin: "These ladies were my mother and myself. God knows it was little we could do."

What she did do, a few weeks later, was offer to care for a dying Union prisoner inside her own home. Calvin Huson, who was married to the niece of US secretary of state William Seward, had been captured at Manassas and sent to Ligon's, where he contracted typhoid fever. Huson, it was rumored, came to Manassas hoping to be made governor of Virginia after a Union victory ended the war, and the Southern press reported his current predicament with glee. "Poor Calvin Huson," wrote the *Charleston Daily Courier*, "who came out to Manassas to 'see the fun' and who fell into the hands of the funny rebels." Again Winder granted Elizabeth's request, and when the closed carriage rumbled up Grace Street her neighbors all knew a Yankee was inside. But if any of them followed Elizabeth's patient to the door, they would spy an oversize Confederate flag, hanging boldly on her entry parlor wall.

LITTLE REBEL HEART ON FIRE

THE SHENANDOAH VALLEY, VIRGINIA

Belle shut her eyes, arms still outstretched, waiting for the click of a musket. Instead, beneath her mother's soft whimper and Eliza's prolonged and piercing wail, she heard short, jagged intakes of breath. She opened her eyes. The Union soldier was still alive at her feet, blood bubbling from his neck, and she was no longer the focal point of the room. His comrades fashioned a stretcher with their arms and carried him to the surgeon's tent, where he died later that day.

Soon after the burial of twenty-five-year-old Private Frederick Martin, Company K, 7th Pennsylvania Volunteers, General Patterson and several of his staff paid a visit to Belle's home. She watched them approach the door, her breath quickening. They could throw her into prison or try to kill or rape her, a violation that would degrade and declass even a so-called fast girl. Her mother allowed them in, positioning herself close to the door. Once again Eliza had hidden Belle's collection of rebel flags. Belle kept her purse nearby, her pistol stashed inside.

The men questioned Mary Boyd, Eliza, and Belle, all of whom recounted the deceased private's coarse language and threatening gestures and insisted that Belle had no choice but to shoot him. As they interrogated her, Belle confessed—silently, to herself—that she had not "one shadow of remorse" for killing the Yankee, that the blood she'd shed "left no stain" on her soul. She had saved her mother from "insult and outrage," perhaps even from death.

Fortunately for Belle, Washington was still practicing appeasement; Lincoln was intent on keeping the border states of Maryland, Delaware, Missouri, and Kentucky in the Union and didn't want to stir up a revolt among their secessionist citizens. Knowing that the Shenandoah Valley (which offered a potential backdoor route to Richmond) was strategic territory, and reluctant to turn this teenager into a Confederate martyr, Patterson declined to take any action and declared the case closed. Belle would recall a more exuberant exchange, with the general deeming her a "plucky girl" and advising her to "do it again if any more such brutal fellows came around." Hoping to avoid further trouble, he stationed Federal sentries around the Boyd home and ordered them to keep close watch on her.

Belle welcomed the increased attention, waving to the Yankees from her balcony, piquing their interest and horrifying the neighbors. She noticed that the town was growing accustomed to Union control, easing back into routine. Patterson had even issued strict orders to shoot the first man caught stealing private property, and announced that any soldier who insulted a lady on the street would be confined in the guard house. The Union blockade, which stretched along 3,500 miles of Confederate coastline, was intended to deprive the rebel army of goods and supplies but had a direct impact on civilians as well. Coffee and sugar were suddenly scarce, and housewives complained of shortages and rising prices. Publicly, though, Southerners dismissed the blockade as ineffective. "The old ass," one Charleston resident said of Lincoln, "thinks he can starve us out, but he never made a greater mistake." Merchants reopened for business and appreciated the Union army payment in gold and silver instead of shinplasters, the only currency of the rebel troops. A Federal flag hung in the public square unmolested, and women and children, according to one Northern report, "thronged the streets in perfect security with joyous hilarity."

Belle sensed an opportunity in the newly relaxed atmosphere. The war seemed to have a place for any prospective spy, even one more concerned with recognition for her deeds than with the deeds themselves. There was ceaseless movement and confusion, armies scattered across miles, the border between North and South riddled with holes. Even accents couldn't betray allegiances. Rebel spies lurked in the pro-Union counties in western Virginia, and Union spies in the staunchly Confederate counties farther south. They operated on varying levels of importance and authenticity: "walk-in" friends with potentially useful information; couriers who traveled their routes with eyes open; and trusted individuals actually tasked to go and procure information.

Most of the spies Belle knew were of the first, informal variety, women who considered spying as much a part of their patriotic duty as making bandages and stitching uniforms. Some merely stood on front porches, counting Federal troops as they marched (one regimental historian wryly noted that the ladies of Winchester, Virginia, "did a little spying in which they were almost always perfectly safe"). Others, like Belle herself, vied to uncover something that mattered, a piece of the war that would remain wholly theirs even after they passed it along.

At night, as the 21st Pennsylvania's regimental band gallantly honored requests for "Dixie," Belle chatted with her guards, slipping questions between the pauses in the music, etching flattery around the edges of her words. What a relief to see that the Union boys truly meant Martinsburg no harm! Were they all very lonesome, so far away from their sweethearts and wives? How were the Union troops faring over at Rich Mountain? How many thousands were camped there? And was it true that General George McClellan was reorganizing his scouts, having decided that his current ones were careless and useless? She ran her hand along the sleeve of a frock coat, touched a fingertip to a gleaming brass button, retracting all traces of the bold, hard girl who had shot their Yankee comrade dead—although she kept the pistol that killed him with her at all times.

Alone in her room, long after her mother was asleep, Belle lit a candle and transcribed everything she'd been told or gleaned from eavesdropping: Union troop movements, troop numbers, the state of troop morale. There was no sure way to ascertain the value of her information, so it would be up to the generals to sift through the scraps. She gave the notes (or *lettres de cachet*, as she preferred) to Eliza and instructed her to walk to Stonewall Jackson's camp seven miles away, reasoning that no one would suspect mischief from a Negro servant running an errand. Sometimes she enlisted the help of a teenage neighbor, a "lovely girl" named Sophia B., who seemed thrilled to do Belle's bidding. Belle congratulated herself on her cunning and deceit until, one mid-July afternoon, Captain James Gwyn, the third assistant provost marshal of the Federal army, appeared on her doorstep and demanded she accompany him to headquarters.

Belle's initial trepidation dissolved once she reminded herself that she'd been in trouble with Union officials before—and literally gotten away with murder. On West King Street, with Captain Gwyn at her side, she smiled at every soldier she passed ("large teeth," one noted, "and a loud, coarse laugh") and took her time walking along the aisle inside the Berkeley County Courthouse, stopping to pause beneath the stained glass dome, framing herself in a brilliant halo of light. The captain led her to the office of a colonel. Belle missed his name but noticed a piece of paper on the center of his desk. The door shut behind her and the colonel stood, trapping the paper beneath his hand.

He said a letter of hers had been intercepted.

Her mind spun with theories about how that happened. Perhaps Eliza lost a dispatch on the way to Stonewall's headquarters and was afraid to tell her. Or Sophia B. grew jealous of Belle's popularity among the soldiers and betrayed her.

It occurred to Belle, too, that she had crafted every one of her notes in her own handwriting. Several of her neighbors were familiar with her light, looping script, and the ones who sympathized with the North would be all too willing to identify her. The colonel insisted that her offense was very serious and recited a condensed version of the fifty-sixth and fifty-seventh Articles of War, dealing with espionage and treason:

"Whosoever shall give food, ammunition, information to, or aid and abet the enemies of the United States Government in any manner whatever, shall suffer death, or whatever penalty the honorable members of the court-martial shall see fit to inflict."

If Miss Boyd repeated this behavior, the colonel concluded, she would suffer the punishment prescribed.

Belle sat silently, studying the pressed tin ceiling while the colonel read his script. She summoned the bravery and spirit of her Revolutionary War ancestors, telling herself she wasn't frightened. She did not know that Ward Hill Lamon, a bodyguard of Lincoln's who also happened to be an old friend of the Boyd family, had intervened on her behalf, asking Union officials to treat her with leniency.

When it was clear that the colonel had finished, Belle stood, lifted the sides of her hoop skirt, and bowed as if overcome with reverence and awe.

She said, "Thank you, gentlemen of the jury," and showed herself out.

"My little 'rebel' heart was on fire," she wrote, "and I indulged in thoughts and plans of vengeance."

When Patterson's troops departed Martinsburg, leaving behind only the 1st Pennsylvania Regiment to maintain order, Belle and her mother packed up their carriage to visit relatives at Front Royal, about forty miles south. They rejoiced when the Confederates won the crucial engagement of Manassas, thwarting the Union plan to push forward and capture Richmond. There were rumors, whispered by the ladies of Martinsburg with either pride or disgust, that a rebel woman, Rose Greenhow, had contributed to the victory. Belle was familiar with Rose's name and reputation as a prominent Washington hostess—her parties during the winter of

1860–61, when Belle made her formal debut, were the most coveted invitations in town—and she knew personally of Rose's emissary, Bettie Duvall: "Miss D. was a lovely, fragile-looking girl . . . remarkable for the sweetness of her temper and the gentleness of her disposition. . . . [She] had passed through the whole of the Federal army." Belle admired the girl's "intrepidity and devotion," and saw no reason why she, too, couldn't facilitate a major victory for the rebels. She yearned for grander, more official involvement than Eliza's occasional run to Stonewall's camp, and determined to find herself a place with the Confederate secret service.

Ignoring her mother's tearful protests, Belle visited her uncle, a lieutenant in Stonewall Jackson's 12th Virginia, and told him that she wanted to be a courier and spy for the rebel army. Her first cousin, William Boyd Compton of the 31st Virginia Militia, was a spy himself; he would keep an eye on her and, she hoped, use her for missions. She believed she had the requisite skills for the job, having been, she said, "at home on a horse's back from my earliest girlhood" and "a good deal of boy myself," spending her days galloping far and away over miles of rough country, chasing foxes with the neighborhood boys. She knew her way around the Valley, from the Blue Ridge to the Appalachian Mountains and all of the gaps in between, every spot where turnpikes or railroads sliced through, every stream and rocky ridge, every broken range and scattered spur. She had hidden inside the Grand Caverns, where Confederate soldiers engraved their names beneath the stalactites. She even trained her beloved horse Fleeter to kneel on command, so that she might evade detection by Union patrols—the only instance in her life when she wished to remain unseen.

Belle's uncle mentioned her name and ambition to Turner Ashby, Stonewall Jackson's cavalry commander and head of Confederate military scouts, and she began riding as a courier between Generals Jackson, Beauregard, and J. E. B. Stuart, carrying commands for the movements of the army back and forth, charging alone through country infested by Yankee scouts and guerrillas.

Sometimes the commands were written, either in plain text or in a substitution cipher, not unlike the one used by Rose's ring; on other occasions she was told the gist of the message and expected to deliver it orally. During these excursions she allowed but one thought to possess her mind: "that I was doing all a woman could do in her country's cause."

Her Confederate relatives taught her the signs and countersigns required to pass through the lines. These changed frequently and constituted simple word exchanges—the challenge being "Stonewall" and the response "Jackson," for example—or a series of intricate physical maneuvers requiring not only memory but coordination. "We have the same old signal," wrote a private in the 5th Virginia Infantry. "Halt anyone, throw up the left arm. He whom you halt must then take off his hat or cap and pass it down below his face. If he fails to do this, fire. This is the day signal. The night signal is the sentinel strikes his leg two or three times with his hand. The person whom he halts has to cough two or three times or clear his throat. If he fails to do this, fire at him." Belle proved equally adept at coaxing Union sentinels into revealing their own secret signals.

She was thrilled by the danger of the job, by the reports in the papers—daily, it seemed—of "bearers of dispatches" being arrested; one, a boy exactly her age, had been caught with papers outlining a plan to capture Union steamships in California, and his fate was yet uncertain. She waited for something to happen, *willed* it to happen, and one night, after she'd leaped over a gloomy and precipitous ravine, it did.

Seven Union men stood facing her in the road, fanning out and blocking her path. The captain ordered her to dismount. She obeyed without protest, and reminded herself that she had prepared for such an encounter.

"Where are you going?" he asked. "Do you have papers?"

She did, hidden inside her bodice—exchanges detailing a coordinated effort among the Confederates to repair nine miles of railroad track above Martinsburg on the route to Richmond, and to remove a considerable portion of the Union telegraph wires in the same vicinity.

"I have no papers," she answered, "and I'm going home."

"Then we will search you." He started toward her.

She spoke her line softly, as if sharing a shameful secret: "Captain, I've but one little paper that my father told me never to give up except to save myself from death in dishonor."

She handed him a chamois bag. He opened it and found a Knights Templar cross wrapped in paper, which stated she was a Knight's daughter. Belle waited while he read the message, turning her eyes skyward to study the moon, silently praying he was a Mason.

He looked up at Belle, down at the paper, and back at Belle. He nodded once and said, "Don't let me ever see you again around here."

She went out again the very next night with new lines memorized, a new lie to tell.

ADMIRABLE SELF-DENIAL

WASHINGTON, DC

The memories of Bull Run implanted themselves in Emma's mind, a slide show of carnage without pause. She couldn't unsee the heaps of severed limbs, all those wizened palms turned upward in supplication, or dispel the taste of the bloodied bandages clenched between her teeth. "One case I can never forget," she wrote. "It was that of a poor fellow whose legs were both broken above the knees, and from the knees to the thighs they were literally smashed to fragments. He was dying; but oh, what a death that was. He was insane, perfectly wild, and required two persons to hold him." She learned quickly to let her body overtake her mind, becoming "simply eyes, ears, hands, and feet." She sent locks of hair to grieving mothers and wives and hauled corpses to the dead house for storage before burial. "I was not in the habit of going among the patients with a long, doleful face," she said. "Cheerfulness was my motto."

In truth, she had not felt so hopeless or helpless since her days of living at home, subject to her father's stern orders, to his capricious and terrifying moods. Washington presented a picture of military life in its most depressing form, she thought, with every saloon and gambling house overrun with officers, and the army—*her* army, which she had risked everything to join—a shambles, undisciplined and uninspired. Drills were forgotten, camps in disarray, men un-

accounted for and deserting left and right. *New York Tribune* editor
Horace Greeley removed the "Forward to Richmond!" banner from
the masthead of his newspaper and sent a private note to Lincoln:
"On every brow sits sullen, scorching, black despair. If it is best for
the country and for mankind that we make peace with the rebels at
once and on their own terms, do not shrink even from that." Lincoln ignored this advice and instead pushed the blockade, strengthened forces in the Shenandoah Valley, called for the enlistment of
a hundred thousand additional troops, and recruited a new man to
lead his army in Washington.

Emma loved Major General George B. McClellan as soon as
he took command, finding him young and gallant and imbued with
"grace and dignity." She was eager to do whatever he asked—an
unexpected reaction, she realized, for someone whose love of independence was equaled only by her "hatred of male tyranny."

General George McClellan,
circa 1861.

McClellan—"Little Mac," as the soldiers called him—was a thirty-four-year-old military wunderkind: second in his class at West Point, hero of the Mexican War, author of manuals on military tactics, brilliant strategist, skilled engineer. Just before Bull Run, he won a minor battle at Rich Mountain in western Virginia, a victory that established him as a national hero. Everything about him seemed painstaking and spit-polished, a physical presentation that implied supreme internal order and control. His fiercely parted dark hair revealed a shining strip of scalp. The bones of his face connected in perfect and pleasing symmetry. His voice was stentorian and his speech unmarred by skips or pauses, each word sounding like the surest ever spoken. He was, according to lore, able to bend a quarter between his thumb and forefinger and lift a 250-pound man above his head. He believed that God himself had elected him to save the Union. "By some strange operation of magic I seem to have become *the* power of the land," he wrote to his wife. "I almost think that were I to win some small success now I could become Dictator or anything else that might please me—but nothing of that kind would please me—*therefore* I *won't* be Dictator. Admirable self denial!" When posing for portraits, he slipped his right hand inside his frock coat in perfect imitation of Napoléon.

Emma noticed a change among the troops right away, the sense of foreboding and gloom loosening its grip, a tentative uptick in morale. "The army under McClellan began to assume a warlike aspect," she observed, "perfect order and military discipline." Little Mac believed, if not in them, in how he could change them; here was an entire army waiting to be remade in his likeness and to his liking. He anointed them with a new name, the "Army of the Potomac," and examined each camp from atop his glossy black stallion, hollering about "perfect pandemonium" and restoring order from chaos. Emma was especially pleased that the general's morals matched her own. He declared that unless the enemy attacked, all work would be suspended on the Sabbath, and on that day his troops were expected to conduct themselves with "the utmost decorum and quiety." Anytime he discovered soldiers with liquor, he ordered them to spill it on the ground.

For the first time since basic training Emma looked forward to the drills, totaling eight hours per day: squad drills before breakfast, regimental drills before lunch, and constant artillery practice, the big guns booming from every entrenchment, the sound oddly soothing to her ears. She reveled in how splendid they looked during McClellan's "Grand Reviews," ostentatious parades down Pennsylvania Avenue, with each regiment vying to outdo the others—beating drums louder, waving flags higher, marching with crisper steps. Sometimes she spotted Lincoln, his wife, and their two youngest sons, Willie and Tad, watching the festivities from a carriage. Nathaniel Hawthorne, on assignment for the *Atlantic Monthly*, witnessed the soldiers' admiration for their new commander. "They received him with loud shouts," he wrote, "by the eager uproar of which—now near, now in the center, now on the outskirts of the division—we could trace his progress through the ranks . . . they believed in him, and *so did I*."

Rose Greenhow, too, stood amid the throngs on Pennsylvania Avenue, fanning herself in the heat, jotting down impressions of the new Union commander. "Under the auspices of the 'Young General' the military are put in motion," she wrote of McClellan. "Hither and thither they are marched and counter-marched, mysterious movement being his forte. He, however, set himself energetically to the task of reorganizing and disciplining the demoralized rabble he was called upon to command."

She didn't yet know that Union officials were also watching her.

As much as Emma enjoyed the drills and parades, she was growing impatient. Every morning she awakened to the sight of the rebel flag flying high over Munson's Hill, just ten miles away across the Potomac, a reminder that they had not yet avenged the loss at Bull Run. The ignorant recruits, herself included, were soldiers now, and she feared battle less than she feared discovery. Every night, eating rations of salt pork and hardtack (often infested by maggots and weevils), they all debated when they might advance on the enemy.

On paper, at least, McClellan's strategy seemed foolproof. Union forces would attack the rebels on three fronts: one army would march into Virginia and take Richmond; another would drive into Tennessee and the border state of Kentucky, and then push into Mississippi, Alabama, and Georgia; and the navy would reinforce its positions along the Southern ports and starve the Confederacy of supplies.

Republicans in Congress grew equally anxious for McClellan to implement his plan. Missouri, another key border state, was in upheaval after Major General John Charles Frémont—without consulting Washington, and overstepping the bounds of his own authority—declared martial law and issued a proclamation freeing all slaves owned by Missouri secessionists. Lincoln was enraged. Such a mandate would also threaten the precarious situation in Kentucky and generally embolden the rebels. Southern newspapers editorialized that the true reason for the war had been unmasked; the president was less interested in preserving the Union than in abolishing the institution of slavery. The incident could also have international repercussions. If the North appeared disorganized and vulnerable, Europe might be persuaded to recognize the Confederacy.

Lincoln strove to remain patient with his new commander, but his cabinet and Congress pressured McClellan to act at once. Instead, the general griped to his wife that "the President is nothing more than a well-meaning baboon" and continued to do nothing.

The members of the 2nd Michigan passed the time in ways both dutiful and debauched. On Sundays the more pious recruits, Emma included, attended church services, while others took a day pass in search of the city's hedonistic offerings. "If the men pursue the enemy as vigorously as they do the whores," one private said, "they will make very efficient soldiers." Despite her best efforts and fervent prayers Emma struggled to separate the sinner from the sin, knowing that if her own sex was discovered she could be arrested for prostitution—a default reaction to finding a woman within the ranks. (Emma had already heard of one instance. A woman named Nellie Williams, serving as a private with the 2nd Iowa Infantry, was arrested in Louisville; afterward, her commanding officer dismissed her as "one of the inmates of a disreputable house on Seventh Street.")

After all, Emma slept next to these men, was privy to their every laugh and tear and exhalation of breath. They subsisted on the same stale bread and bacteria-ridden water, smelled each other's skin beneath uniforms that were never washed, awakened each morning knowing these faces might be the last they'd ever see. She'd grown up thinking of men as "the implacable enemy" of her sex, believing that the only way to escape their treachery was to become one of them. But yet she had come, bit by reluctant bit, to respect her comrades and the curious, bedraggled family they'd made.

How disappointing, then, to witness their excitement at "going down the line," as the soldiers called a trip to the whorehouse, and to hear the details about each establishment: Fort Sumter, the Ironclad, Headquarters, U.S.A., the Devil's Own, the Wolf's Den (managed by Mrs. Wolf), the Haystack (managed by Mrs. Hay), the Blue Goose, Madam Russel's Bake Oven, Madam Wilton's Private Residence for Ladies, and an establishment kept by the Light family—a mother and father who procured clients for their three prostitute daughters. An estimated fifteen thousand white, black, and mulatto streetwalkers also strolled Pennsylvania Avenue, beckoning troops on the picket lines or loitering outside Willard's Hotel.

The men didn't even have to leave camp to indulge their baser urges: "camp followers," women who attached themselves to regiments without having any obvious affiliation to a soldier, were always available. The majority of these women cooked and cleaned and tended to the wounded, but some made it clear they were willing to "handle a gun," military slang for a man's genitals. A portion of

this last group desired to wed and worked to cultivate respectable fa-
cades, acting, as one journalist wrote, "truly wife-like in their tented
seclusion," while others held no such high-minded goal. "Almost all
the women are given to whoredom," a private named Orville C.
Bumpass complained to his wife, "& are the ugliest, sallowfaced,
shaggy headed, bare footed dirty wenches you ever saw."

Emma's own favorite pastime was to watch the pickets on duty,
a position that was by turns monotonous and deadly. For hours
on end, until their eyes refused to stay open, Union and Confederate men stood facing each other, sometimes from mere yards away,
guarding their lines. They were the first to feel or sense any major
enemy movement, the first to be wounded or captured or killed.
They had to remember every new sign and countersign, to be wary
of enemy spies who discerned the magic words. Hours crawled by
without even a whisper. They had to remain vigilant through heat
and cold, rain and snow, in the depths of the forest or in the open
plains or in water knee-deep, watching and listening for the slightest twitch of movement or burst of sound, every nerve and faculty
of their minds strained beyond all sensibility, knowing that at any
moment a foe might spring from the darkness, or that a sharpshooter's bullet could find them.

For Emma, the draw was that picket duty was personal in a way
battle was not; this was war stripped down, thin-skinned and vulnerable. "Oh," she wrote, "how my heart has ached for those men."
Generals on both sides forbade pickets to interact with one another,
an order that proved nearly impossible to obey. Taunts were frequent and merciless. Union soldiers raised flasks, sipping whiskey
and yelling "Here's to Jeff Davis!" and the rebels responded with
quips about their victory at Bull Run. A group of them once, with
seeming sincerity, asked a Union general if he "was in the habit
of eating rebels fried or stewed, for breakfast." They boiled to be
at each other, an urge that occasionally prevailed. One side would

shoot and the other would respond, bullets plunging back and forth, dozens of casualties on both sides. The rebels didn't always fight fair. Their favorite trick was the "cow-bell dodge": ring a cowbell, and a Federal soldier hoping to fetch milk for his coffee would instead find himself surrounded by a half dozen armed enemies. There were also unexpected moments of grace, blue and gray both laying down their arms, trading tobacco and newspapers and buttons from their coats, boasting about their women back home, letting politics and prejudices fall away.

One day Emma accompanied the 2nd Michigan's chaplain, M. Bindell; his wife, "Mrs. B."; and the regiment's doctor to watch the pickets on Munson's Hill. They rode in silence on a narrow dirt road, veering abruptly left and climbing seventy-five feet, the terrain free of vegetation save patches of well-worn grass and clusters of stunted oaks. At the summit a spectacular view unfolded for miles: the Potomac River catching and tossing the light; an uneven, rambling line of breastworks, made of earth and wood; the city of Alexandria, with its dull red necklace of row houses and, close to Emma's heart, the Theological Seminary, a Union flag soaring from its spire. As they neared the Federal rifle pits a group of rebel pickets fired at them, sending the signal that there would be no genial swapping of stories or goods; on this day, for whatever reason, they were enemies and nothing more. A spray of bullets whizzed past, grazing the air above their heads.

With Emma leading the way, they turned and rode back toward a thicket of trees, dismounted, hitched their horses, and proceeded the rest of the way on foot. The rebel pickets found them and fired again. They dropped on hands and knees and began crawling. When they finally reached the rifle pits they spotted a rail fence riddled with minié balls. They leaned against it and watched the performance, passing around a pair of opera glasses. A ball launched in their direction and struck the railing an inch from where Emma rested her head. Another followed, tearing through a Union soldier's arm, mangling muscles and tendons and nerves, dragging with it bits of uniform and dirt. He screamed, the loudest scream Emma had ever heard, and she had to stop herself from running to him.

After the firing ceased she had her chance to rescue him and take him back to camp. He was quiet now, in shock, slick with his own blood. Emma knew he would lose either that arm or his life.

She hoisted him to his feet, wrapping his good arm around her neck, half wearing him like a blanket. He would be okay, she told him again and again, the words louder with each repetition, and in that moment she felt as if nothing or no one could hurt her.

The Birds of the Air

WASHINGTON, DC

Rose sensed she was being watched. Strange men lingered on Sixteenth Street outside her home, derby hats slouched low, taking note of everyone who came and went. They trailed her across the grounds of the unfinished Washington Monument, wading through the herds of grazing cattle and passing by the slaughterhouse situated at its base, where offal was rotting three feet deep; the stench was so bad that Lincoln had taken to spending long stretches of time at a retreat several miles north of the city. They stood paces away as she watched McClellan drill his soldiers on Capitol Hill, jotting notes in her brown leather diary and scribbling in the margins of Yankee newspapers. Sometimes she reversed course and became the pursuer, chasing shadows around corners, finding the whole situation amusing. She suspected she was the victim of private enterprise—an overzealous Northerner who knew of her work, perhaps, or a misguided secessionist offering protection. She recalled a letter from one reputed lover, US senator Henry Wilson, in which he warned that his every movement and act was being observed with "hawkeyed vigilance," and she wondered exactly how much the strange men had seen.

Undeterred, she summoned her accomplices for a meeting, recounting the teenage courier Bettie Duvall's mission and toasting the victory at Manassas. "The Southern women of Washington are the cause of the defeat of the grand army," she boasted. "They have told Beauregard when to strike!" During a feast of terrapin, oysters, and wild turkey, she regaled them with stories about "Ape" Lincoln and his vulgar wife. At his first White House reception, a servant asked Lincoln which wine he'd prefer. He regarded the servant with the most "touching simplicity" and said, "I don't know: which would you?" Mary Lincoln, at the same ceremony, wore artificial flowers in her hair and was mistaken for a servant.

Rose once ran into Mrs. Lincoln while shopping at the market, spotting "a short, broad, flat figure, with a broad flat face, with sallow mottled complexion, light gray eyes with scant eyelashes, and exceedingly thin pinched lips." The First Lady was bargaining for black cotton lace, much to the disgust of the shopkeeper, and her dress was inappropriate for a casual outing: the gown of rich silk emblazoned with gaudy flowers; the white hat adorned with feathers, flowers, and tinsel balls; the white gloves and white parasol lined with pink. "I don't think," Rose concluded, "that Mrs. Davis would have selected it for that hour and occasion."

After dinner the women parted red gauze curtains and settled in Rose's parlor, where they sipped glasses of port and brandished needles and balls of yarn. Little Rose was permitted to say hello to her mother's guests, curtsying and trilling a few stanzas of her favorite song, "Old Abe's Lament":

Jeff Davis is coming, Oh! dear, Oh! dear,
Jeff Davis is coming, Oh! dear;
I dare not stir out for I feel very queer,
Jeff Davis is coming, Oh! dear.
I fain would go home without shedding a tear
About Davis in taking the president's chair;
But I dare not attempt it, Oh! dear, Oh! dear,
I'm afraid he will hang me, Oh! dear.

Mary Chesnut, the wife of South Carolina senator James Chesnut Jr. and a close friend of the Davis family, witnessed one of Rose's gatherings: "It gives a quaint look, the twinkling of needles, and the everlasting sock dangling. A Jury of Matrons, so to speak, sat at Mrs. Greenhow's. They say Mrs. Greenhow furnished Beauregard with the latest information of the Federal movements, and so made the Manassas victory a possibility. She sent us the enemy's plans. Everything she said proved true, numbers, route, and all." Chesnut also added a pointed assessment of Rose's methods: "She has all her life been for *sale*."

Sitting at her rosewood piano, Rose surveyed her group. The network was running smoothly despite the unsettling presence of the strange men. The Confederate government had come up with $40,000, likely in donations from private citizens, to assist rebel inmates in the Old Capitol Prison, a fund for buying clothing and writing supplies and bribing guards. Her team of operatives constantly worked the streets of Washington, checking on reports of new fortifications and trailing persons of interest.

As the women tittered and darned socks for the soldiers, Rose outlined their next steps. She'd scored an intelligence coup, meeting secretly with the Union officer charged with alerting Washington in the event of a Confederate invasion. The signal, he confided, was three gunshots from the provost marshal's office followed by the tolling of church bells. The city's secessionists needed to be equally prepared. They must cut all telegraph wires connecting military posts with the War Department and spike the guns of Fort Corcoran and Fort Ellsworth; aided by information from her scouts, she had sent detailed drawings of both to General Beauregard. Ideally, they would also take General McClellan and other top officials prisoner, "thereby creating still greater confusion in the first moments of panic."

After the meeting she sent an update to Beauregard, warning that McClellan "expects to surprise you, but now he is preparing against one," and dashed off a note to Senator Wilson, scolding him for missing a scheduled rendezvous. "Tonight," he responded, "unless *Providence* has put its foot against me I will be with you, & at as early an hour as I can. That I love you God to whom I appeal, knows."

She filed this letter with others from "H," tying the bundle with a bright yellow ribbon and leaving it where it could be easily found.

Rose's surveillance of McClellan had so far failed to uncover one of his most significant acts: the hiring of detective Allan Pinkerton, who had achieved national renown for investigating a series of sensational train robberies. Pinkerton had already served the new administration well, managing security as Lincoln traveled to his inauguration, foiling an alleged assassination conspiracy along the way. He then contacted the president directly, offering to obtain "information on the movements of the traitors, or safely conveying your letters or dispatches." Pinkerton, whose capacity for self-aggrandizement rivaled McClellan's own, called himself "Chief of the United States Secret Service" even though his unit was technically nameless.

Working as a civilian under contract to the army, Pinkerton assembled a staff of agents, male and female, who would observe Confederate military preparations and rout out rebel spies. "In operating with my detective force," Pinkerton wrote to McClellan, "I shall endeavor to test all suspected persons in various ways. . . . Some shall have the *entrée* to the gilded salon of the suspected aristocratic traitors, and be their honored guests, while others will act in the capacity of valets, or domestics of various kinds." Pinkerton himself planned to go undercover, using the nom de guerre "Major E. J. Allen," since his true name was so famous it had, he said, "grown to be a sort of synonym for 'detective.'" One of his first assignments was to conduct surveillance on Washington matron and suspected spy Rose Greenhow.

Detective Allan Pinkerton with Lincoln.

One evening Pinkerton dispatched a detective to the home of Horatio Nelson Taft, an examiner for the US Patent Office. Taft's teenage daughter Julia found a "bland gentleman with distinguished black whiskers" standing in their parlor. He asked if she knew Rose Greenhow, and Julia replied that she did. In fact, Mrs. Greenhow often visited with her daughter, Little Rose, and the family liked them both. Julia had two younger brothers, Bud and Holly, who were playmates of both Little Rose and President Lincoln's sons, Willie and Tad.

The man nodded, confirming something in his mind. He consulted a list of questions: Did Mrs. Greenhow seem glad to meet the officers who came by? Did she ask Julia, Bud, and Holly for details about their excursions to the White House? Yes and yes, Julia said. Horatio Taft joined the discussion, recalling that Rose demanded details about each new Union regiment that arrived in the city.

"What's wrong with Mrs. Greenhow?" Julia asked.

She watched the detective pull her father aside. The men spoke in whispers. "Very well," her father said, and cautioned Julia and her brothers never to mention the detective's visit, to Mrs. Greenhow or anyone else.

The strange men had entrenched themselves in every aspect of her life. They watched her drift from cart to cart at the Center Market, her maid in tow, sifting through fruits and vegetables, shad and beef, swatting away buzzing clouds of flies. They waited outside Mrs. Arth's Millinery Shop while she tried on hats, staring back at them from beneath the brims. They stood behind her in line at the sweet-cake and ginger pop stands on Capitol Hill. They sipped whiskey in the lobby at Willard's Hotel, peering at her through the crowd of office seekers, wire-pullers, investors, artists, poets, generals, statesmen, orators, clerks, and diplomats, noting every ear she whispered into, every man she beckoned with her gloved hand. They followed her into Senate hearings and sat nearby in the gallery, eavesdropping on her conversation. During one such outing Rose turned to a friend and made a disparaging remark about the Federal army, prompting a young Union lieutenant colonel to confront her.

"That is treason," he warned. "We will show you that it must be put a stop to. We have a government to maintain."

Rose leaned forward. She looked him up and down, taking her time, tracking from the brim of his kepi to his folded arms to the scuffed tips of his boots. She spoke her words slowly, enunciating each syllable as if speaking to a child: "My remarks were addressed to my companions, and not to you. If I did not discover by your language that you must be ignorant of all the laws of good breeding, I should take the number of your company and report you to your commanding officer to be punished for your impertinence."

The Senate doorkeeper, recognizing Rose as a gallery regular, rushed to her side. "Madam," he said, "if he insults you, I will put him out."

Rose smiled, holding her shut fan over her heart. "Oh, never mind," she whispered. "He is too ignorant to know what he has done."

The next time she saw the Union officer, she swiped a finger across her throat and mouthed the word "beware."

She sent another message to Beauregard, this time on her mourning stationery, heavy white paper with a black border—a visual that mirrored her warning about the strengthening of fortifications around the Capitol. "McClellan is vigilant," she concluded. "There is little or no boasting this time."

She continued to take her daily walks. Every eye in the city seemed fixed upon her, every gesture weighted with meaning. There was something strange about the tilt of the washerwoman's basket and the way the young man stood on the corner, twirling his cane. She no longer found the surveillance humorous; even an elderly gentleman sitting on a park bench roused her nerves. "To me," she said, "his manner of polishing his glasses, or the flourish of the handkerchief with which he rubbed his nose, was a message."

She received a peculiar note—peculiar because it came from the city post office, an unlikely route for her Confederate connections to use. The note began: "Lt. Col. Jordan's compliments to Mrs. R. Greenhow. Well, but hard-worked . . ." The rest of the page had been torn away. To protect herself, she showed it to Senator Wilson during his next visit and hid it inside a vase on her mantel for safekeeping.

She prepared another report for Beauregard, noting that "McClellan is very active and very discreet," and sent it by way of a courier named Bettie Hasler, who never questioned Rose and had no idea of the impending danger.

One morning, an old friend and sometime informant, Anson Doolittle, appeared at her door. He was a relative of Wisconsin senator (and friend of Lincoln) James R. Doolittle and, according to Rose, "an occasional and *useful* visitor to my house." Doolittle handed her a letter addressed to a colonel in Richmond and asked if she would send it.

It was not an uncommon request. The Confederate States of America had established a Post Office Department back in February, but the Federal government initially maintained service throughout the Southern states. In May, the Confederate postmaster general assumed control but encountered immediate difficulty. Pickup and delivery were thwarted by Lincoln's blockade, the presence of Union troops, and the increasing scarcity of postage stamps. While Doolittle could easily mail a letter in Washington, DC, he had little guarantee of proper delivery to Richmond. His most reliable option was to entrust letters to those able to cross Union lines: officers on furlough and sick leave; slaves who could hide mail inside baskets of food or clothing; or well-connected individuals like Rose, whose network facilitated daily communication between Washington and the surrounding Southern states.

Doolittle sweated in the heat of her parlor, waiting for a response. For the first time during their long acquaintance Rose wondered who and what he was: another "friend" whose relationship with her took precedence over ties to the Union? Or maybe, just maybe, a double agent who found her company as useful as she found his?

She finally said, "McClellan's excessive vigilance has rendered communication almost impossible, but you might leave it and trust it to chance."

Doolittle thanked her and showed himself out. Over the next several days he called to see if she'd sent the letter, but Rose was "always very sorry that no opportunity had occurred."

In her next missive to Thomas Jordan she warned that General Hiram Walbridge, a former New York congressman and friend of Lincoln, was a spy, and indicated that she was under surveillance: "Do not talk with anyone about news from here as the birds of the air bring back. But I wish I could see you as I know much that a letter cannot give. Give me some instructions. You know that my soul is in the cause. . . . Tell Beauregard that in my imagination he takes the place of Cid," a reference to El Cid, the military hero in medieval Spain.

She signed off with "Always Yours, R.G.," an incendiary choice of words in an era when "Ever your friend" was a customary closing among the betrothed.

She was certain, now, that the strange men were Yankee detectives. "I was slow to credit," she confessed, "that even a fragment of a once glorious government could give to the world such a proof of craven fear and weakness as to turn the arms, which the blind confidence of a deluded people had placed in their hands, for the achievement of other ends, against the breasts of helpless defenseless women and children."

She began urging secessionist friends, couriers, and sources—including John C. Breckinridge, the former US vice president, and William Preston, former minister to Spain—to leave the city before the Yankees could apprehend them. She wasn't going anywhere, at least not yet; there was still much work to do and fewer people to do it. She now had to undertake missions herself, meeting her spies on selected street corners to deliver information, which they would then pass on to Beauregard. But she took new precautions, practicing how to hide information in places she hoped the Yankees wouldn't dare look. She spent hours at her Singer "Grasshopper" sewing machine, fastening a pearl-and-ivory tablet on a silver chain and other contraband for Confederate soldiers into the voluminous quilted underskirt of her gown. She stitched maps of fortifications into the lining and cuffs, and slipped the latest about General McClellan behind the laces of her corset, pulling the notes tight against her body. She sat before her mirror and shook loose her hair, hiding secrets inside the twists and folds. The words from the message she'd received just before Manassas played inside her mind: *Let them come.*

She was ready for them.

The Secret Room

RICHMOND

Elizabeth could hardly conceal the fact that a prominent Union prisoner was dying in her home, but she assured her neighbors and sister-in-law that she was motivated by Christian benevolence, as any proper Southern lady would be. Calvin Huson was only thirty-nine years old, a married father of five. His wife would be widowed at age thirty-six, with dismal chances of remarrying, and his children, ranging in age from ten to one, would never see him again—how could she refuse to extend charity in such a case? They would wish for the same should, God forbid, one of their sons, brothers, or husbands be captured by the Yankees. Still, Elizabeth feared they remained unconvinced, and heard furious whispers about her shameless "aid and comfort" to a "Black Republican enemy," a slur against white supporters of Lincoln.

She focused on her patient, consulting medical books and summoning the prison physician. They tried everything: iodide of potassium every six to eight hours; anal injections of cold water, helped along by swigs of laudanum; enemas consisting of oil of turpentine and the yolk of an egg. She talked him through prolonged periods of agitation, violent ravings, frightening delusions. His fever soared above 100 degrees and his tongue grew a strange brown fur. When his breathing turned hard and raspy she hurried in her carriage to the prison. In tears, she told Congressman Alfred Ely that his colleague was near death, and he should come visit at once. Twenty minutes later, as a guard prepared to escort the congressman to Elizabeth's mansion, a messenger approached and said that it was too late.

She urged Ely to be cautious and hold a small, quiet service at the Church Hill cemetery. Ely agreed, but the neighbors still witnessed pallbearers carrying a walnut-wood coffin past Elizabeth's front door, and the slow procession of the hearse and four carriages down Grace Street. They heard that Elizabeth placed a bouquet of roses upon the Yankee's grave. Soon afterward, during a walk, she sensed someone following her, keeping pace as she quickened her steps. He trailed her past Franklin and Main and Cary Streets to the sludgy bank of the James, and westward to the Tredegar Iron Works, where slaves and poor immigrants worked seven days a week making warships and cannonballs for the rebel army. She stopped to catch her breath and heard a gruff, low voice by her ear: "You dare to show sympathy for any of those prisoners. I would shoot them as I would blackbirds. And there is something on foot up against you *now*!"

She waited until his footsteps receded to turn around and head home, checking over her shoulder all the way.

The Van Lew mansion, Church Hill, Richmond.

Elizabeth hoped the neighbors were watching just as closely when Confederate captain George Gibbs, the new head of the tobacco prison complex, moved into her home with his wife and three children. As soon as she'd heard that the Gibbs family, who came to Richmond from their plantation in St. Augustine, Florida, needed temporary lodgings, she offered rooms at the mansion, reasoning that no one could accuse her of disloyal activity while she had a rebel officer eating at her table and sleeping under her roof. She planned an elaborate Southern welcome, asking her cook, Caroline, to prepare a feast: duck soup, French chicken pie, veal olives (slices of seasoned veal rolled up and cooked on skewers), sweetbreads with cauliflower, fried artichokes, scalloped tomatoes, pineapple pudding, whiskey, and champagne. The menu was especially extravagant, considering the blockade's steady effect on prices. Salt, which used to sell for 80 cents a bushel, was now $1.25, and bacon, normally 12 cents a pound, had risen to 30 cents. In some parts of the South people decided meat was too expensive to buy, and placed ads seeking hunting dogs so they could kill their own.

The servants helped Captain Gibbs and his family to their rooms and prepared the table, setting out the Van Lews' finest china and silver, the antique spoons all stamped with figures of lions and panthers. Elizabeth tried to keep an open mind about the captain's wife and children but found Gibbs himself abhorrent, "not a man of much intellect and untidy," and cringed at his declarations of devotion to the Confederacy. "I would give my right arm," he told her, jabbing his fork in the air for emphasis, "for one word of commendation from Jeff Davis."

Elizabeth smiled and nodded, hoping her silence didn't betray her disgust. Meanwhile Mary, her brother John's wife, offered the captain more explicit support. As she agreed with and toasted nearly everything he said, John caught Elizabeth's eye across the table. She worried about her younger brother, whose marriage had never been an ideal match; Mary was a blue-blooded Virginian, a first cousin of Thomas Jefferson, while the Van Lews were a nouveau riche family with Yankee blood. It soon became clear, though, that their problems were rooted in cultural, not class, differences.

John Van Lew.

Shortly after the couple married, in 1854, they moved north to Philadelphia. Mary felt uncomfortable in the big city, especially without her longtime personal slave. Margaret was loyal and devoted and, above all, knew her place, and Mary didn't want to risk introducing her to the radical abolitionist ideas of the North. Three years later they moved back to Richmond, one daughter in tow and another on the way. Once they settled in the Church Hill mansion, every old, latent tension crept to the surface. When John insisted, per Van Lew family tradition, that Margaret be put on the payroll so she could earn her freedom, the tension developed into a power struggle that poisoned everything and everyone around them.

Mary began calling all the Van Lew servants "niggers," whether or not John was there to hear it. Whenever he returned from a business trip to Petersburg or Fredericksburg, the servants reported Miss Mary's crude language and poor manners. John would confront her with Elizabeth by his side, both of them reminding her that the Van

Lew servants were not slaves and would not be mistreated in their own home. "The Negroes have black faces, but white hearts," Elizabeth told her, hoping she might relent, but Mary's behavior only worsened.

She was especially cruel to Mary Jane Richards, a twenty-one-year-old whom Elizabeth treated like a member of the family, even taking the unusual step of having her baptized in St. John's Episcopal Church, the city's preeminent (and primarily white) place of worship, which had been built by Mary's great-great-great-grandfather. A few months prior, in April, as Virginia debated secession and Fort Sumter fell, Mary Jane had married another Van Lew servant, Wilson Bowser, in the same church, an event that further insulted Mary's sense of tradition and gave Elizabeth a keen and secret pleasure.

Mary and John argued constantly, long, rancorous battles that escalated after the onset of the war. He did not want their daughters—Annie, now seven, and Eliza, four—witnessing her mistreatment of the servants or internalizing her secessionist opinions. She accused him of deferring to Elizabeth, always choosing his sister over his wife. They began sleeping in separate chambers, a decision that, while not uncommon for the time, broke a long-standing Van Lew family custom of couples sharing a marital bed.

Now, at the dinner table, Mary insulted the servants with impunity, confident that the Van Lews wouldn't chastise her about such matters in front of a Confederate officer. Elizabeth knew that her sister-in-law was a potentially dangerous foe, even more so than the neighbors. Mary might find her diary, filled with disparaging entries about rebel officials and the Confederacy in general. She could report every so-called traitorous conversation held within her home. She could even say that Elizabeth's disloyal thoughts had evolved into disloyal activity, which—at this point, at least—would be a blatant, and possibly deadly, lie.

Captain Gibbs began regaling the table with tales from the tobacco prisons, a topic that seized Elizabeth's attention. Recently, he said, he had to contend with the escape of eleven prisoners, a catastrophe he blamed on drunken guards. It seemed they all hit the saloons before reporting for duty, and drank whiskey from their canteens on the job. He petitioned General Winder to close every grogshop by the prisons and supply him with balls and chains, since handcuffs alone failed to deter recaptured inmates from attempting another escape.

Elizabeth feigned sympathy, but an idea had entered her mind and lodged there.

As long as Captain Gibbs boarded at the mansion, Elizabeth enjoyed unfettered access to the prisoners. Sometimes she sent a servant, as in the case of a private from the 14th New York who was wounded at Bull Run, stabbed with a bayonet more than fourteen times even after he lay motionless on the ground. Elizabeth had heard about the prisoners' pitiful rations—scanty servings of moldy bread and meat, usually horse or mule meat—and she packed baskets of fresh vegetables from her family farm, located a quarter mile below the city near the James River. "I should have perished for want," the private said, "but a lady named Van Lew sent her slave every other day with food."

Occasionally the guards refused her entry, even with permission from Captain Gibbs and General Winder. On one such occasion she appealed to Colonel A. G. Bledsoe, the Confederacy's assistant secretary of war, sending her request and some custard that she hoped he might deliver to the prisoners. Instead Bledsoe ate the custard himself and wrote a thank-you note to Elizabeth: "The custard was very nice, & many thanks to you. I borrowed some cups from an eating house nearby, & bought some crackers, so that it was eaten in fine style."

On the back of the note Elizabeth scrawled, "God help us."

She kept track of every escape and attempted escape: two volunteers, one from Michigan and one from Rhode Island, somehow slipped past their guards and made it to Washington; several soldiers from New York conspired to overpower their guards and were separated when the plot was discovered; and, most dramatically, a pair of recruits from Kentucky and Ohio crossed the Potomac by boat into Union territory, and shared their adventure with the press.

They'd noticed that the prison surgeons, distinguished by a bit of red ribbon pinned to their sleeves, were permitted to pass in and out as they pleased, and that the sentinels changed shifts every two hours. The first prisoner tore a bit of red flannel from one of his shirts, affixed it to his coat, and passed without incident. The second prisoner, red ribbon in place, escaped during the next shift change and found his partner at a designated corner a few blocks away. By a previous and unspecified arrangement they acquired a pocket compass and map of Virginia. They set out, following the Union turnpike, narrowly avoiding a tollgate guarded by rebel soldiers, hiking through miles and miles of fields, sleeping during the day and moving at night, eating corn and potatoes they found along the way. They crossed the Chickahominy on a milldam, the Pamunkey on a homemade raft, and the Rappahannock on a small boat; waded through a swamp; and at the mouth of the Potomac told a party of Negroes they were Confederate officers with dispatches and needed a ride across the river.

One evening, after returning from a trip to the prison, Elizabeth climbed the stairs to the top floor of her mansion and walked the length of the eighty-foot hall. There it was: the secret room, five feet high and extending the entire length of the house, its roof tilting to the south, its entrance only two feet square. She estimated it could hold fifty to seventy-five men at a time. Its door was made to fit flush against the wall and fastened on the inside. If she had a servant apply a coat of whitewash and place an old dresser against it, no one would ever detect what lay beyond.

She wanted to nudge things along beneath the surface of this mad and maddening city, where seemingly respectable women collected the bones of Union corpses and taught their daughters to "dread and fear the Yankee above all tame or wild animals." She had no vote, no public forum, no way to make McClellan advance and attack, and she longed to grasp a bit of control. She would be the antithesis of Rose Greenhow, discreet and cautious, as courteous to the enemy as the rebel spy was disdainful. But the risks were considerable. Unlike Rose, she wasn't a widowed mother; the Confederate government would sooner hang her, an old spinster, than someone raising a young girl. She needed to expand her network, to find every last Unionist in the city. A widespread underground movement was ripe to happen, disparate forces ready to rise and converge.

Elizabeth knew what she had to do, even if her sister-in-law was watching.

Mary became even more intolerable after Captain Gibbs and his family moved out of the mansion. The servants could do nothing right: laundered clothes were still dirty and wrinkled, furniture was still dusty, food was not fetched quickly enough, bathwater was too cold. Mary Jane's husband could no longer tolerate the abuse, and Elizabeth offered a solution: he could leave Church Hill and work on the family farm, where they would certainly need extra hands during the coming fall harvest.

Which left the question of what to do about Mary Jane. The answer came in the form of an ad in the *Richmond Dispatch*, mentioning that the Confederate First Lady, Varina Davis, was seeking qualified servants. She reportedly felt the strain of keeping up appearances, and found it difficult to "get suitable servants, everything double the price." Elizabeth would pay her a social call and offer Mary Jane for the position. It was the customary thing to do among the city's better placed, an overture that conveyed status as much as it did decorum, and would give Elizabeth the chance to ingratiate herself with another Confederate official. The situation could prove beneficial in other ways, as well. No one, not even sister-in-law Mary, knew that Mary Jane was highly educated and gifted with an eidetic memory, capable of memorizing images in a glance and recalling entire conversations word for word.

The entrance to the secret room.

STAKEOUT

WASHINGTON, DC

On the evening of August 21, Allan Pinkerton set out for Lafayette Square with three of his best detectives. No one stirred at the house of Secretary of State Seward on the east side of the square, and the doors were locked at St. John's Episcopal Church, where every president since Madison had worshipped. At the nearby residence of General George McClellan, the private upstairs rooms were dark but the lower parlors, which housed the telegraph office, hummed with activity, dozens of staffers writing and reading the papers and smoking Cuban cigars. A storm skulked between the clouds and the wind picked off leaves one by one, with sniper precision. It began to drizzle as the group arrived at the corner of Sixteenth and K Streets, just north of the square, stopping at Rose Greenhow's home.

Pinkerton strolled its perimeter, noting the structure and style: three-story brick building with basement, parlors elevated high off the ground, a flight of stairs sweeping up to the main entrance. Venetian blinds adorned two large rectangular windows to the right of the front door. He concluded that no one was home but decided to wait. The rain began falling in sleek silver sheets, soaking through the detectives' clothes. Pedestrians hurried to their destinations, guided by the feeble glow of gas lamps. No one paid attention to the four men standing idly in the downpour.

Pinkerton had compiled a dossier on Rose and her rise to power in Washington, determining that she was a woman of "pronounced rebel proclivities" who considered the Union flag a "symbol of murder, plunder, oppression and shame." He cataloged her reported political paramours, focusing especially on Senator Wilson, and noted that she was "using her talents in procuring information" and communicating it to the rebel government at Richmond. "Mrs. Greenhow," he concluded, "had occupied a prominent position in the social circles of the capital, and was personally acquainted with all of the leading men of the country, many of whom had partaken of her hospitality and had enjoyed a social intercourse that was both pleasurable and fascinating."

One of the ground-floor windows was suddenly etched in light. Someone was home. Pinkerton lifted himself on tiptoe, but the parlor windows were too high. He waited until the street was empty and summoned two of his detectives. Feeling ludicrous, he kicked off his boots. His stockinged feet sank in the mud. The men bent and cupped their palms and hoisted Pinkerton up, adjusting their positions until he had one foot on a shoulder of each man. Rain spilled over the brim of his hat. His red beard dampened into a point. Gently he flipped the slats of the blinds until the parlor slowly revealed itself. He noted the plush damask settee, the rosewood piano, oil paintings of famous statesmen dispersed along the wall, the gilded figurines standing guard by the staircase—but no Rose Greenhow. He was about to complain to his detectives when one of them whispered, "Sh!"

A figure strode up Sixteenth Street and turned toward the house. Pinkerton dismounted from his perch and hid under the stoop. His men flanked him, lying prone, chins half planted in the sodden ground. They heard ascending footsteps, the faint chime of a bell, the creaking and closing of a door. Pinkerton resumed his position atop the detectives' shoulders and peered into the parlor.

The visitor, he saw, was a "tall, handsome man of a commanding figure and about forty years of age." Upon closer inspection

Pinkerton recognized him as Captain John Elwood, an officer in the Union army who commanded one of the stations of the provost marshal, charged with preserving order and, ironically, identifying rebel spies. The captain removed his cloak, revealing a blue uniform. He appeared nervous, fidgeting in his chair and shifting his eyes, and relaxed only when Rose descended the stairs, wearing a mourning dress, her hair coiled into a knot. He stood and bowed deeply, and his face "lighted up with pleasure as he gazed upon her."

Pinkerton watched their lips move, straining to discern their words; he heard "disposition" and "troops." After a few moments Captain Elwood reached into his pocket and unfolded a large square of paper, holding it up to the light. Pinkerton realized it was a map—a map of the fortifications in and around Washington, no less. The captain pointed to certain spots, his finger pausing over them. Rose leaned in close, her shoulder touching his, nodding as he spoke.

The detective was enraged: "My blood boiled with indignation as I witnessed this scene, and I longed to rush into the room and strangle the miscreant where he sat, but I dared not utter a word, and was compelled to stand by, with the rain pouring down upon me, and silently witness this traitorous proceeding." Voices and laughter drifted up from the street and Pinkerton quickly dismounted, hiding again with his men under the stoop. Once the pedestrians passed, he climbed back up but saw that the "delectable couple" had disappeared.

For more than an hour he stood as still as he could, knees locked, his detectives' hands clasped around his calves, waiting for Rose and Captain Elwood to return to the parlor. They did, around half past midnight, slipping through red curtains, walking arm in arm. When the captain reached for his cloak, Pinkerton made one last retreat under the stoop. He heard the couple say good night and then "something that sounded very much like a kiss."

Elwood descended the steps, and Rose shut the door behind him. Pinkerton waited for a count of ten and then set off in pursuit,

still bootless, swiping rain from his eyes. At the corner of Pennsylvania Avenue and Fifteenth Street the captain ducked into a non-
descript building. Pinkerton followed. He heard a shuffling and felt
several points of sharp, fixed pressure on his chest. It took a moment
for his eyes to adjust and see the tips of four bayonets, wielded by as
many Union sentries.

One of them shouted, "Halt, or I fire!"

Pinkerton rummaged for an explanation. He was unfamiliar
with the area, he said, and had lost his way. The sentries took him
to Elwood's office, where the captain glared at him for several silent
moments. Pinkerton knew he looked a fright, with his drenched
clothing clinging to his skin and his socks slathered in mud, and
he struggled to suppress a laugh. "One might more readily imagine
that I had been fished out of the Potomac," he thought, "than that I
was the chief of the secret service of the government."

Keeping his eyes fixed on Pinkerton, the captain pulled two revolvers from a drawer and set them on his desk. He stroked the
barrels, one with each hand.

"What is your name?" Elwood asked.

Pinkerton used his alias: "E. J. Allen."

"What is your business?"

"I have nothing further to say," Pinkerton replied, "and I decline to answer any further questions."

"Ah! So you are not going to speak. Very well, sir, we will see
what time will bring forth." Elwood called for a sentry. "Take this
man to the guard house," he ordered, pointing at Pinkerton, "but
allow no one whatever to converse with him."

Pinkerton bowed in the same courtly manner in which the captain had bowed to Rose. He was tossed into a small pen with a few
dozen men, most of them drunk, singing slurred refrains or passed
out cold on the floor. But the guard was surprisingly friendly, so
Pinkerton took a risk and asked a favor: when he went off duty,
would he mind delivering a note?

The guard agreed, and Pinkerton found a pencil and a damp

scrap of paper in his pocket. He scrawled a few lines to assistant secretary of war Thomas A. Scott, informing him of his imprisonment and requesting that he secure his release as soon as possible—but in a manner that wouldn't arouse the suspicions of Captain Elwood. The guard left at 6:00 a.m. and returned an hour later.

Pinkerton waved him over. "How is the weather outside?" he asked.

"All right, sir!" came the reply, confirmation that the message had reached its destination. A few moments later, the guard's sergeant escorted Pinkerton back to Captain Elwood's office. He had put his guns away.

"The secretary of war has been informed of your arrest," he told Pinkerton, "and you will be conducted to him at once, and then we shall see whether you will remain silent any longer."

Pinkerton nodded and tried to look solemn, privately scoffing at the captain's imperious manner. Still shoeless, he followed the captain to Scott's residence. The assistant secretary of war mocked Pinkerton's appearance as soon as the men were alone, and the detective "joined in his merriment, for a more realistic picture of a 'drowned rat' I never beheld."

Pinkerton detailed the meeting between Rose and Captain Elwood. Scott began pacing around the room, stopping long enough to tap a bell on his desk, summoning the traitorous officer.

"Mrs. Greenhow must be attended to," he whispered to Pinkerton. "She is becoming a dangerous character. You will therefore maintain your watch on her, and should she be detected in attempting to convey any information outside of the lines, she must be arrested at once."

HARD TO NAME

WASHINGTON, DC

When not on the trail of Rose Greenhow, Allan Pinkerton assessed the Confederate forces around the capital, emerging with conclusions that only underscored McClellan's own caution. He told the general that 150,000 rebel forces surrounded the city, armed with a great arsenal of cannon. In reality the Confederate troops numbered around 45,000, and the cannon pointed at the capital were merely mammoth logs painted black to look like artillery, a ruse the Northern public immediately mocked as "Quaker guns." A reporter for the *New York Tribune* likened the situation to the Chinese trick of scaring enemies "by the sound of gongs and the wearing of devils' masks." The revelation embarrassed McClellan but failed to bolster his confidence, and he remained convinced that the rebels intended to invade Washington. Rose took note and sent a message to General Beauregard, stressing that "the panic is great and the attack is hourly expected."

In the absence of marching orders Emma continued her shifts at the regiment hospital, just south of Alexandria. Her patient roster now included a growing number of officers afflicted with venereal disease: syphilis, gonorrhea, and lymphogranuloma venerum, the last of which was manifested in plum-sized swellings in the lymph glands of the groin. Treatment differed from doctor to doctor, and depending on the severity of the disease, Emma assisted in urethral injections of nitrate of silver and sugar of lead; administered pills containing balsam of co-paiba, powdered cubebs, and magnesia; and gave mercury vapor baths followed by the continuous application of a "black wash," a mixture of calomel and lime water.

It was grueling but satisfying work. She felt that Frank Thompson was safe there, one small part of a bloody and boundless machine; no one had time for talk that didn't pertain to the task at hand. So it was by chance that she fell into conversation one evening with a fellow private from the 2nd Michigan.

Emma had seen him before and said hello in passing, forbidding her gaze to linger on his dark eyes and razor cheekbones, his arms like sculpture beneath his sleeves. He had come now to visit a friend who was a patient, and she let her eyes cut to him as she wandered from cot to cot, re-dressing wounds and preparing to give sponge baths, working alongside female volunteers. Their supervisor gave strict instructions on bathing etiquette: "Wash as fast as you can. Tell them to take off socks, coats and shirts, scrub them well, and put on clean shirts." When the nurses "finished them off"—code for washing the most intimate parts of the body—it could be awkward for both patient and nurse, a necessary but flagrant violation of Victorian conventions.

One volunteer confessed that if she'd been asked to "shave them all, or dance a hornpipe on the stove funnel, I should have been less staggered; but to scrub some dozen lords of creation at a moment's notice was really—really—." But she took a deep breath, "drowned my scruples in a washbowl," and "made a dab at the first dirty specimen I saw." The new superintendent of the Department of Female Nurses, Dorothea Dix, would begin recruiting only women over thirty who were "plain-looking" and modestly dressed—no bows, curls, jewelry, or hoop skirts—to reduce the chances of impropriety. Emma noticed her own patients' comfort in her presence, their calm submission to her touch, and took it as further proof that she was doing right both by her adopted country and by God. Her secret protected her patients; her deception was a kindness, not an abomination or a sin.

When the man—no mere boy, he—finished visiting with his friend, he approached Emma and introduced himself as Jerome Robbins. He told her that he was a member of Company I, and that they had been mustered into service on the same day. He was twenty, a year older than Emma, and seemed impervious to the chaos around him, a sluice of calm channeling through the rows of weary men. He had been in college when the war broke out and considered himself to be more of a scholar than a soldier. They both admired McClellan and abhorred the sinful temptations of camp life; he exalted "blissful celibacy" and ranted against trips down the line. "To me," Jerome wrote, "nothing in the whole of human actions is so despicable, degrading, and devoid of all respectability as a man wearing the garb of humanity who visits the city brothels." He kept a journal chronicling his ongoing efforts to perfect his relationship with God.

Jerome Robbins, 1861.

Emma, sponge in hand, blood on her coat, listened to him, rapt, but each revelation, each connection, reeled her back years and miles away, settling on a singular moment that had cleaved her life in two. In a corner of her mind it was 1850 and she was nine years old, living on the farm in New Brunswick, cleaning up after supper. A knock at the door interrupted her work. An old peddler stood on the porch, listing under the weight of his cracked leather bag. Emma's mother took pity on him and offered him a bed for the night.

Before leaving the following morning, he pulled Emma aside and pressed a book into her hands. She drifted a fingertip across the title: *Fanny Campbell, the Female Pirate Captain: A Tale of the Revolution!* She had never read any book but the Bible. It meant something, she decided, that the peddler gave the novel to her instead of to one of her sisters. That afternoon she shirked her chores, leaving the potatoes unplanted and the cows unmilked, and lay facedown in the soil to read. "I felt," Emma said, "as if an angel had touched me with a live coal from off the altar."

The story imprinted itself on her mind. When Fanny's lover, a sea captain, was captured during a mutiny in the Caribbean, she hacked off her curls and hid her figure beneath breeches and a blue coat. Fanny broke wild horses and tangled with panthers and transformed herself into a pirate, eventually finding her lover and rescuing him. In Emma's view, the lone flaw was that the heroine "had no higher ambition than running after a man." At night, by candlelight, Emma reread the book and began planning her escape.

This idea burrowed deeper as her father married off her sisters, one by one. Emma watched them settle into lives where each day seemed longer and plainer than the last: every moment preordained, every thought pragmatic, every chore chafing at the same raw spot on their bones. "In our family the women were not sheltered but enslaved," she wrote. "If occasionally I met [a man] who seemed a little better than others, I set him down in my mind as a wolf in sheep's clothing, and probably less worthy of trust than the rest." After Emma turned fifteen, her father announced she would marry next, promising her to a lecherous neighbor, a farmer dozens of years her senior. When the time came she would invert Fanny's plan, becoming a man in order to avoid one.

And yet here she was, taken by a man's conversation and manner, answering his questions without suspecting any secret motives behind them. She watched his eyes, too, noting their careful scan of her features, the quick and subtle trip down her frame, and a part of her wondered what she'd failed to hide.

Jerome recorded their meeting that evening in his diary. "I had a very pleasant conversation with Frank Thompson on the subject of religion," he wrote. "He is an assistant in the hospital and I think well able to win and repair the hearts of those about him." And yet something—Jerome wasn't sure what—was strange. "A mystery seems to be connected with him," he added. "Hard to name."

Emma looked for Jerome each night and he always showed up, talking with her longer than he did his sick friend. She told him about her childhood in the Anglican church, her belief that a thing had to be right "if there were instances of it in the Bible," and her faith that "God was there," no matter what happened. She shared stories of her time as a Bible salesman, going from door to door in Canada and eventually crossing the border into America, where everyone was talking about John Brown's raid and the looming crisis. She said that she, as a disciple of Christ, could not abide the institution of slavery.

"I visited my friend Thompson this evening," Jerome wrote, "and was highly entertained for the evening in conversing about the subject of religion and human nature . . . he is a good noble-hearted fellow as far as limited acquaintance will allow."

She was thrilled when Jerome announced his appointment as a steward in the hospital, responsible for dispensing medicine and applying bandages and leeches. That afternoon they napped side by side. With his eyes shut she felt invisible, safe to study the planes and ridges of his face. She could have shifted her hand and felt the rise and fall of his chest. She recalled her dates with women back in Canada, excursions meant to assure her friends she wasn't "queer," involving nothing more than witty banter and the kiss of a hand. They all expected her to marry one companion, "a pretty little girl who was bound that I should not leave Nova Scotia without her," but leave, of course, she did. She let her mind wander to a dangerous place, wondering what might happen if she told Jerome the truth, the prospect—both thrilling and petrifying—gnawing at her heart.

Jerome duly recorded the encounter: "I arose greatly refreshed after a good sound sleep on a couch with my friend Frank Thompson."

The next day they strolled together across the Long Bridge into Washington, taking their time. Without Emma's explicit consent her body tweaked the way it moved, finding vestigial traces of the girl she had been. The truth bubbled up inside her, catching in her throat.

"I revere as a blessing the society of a friend so pleasant as Frank," Jerome wrote, "though foolish as it may seem, a mystery appears to be connected with him which it is impossible for me to fathom. Yet these may be false surmises—would that I be free of them for not for worlds would I wrong a friend who so sincerely appreciates confiding friendship."

Later that evening they returned to Washington to watch a fireworks display honoring General McClellan, who had convinced high-ranking Federal officials that the army's seventy-five-year-old general-in-chief, Winfield Scott, was the greatest obstacle to Union advancement. Scott was forced into retirement, and Lincoln named McClellan his successor, so that the young general was now in command of both the Army of the Potomac and the entire US Army. The president voiced concern over the challenges of the dual role, but McClellan had a ready response: "I can do it all."

Emma and Jerome took strolls along the hospital grounds during the day and enjoyed hushed, fervent discussions at night, heads nearly touching. Together they attended prayer meetings, which

Jerome called "a delicious morsel for our thirsty souls." He needed her, she thought—no one else understood the quirks and turns of his mind, or could distract it so easily when he was desperate to go home. She needed him just as much; he reinforced her sense of her own virtue and goodness, defusing the persistent, subconscious fear that nothing about her was real.

"My time is greatly eased by conversations with Frank," Jerome wrote, "which to me bind his friendship more firmly."

One day Jerome received a letter from Anna Corey, his sweetheart back in Michigan. She was, he said, "the only young lady correspondent I have had since I enlisted, and well worthy of the highest esteem of any who appreciates virtue and true nobility." He shared his excitement with Emma. With each word his voice seemed to grow smaller and farther away, receding to a place where she couldn't hear him at all.

The next day she told Jerome she didn't feel well—would he mind covering for her at the hospital? He agreed, and she retreated to her tent and stretched out on her bunk, leaving her boots on. She remembered the tale of Fanny Campbell, and how the heroine did not reveal herself immediately to her lover once she rescued him. Instead she worked alongside him for several weeks, impressing him with her strength and savvy, entrenching herself in his life. Only then did she tell him the truth. Emma had followed the same sure course, earning Jerome's respect and affection, making herself indispensable during these long, taut days of waiting for war. The time had come to tell him.

She found Jerome and said she was much improved, and suggested they go for another walk to the city. It was a cool, clear night, the moon showing only half of itself, and midway along the bridge she stopped him. She forgot about the possible repercussions, both legal and personal, of what she was about to do. Frank Thompson began to speak, unloading each of her lies until only Emma Edmondson was left.

CRINOLINE AND QUININE

ON THE CONFEDERATE
UNDERGROUND RAILROAD

As a rebel courier and spy, Belle rode Fleeter throughout the Valley and to Richmond and Washington, carrying dispatches between Confederate generals and their subordinates, often traveling through the night, a bright moon carving her path. She was now part of the Confederacy's own Underground Railroad, the same route traveled by spies in Rose Greenhow's ring. She used her home in Martinsburg as her personal headquarters, with occasional stays at her aunt's hotel in nearby Front Royal. Mary Boyd had resigned herself to her daughter's choice of career, knowing that Belle would not stand still long enough to listen to her concerns. She waved from the front door and wept after it closed, praying her daughter would find all the safe houses along her route, private establishments—many operated by women—where rebel operatives could exchange information and rest for the night.

The Confederate Secret Service was constantly evolving, attracting new recruits and inventing ways to outwit the enemy. The want-ad columns in newspapers had become a regular medium for the exchange of underground information, with agents announcing arrivals and departures by advertising under previously determined aliases. The Underground Railroad would soon be renamed the Secret Line and expand its scope. A Doctor Line in Southern

Maryland and Washington employed real and bogus physicians, carrying leather bags concealing dispatches instead of instruments, able to travel at all hours without arousing suspicion. A Postmaster Line in the same territory employed mailmen sympathetic to the Confederacy, many of whom were arrested and then replaced by their wives. In Louisville, Kentucky, members of the Knights of the Golden Circle, a precursor of the Ku Klux Klan, met at the home of a Mrs. Jack Taylor on First Street, where they created anti-Lincoln propaganda, identified safe houses for Union deserters, and plotted ways to spy for the South. (The proprietor of a local boardinghouse, Mrs. Long, went a step further, serving Union soldiers fresh coffee seasoned with a dash of poison.) Even Union spies had a grudging respect for their rebel counterparts; one officer with the 6th New York Cavalry admired how Confederate agents "could take hints quickly, adapt themselves to circumstances with readiness, and had their hearts in their business."

Belle vied to distinguish herself from the growing field of operatives, and began acting in oddly conspicuous ways for a girl purporting to be a spy. One rebel officer noticed a little black lapdog scurrying alongside her wherever she went. When she thought no one was looking, she reached into her purse and produced a white dog skin cut to fit precisely over her pet's body. She placed messages atop the dog's back, wrapped him tightly in the false hide, and smiled sweetly at the Union pickets. "Some of the old and ugly ladies make a great fuss about being searched," one guard wrote to his wife, "but the young and good looking ones are a great deal more civil."

She told stories, to anyone who would listen, about killing a Yankee on the Fourth of July, a boast received with some skepticism by Union officers. "Some of the boys met the woman Belle Boyd, a violent rebel," wrote a captain with the 5th Connecticut. "She claims to have shot a Union soldier who insulted her. It is not believed by the boys. . . . She appears to be the only witness." None of the boys, Yankee or Confederate, knew quite what to make of her. One day she'd show up at camp with her face scrubbed free of powder and dressed as a man, donning the gray wool frock coat and butternut kepi of a Confederate private. At other times she wore a rebel soldier's belt cinched around her waist, a velvet headband emblazoned with the seven original stars of the Confederacy, and a gold palmetto tree pin, in homage to South Carolina, affixed beneath her chin. She tried on different personas, playing the part of a demure Southern maiden one moment and an able warrior the next.

Belle Boyd in Confederate attire.

Once in enemy territory Belle wondered aloud about troop movements and numbers, disguising her intentions as curiosity, tying bows around all of her sentences. She had studied politics in school, she explained, and longed to know more. If prodded, she could recite the names of every general in the rebel army, every member of Congress and the district he represented, and the details about every battle since Fort Sumter. She never once prevaricated about her allegiance to the South, and this brazenness became a vital part of her costume: After all, she asked, how dangerous could such an open foe possibly be?

While the boys contemplated their answer, Belle flashed the bowie knife she carried on her belt.

The Shenandoah Valley witnessed a few minor skirmishes during the fall—one after the Yankees seized twenty-one thousand bushels of wheat in a mill near Harpers Ferry—but the war seemed to be on hold until General McClellan made a move. The lull in activity bred a spirit of playfulness among some of the opposing troops, particularly those who had been friends before secession. Belle occasionally delivered messages between Union lieutenant Orlando M. Poe, the new commander of Emma's regiment, and Confederate general J. E. B. Stuart, his old West Point classmate. "My dear Beauty," Poe wrote, using the derisive nickname Stuart had earned at the academy for his "personal comeliness in inverse ratio" to the word. "I'm sorry that circumstances are such that I can't have the pleasure of seeing you although so near you." She copied Poe's note verbatim and filed it in her growing collection of war mementos.

At night, after the last snap of snare drum and game of cards, she crept about Union camps gathering unattended sabers and pistols, and depositing them at a temporary hiding place in the woods, just far enough from the enemy pickets. A network of rebel ladies joined her, weaving arsenals through the steel coils of their hoop skirts, passing each other balls of string to secure the weapons tight. One day the 28th Pennsylvania Regiment, encamped near Harpers Ferry, discovered a cache of 200 sabers, 400 pistols, cavalry equipment for 200 men, and 1,400 muskets, all stashed inside barns and outhouses and buried underground, awaiting transfer to Southern lines. "I had been confiscating and concealing their swords and pistols on every possible occasion," Belle confessed, "and many an officer, looking about everywhere for his missing weapons, little dreamed who it was that had taken them, or that they had been smuggled away to the Confederate camp, and were actually in the hands of their enemies to be used against themselves."

The South was suffering from more than a shortage of arms. Lincoln's blockade kept tightening, interrupting business and choking off the Confederacy's supplies. No one could receive checks or access funds held in Northern banks. Some merchants accepted sewing pins as currency. Palmetto hats and raccoon-skin shoes came into vogue. Atlanta jewelers set coffee beans instead of diamonds in breast pins. Coffee was becoming a luxury good, and newspapers printed suggestions for ersatz brews: take the common garden beet, wash it clean, dice into small pieces, roast in the oven, grind, boil with a gallon of water, and settle with an egg. Above all, medicine was scarce and prohibitively expensive—especially quinine, derived from the bark of a South American tree and used to combat frequent outbreaks of malaria among the troops.

But England, which remained neutral, allowed Southern agents to buy at will, and a blockade-running business flourished abroad. Low, sleek ships like the *Sumter* and the *Robert E. Lee* set sail from Liverpool for the Caribbean, arriving in Havana or Nassau with a million-dollar cargo: cannon, rifles, cartridges, gunpowder, shoes, blankets, morphine, and the ever-valuable quinine. From there, smaller blockade runners called sprinters picked up the goods and broke the cordon, slipping into the port at Savannah or Wilmington. Domestically, two Philadelphia-based, politically connected chemical manufacturing companies, Powers Weightman and Rosengarten Sons, supplied quinine to Union troops, but employees who valued profit over patriotism or sympathized with the South were always eager to make a deal.

Belle began running the inland blockade, smuggling quinine and other necessities. Despite constant patrolling by the Federal navy, she and hundreds of other rebels crossed the Potomac River at Popes Creek, where the water separating Virginia and Maryland was less than two miles wide. A farmer named Thomas A. Jones— who would aid John Wilkes Booth's escape in 1865—lived on the Maryland side. He had calculated a sliver of time, just before dusk, when the sun grazed the high bluffs above Popes Creek and threw a shadow across the river, enabling small rowboats to land and hide without detection.

Jones cooperated closely with Benjamin Grimes, a fellow farmer on the Virginia side, and together they orchestrated at least two crossings a night—some of them conducted by nine-year-old Robert Fitzgerald, who would become the father of the future writer. Boats set sail from Grimes's property, deposited packages in the fork of a dead tree on Jones's shore, and collected packages from the same spot. When it wasn't safe for a boat to cross from Virginia, a Miss Mary Watson, the twenty-four-year-old daughter of a Confederate major, sent a signal by draping a black shawl from a certain window, right over the heads of the troops stationed there.

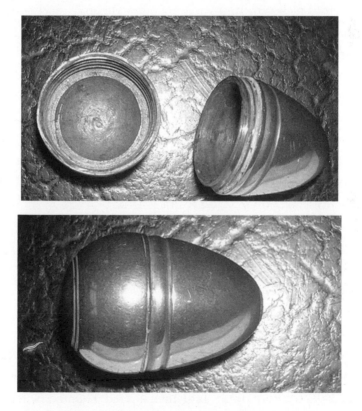

Confederate operatives used acorn-shaped contraptions to smuggle quinine and other goods across the lines.

The blockade runners devised ingenious ways to conceal their contraband. Some male agents used an acorn-shaped brass contraption, just large enough to hold a message or a bit of quinine powder, and hid it in the place least likely to be searched: their rectums. One woman managed to conceal inside her hoop skirt a roll of army cloth, several pairs of cavalry boots, a roll of crimson flannel, packages of gilt braid and sewing silk, cans of preserved meats, and a bag of coffee—the contraband tally for a single crossing. Another found a functioning pistol on a battlefield, took it apart, and dared to smuggle the pieces to a rebel soldier. She hid the two halves of the wooden butt between soft ginger cookies, pushed the barrel into a loaf of bread, and buried the screws in a jar of potted shrimp, also filched from a Union officer. Mothers packed quinine in sacks of oiled silk and tucked them inside the hollowed papier-mâché heads of dolls, instructing their daughters to hold the toys tight and not say a word.

Throughout the summer and fall Belle visited Confederate camps, bearing smuggled quinine and stolen Union guns, offering to deliver letters to Richmond for $3 each. At the new headquarters of General Stuart, a farmhouse near Centreville, Virginia, she cornered her latest love interest, Major John Pelham. He was twenty-three and blond and thin as a banister, six feet tall and 150 pounds, and, according to his fellow soldiers, "as grand a flirt as ever lived." Belle pressed a Bible into his hands and told him to read her inscription:

To John S. Pelham
From Belle
With the sincere hope he will read carefully and attentively
For his own if not for her sake.

A few days later, perhaps because Pelham had rebuffed her, Belle was in a less generous mood. She wandered from Stuart's headquarters to those of General Beauregard, situated nearby, between Fairfax and Centreville. The scene struck her as chaotic and unruly, with pieces of artillery indifferently manned, the cannoneers undisciplined, and the Quartermasters Department deficient in wagons. Many troops wore burlap tied around their feet in lieu of boots or went barefoot altogether. The army was experiencing daily desertions; eleven men had recently fled one company in a single night. It felt like a place where anything could happen, where she could *make* anything happen.

Her chance came when a Confederate soldier approached and asked her for whiskey. Alcohol was at times rationed out as a reward or a stimulant before battles or marches, but in general enlisted men were not allowed to keep supplies of liquor and suffered swift punishment for drunkenness on duty (meanwhile, Beauregard served special guests, like French prince Jérôme Bonaparte, juleps of dark cognac made by his aides in large buckets filled with ice and mint). Belle considered the soldier's request, and decided she wasn't in the mood to give anything away.

Two dollars for a pint, she told him.

The soldier countered, offering one. Belle refused. The man thrust his hand beneath Belle's skirt. He tore a bottle from the hoop, took a long drink, and raised it in victory.

Belle reached into her belt, brandished her knife, and demanded the bottle. The soldier dropped it to the ground, the liquid pooling at Belle's feet. He unsheathed his own knife. About a hundred of his comrades gathered, waiting for what would come next. Belle charged, bellowing and swinging her knife.

Men from Wise Artillery rallied behind Belle, and men from Borden's Guard Artillery, behind the soldier. The melee continued long after she removed herself from its vortex, slipping out to the sidelines. She watched the knives gleam and the blood spill; some of the men defending her hailed from Martinsburg and had known her since childhood. She read newspaper reports calling it a "bloody fracas" and a "fierce conflict" between "rivals for her stimulating donations and sweet smiles." It was the first time her name made national news. Thirty of her beloved rebels were badly wounded, more than twice the number of casualties incurred during a recent skirmish with the enemy.

Dark and Gloomy Perils

WASHINGTON, DC

Massachusetts senator Henry Wilson would be arriving any moment, but Rose's mind was on another paramour and source, Union captain John Elwood. She'd heard what happened to him that rainy night after he left her home and arrested Allan Pinkerton. The detective brought him to assistant secretary of war Thomas Scott, who confiscated his sword and sent him to Fort McHenry, where he would slit his throat with a penknife. Rose felt sympathy for the poor, dim captain, described in his obituary as "by no means a person of sharp and quick intellect," and for his two young children and widow, especially since, according to the press, the widow shared Rose's secessionist sympathies. Had Rose known earlier, she might have spent as much effort recruiting Mrs. Elwood as she had seducing her husband.

At the sound of the doorbell Rose's maid, Lizzie, jumped to admit her visitor. The hall lamp briefly lit Senator Wilson's stern face, its jowls spilling over the rim of his collar. Lizzie disappeared on the other side of the red curtain, leaving them alone in the parlor. Rose let Wilson kiss her cheek and poured him a tumbler of brandy. Settling beside him on the settee, she directed the conversation, nudging him with pointed questions, murmuring her encouragement during every pause. He spoke of Union troops being sent to fortify Maryland and new batteries being erected around Washington. He mentioned McClellan's apprehension about rebel activity in the Shenandoah Valley. And he confided that he himself planned to join the general's staff as an aide and adviser, a development Rose hoped might work in her favor. At least, she thought, she would enjoy even greater access to the secrets of Lincoln's cabinet and War Office.

She walked Wilson to the door, lingering until his carriage turned the corner. For once she failed to spy the pair of strange men under the streetlamp, revolvers peeking from belts. Allan Pinkerton watched her retreat into her home and made a notation in his file: "A number of prominent gentlemen were received by the fascinating widow, and among the number were several earnest and sincere Senators and Representatives, whose loyalty was above question, and who were, perhaps, in entire ignorance of the lady's true character."

He had seen enough. It was time to apprehend her.

Two days later Rose encrypted a message to Beauregard, including the details of her conversation with Wilson and ending with a plea: "Tell me what to send you, as I know nothing from you of your wishes, and I may be wasting means in sending you what is of no use." She checked on Little Rose in the garden, playing hopscotch, stomping inside squares etched roughly into the ground. The child's bonnet had slipped from her head and bounced against her back with the force of each jump. Lizzie would keep an eye on her while Rose made her delivery; one of her scouts was scheduled to find her. She tucked the dispatch into the pocket of her mourning dress and started on her walk.

Crossing Sixteenth Street, she noticed two men, one in a Federal army uniform and one in civilian clothes, angling themselves in her direction. She rushed over to a neighbor and made casual conversation, asking about the health of her children, shifting her eyes left and right, keeping the men in view. The air was blurry with heat. Sweat coursed down the length of her silk sleeves, and her lace collar scratched at her neck. Her scout—"a distinguished member of the diplomatic corps"—approached, right on time, but it would be impossible to slip him the note.

As he passed, she whispered: "Those men will probably arrest me. Wait at Corcoran's Corner, and see. If I raise my handkerchief to my face, give information of it." The scout didn't break stride, whistling as he walked—confirmation that he had heard her. Without any explanation to her neighbor, Rose crumpled the note, crammed it into her mouth, and swallowed it.

Ignoring her neighbor's puzzled expression, Rose bade her good-bye and strolled back home. She felt her heartbeat in her throat, heard its echo in her ears. She told herself: "The fate of some of the best and bravest belonging to our cause hangs upon my own coolness and courage." She forced her steps to stay light and slow. Her face composed itself. She heard them coming behind her, panting at her back.

She turned, stopping them. After all that time watching her, they now seemed surprised to stand mere inches away, eye to eye. One wore an untidy red beard and introduced himself as "Major E. J. Allen," but she knew him as Allan Pinkerton. She had started her own dossier, describing Pinkerton, the epitome of a dour Scotsman, as "a German Jew, possessed [of] all the national instincts of his race." His partner had a mass of curly brown hair leading to whiskers shaped like wooden spoons. She recognized him as part of the group that had followed her around Washington.

"Is this Mrs. Greenhow?" Pinkerton asked.

"Yes," she replied, and decided to play coy. "Who are you, and what do you want?"

"I have come to arrest you."

"By what authority?"

"By sufficient authority."

"Let me see your warrant," Rose demanded. She reached into her drawstring purse for a handkerchief and patted her cheek.

Glancing back at the street, she confirmed that her signal had been understood. The entire city—the entire country, both North and South—would soon know of her arrest. The men each clutched an elbow and escorted her inside. She yanked herself free, turned to them and rasped, "I have no power to resist you, but had I been inside of my house, I would have killed one of you before I had submitted to this illegal process."

Pinkerton seemed taken aback. "That would have been wrong," he responded, "as we only obey orders and both have families."

"What are you going to do?" she asked him.

"To search."

"I will facilitate your labors." She reached for a vase on the mantel and extracted a fragment of an old letter—the one that she had suspected was an attempt to entrap her; it had come from the city post office, a route her sources knew better than to use.

Tossing the note at Pinkerton, she said, "You would like to finish this job, I suppose?" He pocketed the note but discarded the city envelope in which it had arrived. Silently Rose congratulated herself on her "cool and indifferent manner," her refusal to commit "some womanly indiscretion by which they could profit," and in the very next moment she heard Little Rose scream. Fear struck and kindled inside her chest. Another scream and the words: "Mother has been arrested! Mother has been arrested!" She picked up her skirts and ran toward her daughter's voice. From the back door she spotted Little Rose dangling from a tree, her body folded in half over a branch. Two men—more Pinkerton detectives—reached up, clutching at her spindly little calves, and the girl kicked at their heads, pinwheeling her feet. She was crying now, long, reverberating wails that seemed to come from someone three times her age and size, the worst sound Rose had ever heard. She lurched forward to save her but her arms wouldn't move. Someone had trapped her, twining her wrists at the small of her back. She heard Pinkerton say, "Take charge of this lady, detain her in the parlor."

There was an army of them now, uniformed men and women spreading to every corner, filing upstairs to invade Gertrude's room. Rose had not touched the room since her daughter's death—the toiletries were still aligned on her shelf, the dresses still hung in her wardrobe—and she watched as detectives confiscated every last thing the girl had owned. Even the bed where Gertrude had died and lay in her winding-sheet was desecrated by Pinkerton's men.

They invaded Rose's room next, rummaging through her hamper, shaking out soiled crinolines, corsets, stockings, and underpinnings to see what might fall from the pockets and folds. They even seized Little Rose's unlettered scribblings, suspecting correspondence with the enemy. They tunneled their hands into vases, smashed picture frames to check behind photos. They uncovered Henry Wilson's thirteen love letters to Rose and several others from Oregon senator Joseph Lane, concerning his "feebleness" and desire to visit as soon as possible. Pinkerton made a notation on the last missive: "Wants to see Mrs. G very much—but is engaged with Committee on Military Affairs. He is mostly sick because he can't see her." Each item was classified and filed into a parcel marked "Highly Important," "Legal," or "Political."

A s instructed, one of Pinkerton's detectives—the one with the spoon-shaped whiskers—sat in the parlor with Rose while guards searched the house. Rose sat next to him on the settee, fanning herself. Wisps of hair escaped from her braided bun and swung with each flutter. She assumed that the detective, like all men, was his own favorite subject of conversation, and she would let him think she was enthralled with every revelation. After saying that his name was Pryce Lewis, and that he had immigrated to the United States from England five years earlier, to her surprise he asked questions in turn. She was incapable of censoring herself. He found Rose "a beautiful woman . . . careless and sarcastic and, I know, tantalizing in the extreme." He was young enough to be her son.

Rose tilted her head and let her voice go soft and Southern, stretching syllables like taffy. "May I go upstairs for a few minutes?" she asked. The parlor had gotten so unbearably hot, she explained, and she wished to change her dress.

Lewis, charmed by the "winning way" of her manner, agreed.

He followed her upstairs, one step behind. She shut the door in his face and began sifting through dresser drawers, searching for evidence not yet found: notes on McClellan's army, copies of maps she'd drawn or acquired from sources, drafts of her dispatches to Beauregard, her cipher and its key. She tore them to pieces and sprinkled the remains in her iron stove, hoping the detective couldn't hear the sound of the paper shredding, the clank of the metal. She unbuttoned her dress, kicked off her crinoline. There was a shuffling outside her door.

"Madam, madam," a voice called, and without further warning Pryce Lewis barged inside.

He stared, stunned, at a half-naked Rose. She took advantage of his embarrassment, grabbing her revolver from the mantelpiece and aiming it directly at his head.

"If I had known who you were when you came in," she said, "I would have shot you dead!"

For a moment they stared at one another, waiting. Finally Lewis smiled and said, politely, "The revolver has to be cocked before it will go off."

Rose glared at him as he peeled it from her hands.

Lewis excused himself when a female detective entered the room, introducing herself as "Ellen." Rose sized her up, noting that she resembled "one of those India rubber dolls, whose expression is made by squeezing it," and that her weak gray eyes seemed perpetually on the verge of tears. She ordered Rose to finish undressing in preparation for a strip search.

Rose handed her garments to "this pseudo-woman" until she wore only her stockings, linen, and shoes. Ellen examined each piece, skimming her fingers across the silk, and then allowed Rose to dress herself. They were halfway down the stairs when Rose heard a smothered scream—the voice of a former servant's sister, who had been taken prisoner because she happened to pass the house. "I began to realize," Rose wrote, "the dark and gloomy perils which environed me."

As Ellen brought her back downstairs to the parlor, the front door swung open. Rose's friend and scout, twenty-two-year-old Lily Mackall, pushed her way past two guards, and the women embraced.

"I did not know what they had done with you," Lily whispered.

Rose pressed her mouth against her ear and replied, "Oh, be courageous, we must whip these fiends."

Pryce Lewis lowered an arm between them, breaking them up, and told Lily that she, too, was under house arrest.

Pinkerton strode into the parlor and barked orders to his men: remain inside the house, prevent anyone from leaving, arrest all visitors. He was heading out for the evening to report to the War Department.

As soon as he left, his men began helping themselves to Rose's brandy. She let them imbibe without protest or comment. They began stumbling about the parlor and slurring their words. She followed them into her dining room, where they dropped heavily into chairs. She adopted a scolding tone, telling them to take a good look at themselves—their hygiene was deplorable, their manners were atrocious; didn't they know better than to sit in shirtsleeves at the table? When was the last time they'd washed their hands? They were all mindless "slaves of Lincoln," at the mercy of that "Abolition despot." They were lucky to be in the same room with her, a lady on a par with Marie Antoinette or Mary Queen of Scots, and they should muster any scrap of propriety they had and treat her with respect.

The men received her monologue without comment. One, an Irishman, took a long swallow of brandy and landed his logy eyes on her bodice. He was looking forward, he said, to the "nice times" they'd all have with her later. The others laughed and raised their glasses to the thought.

While they drank, Rose crept upstairs to her library and reached for a folio slipped deep between Robert's books, another hiding spot that had so far eluded the detectives. The pages fluttered to the floor, and she began tucking them into the deep folds of her dress. She found another folio, and another. There were more documents than she remembered; disposing of them alone would be difficult. She recalled that during her strip search the detective had neglected to check her stockings and boots. If Rose gave the papers to Lily Mackall, her friend could easily conceal them. She tiptoed back downstairs, checked on the drunken guards, and Lily agreed to her plan. Should the detectives discover the papers, they vowed to set the house on fire. Around 4:00 a.m. the detectives, still inebriated, ignored Pinkerton's orders and permitted Lily to leave. She took the papers with her, but detectives brought her back the next day.

Throughout the week the detectives continued to search Rose's home. She watched them haul out boxes of fine china and crystal, dismantle her Duncan Phyfe tables to check the sockets of the legs, plunge their hands beneath her mattress. They panicked when a sprig of jasmine tumbled from one of Rose's undergarments, certain it had been smuggled to her as a hidden message. They finally uncovered the message fragments stashed in her stove, partially singed scraps of treasonous origami they spread across her tables and studied for hours at a time. Rose delighted in their confusion. "One by one," she gloated, "they had allowed the clue to escape them."

The guards began treating her with a fawning and suspicious kindness. The Irishman who had threatened to rape her sat across from her in the parlor and chatted as if they were old friends; he shared her Catholic faith, he told her, and he did not fathom how "so noble a lady should be treated as a common malefactor." He seduced her maid, Lizzie, sneaking her to an upstairs chamber, no doubt seeking incriminating evidence against Rose. Lizzie willingly went along—"I led Pat a dance," as she told Rose—and said she would continue the dalliance to see what she could learn.

A Scottish guard tried to engage Rose as she ate, speaking somberly of the "sublime fortitude" she had shown throughout the ordeal. He riffled through his coat pocket and produced a document, lowering it next to her plate. This belonged to General Mc-Clellan, he said, and he would be most honored by her autograph.

"Madam," he said, "there is no telling what may happen, and I would like to look at your name and know that you had forgiven me."

Rose found his manner "touchingly pathetic" and ignored him. He tried again: Would she like him to sneak letters out of the house for her? He promised they would safely reach their destination, and he would not tell a soul. It was the least he could do, after the suffering she'd endured. Rose spooned her beef stew and declined.

Instead, after dinner, she went to the room she shared with Lily and Little Rose. She knelt by her daughter's bed, and clasped her hands as if in prayer. She considered what she was about to ask of her eight-year-old child, her namesake, her last connection to her husband. Little Rose had always been *her* child—she had never known her father—and she worshipped her mother, internalizing Rose's opinions, mimicking her words. She talked about how she "hated" the Yankees and memorized a repertoire of rebel songs: "The Bonnie Blue Flag," "God Save the South," "Old Abe's Lament." She was a precocious girl, more accustomed to the company and habits of adults than children, and Rose hoped she was up to the task. She told herself her daughter would be in greater danger if she gave up entirely, if she stood idly by and let the North win the war.

She lowered her mouth to Little Rose's ear and whispered it like a story: Friends would be coming to visit soon. She would be outside playing, and she would see them approach. They would bring candy, maybe peppermint sticks or Necco wafers, wrapped inside a very important note. She, too, would have a very important note, written by Mother and hidden in her pocket. She would trade, taking the candy and giving the note. She would say nothing but "Thank you." She would hide the friend's note in her pocket and bring it to Mother as soon as she could. The Yankees would be watching and listening. She would be her mother's "Little Bird," carrying news just like the homing pigeons. This was their secret, the most important secret they would ever share, and she had to promise to keep it.

Little Rose promised, her eyes peeking over the hem of her blanket, the Yankee guard just outside the door.

A few days after Rose's arrest Pinkerton brought in a detachment of Sturgis Rifles, the same company that served as bodyguards for General McClellan. Rose was left alone only when bathing or using the chamber pot. When she slept, a guard sat on a chair inches from her bed. When she changed her dress, she was ordered to leave the door open, the lewd guard invariably peering inside.

Even under constant surveillance Rose continued her espionage. Little Rose understood exactly what was expected of her when she went out to play in front of the house, making quick and surreptitious exchanges with scouts, reporting back to her mother with candy in her palm and a note in her pantalettes. In this way Rose learned of Union general Ambrose Burnside's plan to send an armada and twenty thousand men to ports in North Carolina—part of the Union's overarching "Anaconda Plan" to surround the Confederacy and cut off supplies. She told her guard she had to use the chamber pot and prepared an enciphered dispatch, to be delivered by Little Rose, warning Confederate officials to fortify the South's coastal defenses; at the moment, only eight modified workboats known derisively as the Mosquito Fleet patrolled the waters. She devised a system of communicating by needlework, knitting tapestries in specific patterns based on the Morse code, a precise vocabulary of stitches and colors. Another opportunity came on Sunday mornings, when Lizzie took Little Rose to Mass at St. Patrick's Church. Although a guard sat nearby, certain parishioners managed to whisper information about troop numbers and military plans during the sign of peace. If Little Rose was napping, or if Rose sensed a guard was suspicious, she communicated with the outside by writing letters with hidden meanings, all of them scrutinized before they were mailed.

"Tell Aunt Sally that I have some old shoes for *the children*," one read, "and I wish her to send someone *down town* to take them, and to let me know whether she has found any charitable person to help her to take care of them." Only the intended recipient would understand the translation: "I have some important information to send across the river, and wish a messenger immediately. Have you any means of getting reliable information?" The detectives, Rose quipped, came to the conclusion that, "for a clever woman, Mrs. Greenhow wrote the greatest pack of trash that was ever read."

Once, and only once, did Rose bribe a guard to deliver a message intended for Thomas Jordan. She told him that "artillery is constant and severe," with about sixty thousand troops surrounding the city, and that they were badly equipped, asking for private contributions of blankets. She reported that her scouts were still in place and enclosed a list of forts. "I have signals," she said. "Take care that the cipher does not fall into their hands. I destroyed it and every paper of consequence, but oh my God, with what danger, with twenty detectives following every step . . . spies are employed in every class, and more is spent in the secret service by McClellan than ever before."

She slipped the note to Lieutenant Sheldon, the only guard who had showed her any kindness, and hoped he wouldn't betray her.

One afternoon Rose heard a rustling and familiar voices at the front door. She looked up to see members of her spy ring: Eugenia Phillips, the wife of former Alabama congressman Philip Phillips; her sister; two of her daughters (whom Eugenia had trained to spit on Union officers); and Bettie Hasler, the courier who had delivered messages entirely unaware of their content. They had each been arrested and charged with corresponding with the enemy. Rose nodded at them in silent greeting and tried to catch Bettie's eye, but the woman was weeping into her handkerchief, tattered points of lace sprouting between her fingers.

Forbidden to talk with her fellow captives, Rose passed the time knitting, seemingly absorbed in the stab and click of her needles as she eavesdropped on the guards. One mentioned that the government was offering a handsome sum for the key to her cipher. Another reported that President Lincoln and General McClellan had convened a Cabinet Council, summoning Senator Wilson and other Republican officials to explain themselves after they'd been implicated in her intercepted correspondence. She heard that her enciphered dispatches containing drawings of Union fortifications were, in her words, "complimented as being equal to those of their best engineers—*as well they might*." She learned that Union officials had arrested several suspected accomplices: a former minister to Brazil; the current minister to France; a clerk in the post office; a lawyer named Charles Winder, who happened to be the brother of General John Winder, provost marshal of Richmond; and even Washington, DC, mayor James G. Berret, who, although a Democrat and suspected Southern sympathizer, had never been involved in Rose's spy ring. Berret was sent to Fort Lafayette and released after taking an oath of allegiance.

She was gratified to learn her arrest had made immediate national news. Newspapers across the country dubbed her home "Fort Greenhow" and pondered how to contend with female traitors, a situation that had seemed unthinkable before the war began. War, like politics, was men's work, and women were supposed to be among its victims, not its perpetrators. Women's loyalty was assumed, regarded as a prime attribute of femininity itself, but now there was a question—one that would persist throughout the war—of what to do with what one Lincoln official called "fashionable women spies." Their gender provided them with both a psychological and a physical disguise; while hiding behind social mores about women's proper roles, they could hide evidence of their treason on their very person, tucked beneath hoop skirts or tied up in their hair. Women, it seemed, were capable not only of significant acts of treason, but of executing them more deftly than men.

Rose's acquaintance, Mary Chesnut, viewed the shift in public perspective as a titillating game, quipping, "It is so delightful to be of enough consequence to be arrested," but men were unnerved by this flagrant violation of social mores. "The 'heavy business' in the war of spying is carried on by women!" declared the *Albany Evening Journal*. "Is it not about time that an example were set which will prove a terror to these artful Jezebels? Is it not time to impress them with the conviction that they may presume too much upon the privileges of their sex and the gallantry of those in authority?"

There was talk of trying Rose for treason, along with calls to condemn her to the ducking stool for the remainder of the war. She ran the possibility of such punishments through her mind and decided that the authorities wouldn't dare. Persecuting a widowed mother, especially one of her political and social stature, would only establish her as a Confederate martyr.

"Let it come," she challenged one guard. "I will claim the right to defend myself, and there will be rich revelations."

During the next few months, into the fall and winter, Pinkerton and his detectives tried to break Rose, employing the cruelest methods possible under the circumstances. They sought to isolate her by releasing all of her fellow prisoners. "I felt now that I was alone," Rose wrote. "The wall of separation from my friends was each hour growing more formidable." They refused to allow her any fresh air or exercise. They brought in another prisoner, a Chicago prostitute turned spy by the name of Mary E. Onderdunk, and insulted Rose by placing Mary in Gertrude's bedroom. They sent away Catholic priests who tried to visit and debated nailing boards across her windows to deprive her of sunlight. They replaced catered meals with meager servings of cheese and crackers; many nights Little Rose cried herself to sleep from hunger. They refused to let the family physician treat Little Rose when she got sick, offering instead to send an army doctor. Rose declined, preferring instead to "trust her life to the good Providence which had so often befriended me."

They let Rose read select letters from Lily Mackall. Rose's friend was gravely ill and wanted desperately to see her, unsuccessfully imploring every member of Lincoln's cabinet before trying the president himself. Lincoln, according to Rose, told Lily that "she had had too much of my teachings already—that I had done more to damage, and bring his Government into disrepute, than all the rest of the darned rebels together; and by God she should never see me again, if he could help it." Lily died on December 12.

Every day Pinkerton's team interrogated Rose for hours: Who were her sources? Who were her accomplices? Who in the rebel government was still plotting treasonous activity? They contacted her suspected informants and hinted that she had accepted a bribe to reveal their identities; the informants might as well confess to save themselves. When this ruse failed, the interrogators suggested to Rose that a "graceful concession" on her part would secure her release. "Your whole bankrupt treasury," she retorted, "could not tempt me to betray the meanest agent of our cause."

"Do you know your life is in danger?" one detective asked. "Probably, to save your neck, you might answer differently."

Rose replied, "The life of anyone is in danger when in the power of lawless scoundrels."

The detectives gave up for the moment, but Rose was emboldened, sending a letter to US secretary of state Seward that was as duplicitous as it was defiant: "In the careful analysis of my papers I deny the existence of a line that I had not a perfect right to have written or to have received . . . you have held me, sir, to a man's accountability, and I therefore claim the right to speak on subjects usually considered beyond a woman's ken."

A Southern sympathizer leaked the letter to the *Richmond Whig*, which deplored the "cruel and dastardly tyranny of the Yankee Government" and the "incarceration and torture of helpless women." Newspapers throughout the North and South picked up the story. Rose reveled in the attention (and in imagining the great embarrassment she caused Seward), but Pinkerton was at a loss. Rose Greenhow was his highest-profile case to date, and he worried that Federal officials would bow to public pressure and release her. In a report to the provost marshal, he argued that Rose was a serious threat to the Union, using her "almost irresistible seductive powers" to aid the rebels. "She has not used her powers in vain among the officers of the Army," he stressed, "not a few of whom she has robbed of patriotic hearts and transformed them into sympathizers with the enemies of the country which had made them all they were."

Rose, meanwhile, received a visit from an officer of the army—Thomas M. Key, an aide-de-camp to General McClellan. Key offered to negotiate her release, asking what conditions she would accept.

She was conflicted. She thought of Little Rose, her stomach growling, crying in the dark. But she didn't quite trust the messenger: "I crushed down the impulse, for I saw that he was watching me very narrowly."

She responded, "None, sir. I demand my unconditional release, indemnity for my losses, and restoration of my papers and effects."

Key left without offering a conclusive response, but Rose believed that the Union had finally tired of her antics and was prepared to exile her to the South for good. If not, she had another idea. One of her scouts—"a party well known to the government"—was preparing to send a Christmas gift: a cake with several Confederate bills baked into the batter and an escape plan hidden between the layers.

Unmasked

WASHINGTON, DC

Standing on the Long Bridge, under the weakening light of the moon, Emma told Jerome about everything: the "severity" of her father; the impending arranged marriage to a vulgar old neighbor; the night she looked in a mirror, hacked off her long dark hair, dressed in a man's suit, and became Frank Thompson—Jerome's good friend, his best friend, the person he knew better than anyone else. She tried to persuade him even as she watched his thoughts register, his features contracting in confusion and expanding in shock, and settling, finally, into understanding, his face closing like a fist, looking as though he had never known her at all.

Separately they walked back to the same place, retreating to their tents, where Emma hid under her blanket and Jerome opened his journal. At the top of the page, in his bold, tidy script, he wrote, "Please allow these leaves to be closed until the author's permission is given for their opening." For once he didn't care about conserving paper, and filled the next several pages with thoughts he could never say aloud.

"My friend Frank is a <u>female</u>," he wrote, the underscore thick from the pressure of his hand. "I won't say that it is not strange to me. . . . How sad is the reaction which often occurs when we think we have friendship in exchange for friendship and find that friend differing so widely from our own natures. . . . I learned that in friends we may be deceived. This have I been with the one here mentioned as he possesses a nature too willful to be pleasant, too jealous to be happy." He switched pronouns and rendered his verdict: "God knows my heart that towards her I entertain the kindest feelings, but it really seems a great change has taken place in her or that the real her has been unmasked. It may be that I am partial to another." He decided to keep her secret, although it was dangerous information to know; he could be court-martialed for helping Emma hide her sex.

Emma did her best to avoid Jerome at the hospital. She couldn't bear the pitying cadence of his greetings and the vaguely smug tilt of his smile. (If only she were privy to journal entries in which he skimmed past the cause of their estrangement: "I fear I have been somewhat deceived by his disposition yet as before intimated the fault may be mine, but certain it is there is not so warm friendship existing between us as there formerly has been.") She hurled herself into work and completed tasks with willful—yes, she *was* willful—efficiency: prepare the injection, dig the graves, tell the soldier that he was dying. She went out of her way to indulge every inconvenient request. One Dutch soldier, suffering from typhoid fever, declared he could eat nothing unless he first had some fresh fish.

"But," Emma objected, "the doctor must be consulted. Perhaps he will not think it best for you to have any fish yet, until you are stronger."

"He dusn't know vat mine appetite ish—the feesh I must have . . . I must have some feesh!"

So Emma acquired a bamboo rod and fashioned a line out of horsehair and walked to Hunter's Creek, a mile and a half from camp. Soon after she cast her rod a monstrous eel struck. She fought it, pulling and leaning, her rod bending in jerks, the eel bucking against the current. She planted her foot, digging in. She would not lose this one. It twisted and danced, the point of its tail skimming the water's surface, relenting in stages. She carried it back to camp like a baby, lying slack across her arms.

"Dhat ish coot," the Dutchman said when she returned, rubbing his hands together. "Dhat ish coot."

For once, camp life was as hectic as work. It was clear that General McClellan had no immediate intention of invading Virginia, and so the entire Army of the Potomac began establishing winter quarters, constructing log cabins, complete with fireplaces and chimneys, to insulate against the coming cold and snow. The most significant military development of the past five months had happened not in the eastern theater but out at sea, when the USS *San Jacinto* intercepted what it believed to be a Confederate man-of-war cruising in the Old Bahama Channel. Instead it was the British mail packet *Trent*, en route to Europe and carrying two Confederate envoys, James Mason and John Slidell. Union authorities took the men into custody, an act that England considered an insult and a violation of its neutral rights. The Confederacy grew giddy with hope that the two countries would go to war over the Trent Affair and that England would recognize its legitimacy.

Closer to home there had been scouting excursions and brief clashes, and one humiliating Union defeat at Ball's Bluff, near Leesburg, Virginia, but McClellan had let the fine fall weather slip by. The once reassuring bulletin "All quiet on the Potomac" had become a mockery. Even optimistic Unionists felt what Walt Whitman described as "a mixture of awful consternation, uncertainty, rage, shame, helplessness and stupefying disappointment." Distracted by discussions about the Trent Affair, President Lincoln urged patience, but the more radical Republicans griped that McClellan had done nothing but stage his grand reviews (the latest of which involved sixty-five thousand men in a spectacle that *Harper's Weekly* called "brilliant beyond description"). They even hinted that McClellan, a Democrat, kept his army idle out of sympathy for the Southern cause.

Politics aside, transporting just one regiment of a thousand men along with horses, artillery, and supplies was a laborious process even under ideal weather conditions, requiring some eighty wagons, fifty passenger cars, and nine locomotives. The government furnished no mittens, and stiffened fingers and numbed hands made it impossible to load and fire a musket with any rapidity.

But Confederate generals remained wary of an attack, especially after receiving Rose Greenhow's latest enciphered dispatch, based on news gleaned from an exchange between Little Rose and one of her scouts in the yard. She warned that Union forces aimed to cut off railroad and other communications and attack from the front. "For God's sake heed this," she concluded. "They are obliged to move or give up." Thomas Jordan forwarded Rose's message to Beauregard along with a note of his own: "No living man ever made such a desperate effort as McClellan will make. Nevertheless I believe he is a coward and is afraid to meet you."

Even if McClellan did order an advance, his army was too sick to comply. Entire regiments suffered fits of coughing, short, barking rasps coming from every tent. McClellan himself contracted typhoid fever, a diagnosis that brought three homeopathic physicians rushing from New York to tend to him. The rival armies would not meet again until spring.

Still, Union troops celebrated Christmas, decorating evergreen arbors with pinecones, singing "Silent Night" and "Deck the Halls," opening boxes from women's relief groups filled with homemade clothes and dried fruit, eating a lavish dinner of chicken, turkey, and cider. Some of the men raised small trees inside their tents and decked them with hardtack and pork. Emma missed all of it. Hoping to avoid Jerome, she had requested a transfer; her new appointment, as a nurse at Mansion House Hospital in Alexandria, began that morning. The women volunteers bustled about trying to create a festive atmosphere, carrying platters of plum pudding topped with sprigs of holly, or sitting cross-legged on the floor bending pine branches into wreaths. Mary Todd Lincoln had donated all unsolicited gifts of liquor that came to the White House, and a pyramid of whiskey bottles towered in one corner. Emma thought, ruefully, about how Jerome would disapprove, and then pushed the image of his face from her mind. She went from bed to bed, reading Bible verses and taking dictation for letters home. She expected to remain there indefinitely, but the command of the Army of the Potomac had other plans.

THE DEFENSELESS SEX

RICHMOND

Elizabeth, like all of Richmond, followed the exploits of Rose Greenhow, but she alone noted the hypocrisy of the city's reaction. The same newspapers that had threatened Elizabeth for visiting Union prisoners now excoriated the Northern government for punishing a "true woman" like Rose. "Nothing," roared the *Richmond Dispatch*, "is so hideous in the tyranny inaugurated at Washington as its treatment of helpless women. . . . None but savages and brutes make war upon the defenseless sex." Richmond's outrage over Rose's predicament was exceeded only by a cold and constant fear: If the Union capital could be infiltrated so easily, might the Confederacy's be just as vulnerable?

That thought gripped Elizabeth's mind as she summoned one of her family's former slaves, Mary Jane Bowser, to discuss potential employment in the Confederate White House. They sat in the parlor of the Van Lew mansion, where Edgar Allan Poe once conducted static electricity experiments and read *The Raven* to terrified Richmond belles, and in view of the Confederate flag now dangling from the wall.

At the moment Elizabeth's sister-in-law, Mary, was out, meeting with other Confederate women to sew winter uniforms for the troops, all of them so devoted to the task that they even knitted in their carriages on the way there and back. Brother John was away on business. His girls, Annie and Eliza, were with their tutor; John and Elizabeth refused to entrust their education to Southern schools, whose lessons now had a decidedly local flavor. (One arithmetic book posed the problem: "If one Confederate soldier kills 90 Yankees, how many Yankees can 10 Confederate soldiers kill?") Aside from the bustle of other servants in the kitchen, Elizabeth and Mary Jane were alone.

Mary Jane had always been special to Elizabeth, more like a daughter than a servant. After Elizabeth's father died in 1843, the rest of the Van Lews had given the slave, then three years old, her freedom. Elizabeth recognized Mary Jane's quick and cunning mind, and on one of her trips North to visit friends she brought the girl with her, enrolling her in a Quaker school in Princeton, New Jersey. When Mary Jane turned fifteen, Elizabeth sent her to Liberia to serve as a missionary and fretted during the four years she was away, routinely mailing money and boxes of necessities, and praying she wouldn't fall victim to illness or warfare. "I will try to do the best I can by her—as I would be done by," Elizabeth wrote. "I do love the poor creature—she was born a slave in our family—& that has always made me feel an awful responsibility." During Mary Jane's journey back to the United States on the ship *Caroline Stevens*, Elizabeth insisted that she travel in a first-class cabin instead of steerage.

When Elizabeth brought Mary Jane home she did so in violation of the law, since a Negro who had left Virginia to be educated in the free states was not permitted to return. In 1860, back in Richmond for five months, Mary Jane was arrested for "perambulating the streets and claiming to be a free person of color, without having the usual certificate of freedom in her possession." It was a serious charge; a free Negro caught without a pass risked being auctioned back into slavery. To protect Mary Jane, Elizabeth's mother paid a $10 fine and lied to the judge, insisting that the woman was indeed her slave and should be returned to her possession.

Mary Jane initially told her captors she was named Mary Jane Henley and then Mary Jones, displaying a talent for subterfuge that would serve her well inside the Confederate White House. She would spend some time getting acquainted with the Davis family and earning their trust. Whenever the First Lady's gowns needed mending, she would bring them to a seamstress across the street, who worked for a family friend; that way Elizabeth could easily arrange a meeting. Mary Jane would be a sleeper agent, waiting to be activated at just the right moment, and many months might pass before they had any contact at all.

Elizabeth considered the risks of engaging her servant in such a mission. Mary Jane, unlike Elizabeth herself or Rose Greenhow, didn't have the protection of race or class or social status. If caught, she might become a public and deadly example, hanged on the gallows at Camp Lee, or she might report Elizabeth for solicitation of espionage in an attempt to save herself. Elizabeth believed that the affection between herself and her servant was mutual, but one could never gauge what might happen under pressure, let alone under threat of death.

After a moment of hesitation, Mary Jane agreed to the plan.

Elizabeth selected her finest walking dress, pinning a calycanthus flower to the collar to ward off the smells of the city, and smoothed the dark blond curls that wreathed each temple. Broad cedar trees rose along Grace Street, shading and dappling the lawns, sketching shadows that shifted with each fetid breeze. She hurried on through Church Hill, the homes strung along like magnificent beads, every other one a different style—Federal, Greek Revival, Italianate, Second Empire, Queen Anne—as if the inhabitants all longed for a different time and place.

It was a fifteen-minute stroll to the Confederate White House, a route that took her past Seabrook's Warehouse, the train depot, the Medical College of Virginia. Every few blocks she turned her head halfway, checking to see if she was being followed. By all appearances this was just a simple social call, one refined Southern lady extending a courtesy to another—a visit Elizabeth needed to make now, while the war was still new and the rebels were holding their own.

The Confederate White House.

The Confederate White House stood at the corner of Twelfth and Clay Streets, a gray-stuccoed neoclassic mansion set high on a hill, overlooking the slave markets in Shockoe Valley. The house was one of very few in Richmond that had a proper water closet; a cistern on the roof collected rainwater to flush the toilet and then emptied waste directly onto Clay Street. The servants' quarters, where Mary Jane would live, were located in an adjoining outbuilding. A diverse group of slaves, free blacks, and immigrants tended to Jefferson and Varina Davis and their four children and organized the daily operations of the home, the cost of which averaged $2,800 per month—this, when the average Virginia farmer earned $137 per year. Katherine, the Irish nanny, supervised play in the nursery and the backyard, where Jeff Davis Jr. shot imaginary Yankees with his miniature cast-iron cannon. Jim Pemberton worked as the president's enslaved manservant, and his wife, Betsy, as an enslaved maid. Another Irishwoman, Mary O'Melia, acted as general housekeeper and devoted companion to Varina, helping the First

Lady dress in the morning and overseeing receptions for Confederate officials. William Jackson, the president's enslaved coachman, prayed daily for his boss's defeat.

Elizabeth navigated her way to the front steps, and Henry, the enslaved butler, welcomed her into the entrance hall, where life-size statues of Comedy and Tragedy clutched gas lamps and loomed in mute judgment. She was led into the drawing room, done entirely in French rococo, red damask adorning each wall and chair. Varina's Parian ware figurines of goddesses and cherubs, a collection that her husband dismissed as "trumpery," cluttered every surface. In the center of the room stood a Carrara mantel with goddesses engraved in the marble. Every evening before bed, Varina's boys, Jeff and Joe, stopped to kiss them.

Elizabeth eased herself into a settee across from Varina Davis. The First Lady was pregnant, due to give birth just before Christmas, and had forgone silk and hoop skirts for a sacque dress, petticoat, and maternity corset, threaded with cords instead of steel for greater flexibility. In another place and time the two women might have been friends. Despite her elevated position, Varina, too, was viewed warily by old Richmond blue bloods. Society ladies, even the ones who professed to like her, whispered about her dark skin and "tawny" looks, calling her a mulatto and a "squaw" behind her back. They thought her too sarcastic, too blunt, too crude. During a summer dinner party, when a guest lamented that the underwear for an entire Confederate regiment had been made with two left legs—most men "dressed right," requiring a more generous cut on that side—she horrified everyone by laughing out loud.

Then there was the troubling matter of her Northern kinfolk and associations: her paternal grandfather was a former governor of New Jersey, and Varina, like Elizabeth, had been educated in Philadelphia (her mother's family hailed from Mississippi, and she jokingly called herself a "half-breed Yankee on one side and Confederate on the other"). Richmond authorities had begun screening mail and arresting people for writing to relatives in the Union, a policy that did not deter Varina in her own correspondence. She trusted, rightly, that her social status and gender gave her immunity from formal investigation, if not from suspicion.

Pregnancy was not a topic of polite conversation, and so Elizabeth and Varina chatted about the holiday season—how lovely that the Davis children were gathering some of their old treasures for the orphans' tree. And the New Year's Day open house promised to be delightful, especially with the unseasonably warm weather—which one newspaper called a "harbinger of gladness in the future"— and high spirits after the recent Confederate victory at Ball's Bluff (a pity, though, that Lincoln seemed poised to release Mason and Slidell and end the Trent Affair). How many callers did President Davis expect? Would the Armory Band play on the lawns? Elizabeth had heard about the carriage accident Varina suffered back in October—how fortunate that her injuries weren't grave.

First Lady Varina Davis.

At the right moment Elizabeth segued into the reason for her visit. She understood that the First Lady was seeking qualified help. Might Elizabeth recommend one of her own staff, Mary Jane, "an excellent house servant who never faltered in the dining room or parlor?" True, Mary Jane was not terribly bright, but Elizabeth found her to be a dedicated, hardworking girl. It was the least she could do to help, especially at this busy time.

Varina accepted, and Elizabeth prepared her servant for the move. If all went smoothly, the Confederate White House would be Mary Jane's home for the duration of the war. She would act the part of the simple, illiterate maid, obsequious in manner and bumbling in speech. No one would notice her slyly weaving herself into the routines and customs of the home. No one would think twice when she cleaned the president's library, lingering as she dusted the desk piled with maps of fortifications and statistics about his troops. No one would guess she could comprehend, let alone memorize, entire discussions between Davis and his generals. She knew how to be invisible, to render herself merely the sum of others' projections, and she would be ready whenever Elizabeth gave the word.

[PART TWO]

1862

Not Your Ideal of a Beautiful Soldier

THE SHENANDOAH VALLEY, VIRGINIA

One January morning Belle donned her dark green riding dress, accessorized with a lieutenant colonel's pair of shoulder straps and a wool felt hat, a feather pinned to its crown. She saddled Fleeter and, riding astride, began galloping through the streets, defying both social convention and Martinsburg's law against traveling faster than at a canter. Confederate soldiers, currently in control of the town, were busy building earthworks fortifications along its perimeter, stabbing at the frozen ground with shovels and picks and piling woven bundles of brush. Waving, she continued, crossing the arched stone bridge over Opequon Creek and plunging into the valley. It unfurled for miles beyond, carpeted alternately by field and forest, bare branches like upturned spiders' legs scrabbling at the sky, snow-dappled mountains fencing her in on all sides, an exquisite but interminable trap.

She was tired of the long, dark slog of winter, spent under her mother's close and fretful watch. After the skirmish over the bottle of whiskey, Mary Boyd—appalled by the incident and concerned about Belle's increasingly dangerous behavior—began imploring trusted Confederate soldiers to chaperone her. Captain John Q. Winfield of the 7th Virginia Infantry obliged, taking Belle to a wedding reception in Martinsburg and delighting in her company. "She is quite a favorite with me," he wrote to his wife, "possessing an originality and vivacity, no-care-madcap-devil-of-a-temperament that pleases. Her mother who is much of a lady and had shown me much kindness asked me to take her daughter under my charge."

Another soldier, Private J. C. Webb of the 27th North Carolina Infantry, was equally charmed. "Not what I call a beauty," he decided, "but a handsome woman" who seemed she would "dare do anything she made her mind up to, regardless of consequences." Belle assured him that she had been much more reserved before the war, and offered Webb a ride in her private carriage. She also had a brief romance with one Dr. Cherry, a wealthy thirty-five-year-old soldier originally from Mississippi, and told everyone they would marry in February—a date she never intended to keep.

Fun diversions, all of them, but Belle longed for spring and the return of the war, with all of its "perils and its pleasures, its griefs and its joys." She feared that the enemy might forget her very existence unless she took measures to remind them. The Virginia papers reported daily updates on General George McClellan, who remained ill from typhoid fever despite the best efforts of his doctor and cheerful serenades from a regimental brass band. He conducted all business from his bed—not that there was much to do, since he had no immediate plans to attack the Confederates and take Richmond. On January 10 an exasperated President Lincoln called a council at the White House and spoke frankly with his aides. "I am in great distress," he confessed. "If something is not done soon, the bottom will be out of the whole affair, and if General McClellan does not want to use the army, I would like to borrow it, provided I could see how it could be made to do something." He was equally frustrated with events in the western theater, save for the capture of Paducah, Kentucky, by Brigadier General Ulysses S. Grant. The Union military command had split into three separate departments—the Department of Kansas, the Department of Missouri, and the Department of the Ohio—and no one could agree on a strategy for operations.

Belle rode northwest, heading toward Morgan County, the frigid wind slapping her cheeks. Her greatest adventure over the winter had occurred only inside her mind. During a ride just like this one, she told neighborhood friends, she happened to run into two Confederate officers, one of them a cousin. There was a sudden crack of a rifle, causing Fleeter to spook and tear off in a panic. She clenched the reins, pulling back and seesawing left and right, ordering him to stop, willing him to be calm with her calm. His defiance infuriated her, and she could do nothing but hang on and let him take her through the valley wherever he wished to go. After his panic subsided, she continued in that direction for several miles, eventually entering Union territory. The rebel officers didn't dare follow.

Without hesitation, she rode straight up to the Union officers in command of the picket.

"I beg your pardon," she said sweetly, "you must know that I have been taking a ride with some of my friends. My horse ran away with me and has carried me within your lines. I am your captive, but I beg you will permit me to return."

One of the officers stepped forward and bowed. "We are exceedingly proud of our beautiful captive," he replied, "but of course we cannot think of detaining you." A pause, and then his tone turned conspiratorial: "May we have the honor of escorting you beyond our lines and restoring you to the custody of your friends? I suppose there is no fear of those cowardly rebels taking us prisoners?"

Belle smiled, although inwardly she seethed: "They little thought how those words, 'cowardly rebels,' rankled in my heart." She smoothed the anger from her voice. "I had scarcely hoped for such an honor," she said. "I thought you would probably have given me a pass, but since you are so kind as to offer your services in person, I cannot do otherwise than accept them." Then she, too, spoke as if they shared a private joke, adding, "Have no fear, gentlemen, of the cowardly rebels."

Belle and two Union officers rode off. As soon as they crossed back into Confederate territory, her rebel friends appeared from behind a cluster of trees. She reveled in all four men's "surprised and embarrassed" expressions, breaking the silence with a laugh.

"Here are two prisoners that I have brought you," she told her friends, and then turned back to the Union officers. "Here are two of the 'cowardly rebels' whom you hoped there was no danger of meeting!"

The Union men were silent for a moment, studying her.

"And who, pray, is the lady?" one of them asked.

She bowed and replied, "Belle Boyd, at your service."

"Good God!" the officer gasped. "The rebel spy!"

"So be it, since your journals have honored me with that title."

The Union men offered no resistance as Belle and her friends escorted them to headquarters, where the officer in command ordered them to be detained. "Let us hope they have profited by the lesson," Belle concluded her story, and announced her motto for the new year: "All was fair in love and war."

A lthough the war remained on pause for the Army of the Potomac, Confederate general Stonewall Jackson went on the march despite the winter weather. He was preparing for his Valley Campaign, the goal of which was to unnerve the Union, prevent Washington from reinforcing McClellan, and keep Virginia in the Confederacy. "If this Valley is lost," he said, "Virginia is lost." Nine thousand Confederate soldiers started at Winchester, Virginia, and moved toward Bath in the western part of the state. The general was intent on disabling long stretches of the Baltimore & Ohio Railroad, tearing up tracks and torching depots, and firing shots and shells at dams on the Chesapeake & Ohio Canal.

Night after night the sky dropped snow and hail, turning the roads so slick that the men had to march in side ditches. Horses were unhitched from the cannon and wagons and sent forward through the woods, leaving the troops to pull the cargo with ropes and chains. Soldiers slipped and fell, firing their guns on the way

General Thomas J. "Stonewall" Jackson.

down. Together they slept on the ground, their blankets freezing fast to the earth; several men never woke up. Belle's father told her that some complained about Stonewall staying in the town hotel, warm in bed—"when, all of a sudden, there was a motion under a blanket nearby, and up rose a tall, stalwart figure . . . he had slept out there through all that sleet storm with his boys."

Belle idolized Stonewall Jackson from the moment her father enlisted in his brigade, harboring fantasies that alternated between the familial and the romantic. She once dreamed that the general entered her bedroom while she slept, his eyes resting "sorrowfully upon me," while her father stood silently behind him. "I was dumb," Belle recalled, "or I should have spoken, for I did not feel alarmed. As I looked upon these two standing together, General Jackson turned and spoke to my father. I remember the words distinctly: 'It is time for us to go.' And, taking my father's hand, he led him away, adding as he did so, 'Poor child!'" On other occasions Belle indulged in more amorous thoughts about the general, and was overheard "giving vent to romantic desires to occupy his tent and share his dangers."

Stonewall Jackson had just turned thirty-eight years old and looked, some said, more scarecrow than human, with eerily bright blue eyes and a mangy brown mass of beard. His preferred uniform consisted of a threadbare single-breasted coat left over from his service in the Mexican War, a battered kepi with a broken visor worn low to conceal his eyes, and an oversize pair of flop-top boots for his size-fourteen feet. His horse, Fancy (whom everyone else called Little Sorrel), stood only fifteen hands high, and Jackson rode him with his feet drawn up so as to avoid dragging them on the ground. He spoke seldom and almost never laughed. On the rare occasions when he did, he tossed back his head, let his mouth gape open, and made no sound whatsoever. Once an injured Northerner, captured by Jackson's men, asked to be lifted to up to catch a glimpse of the general. He stared at Jackson for a moment, and then, in a tone of disbelief and disgust, exclaimed, "O my god! Lay me down!"

Jackson was as idiosyncratic as he was brilliant; his peculiar habits and tendency toward hypochondria became as legendary as his skill on the battlefield. He thought of himself as being "out of balance," and even under fire would stop to raise one arm, waiting for the blood to rush down his body and establish equilibrium. He refused to eat pepper because it made his left leg weak. A partial deafness in one ear often made it difficult for him to detect distant artillery fire or to determine the direction from which it came. Convinced that every one of his organs was malfunctioning to some extent, he self-medicated with a variety of concoctions, inhaling glycerine and silver nitrate and ingesting a number of ammonia preparations. "My afflictions," he told his sister, a Unionist, "I believe were decreed by Heaven's Sovereign, as a punishment for my offenses against his Holy Laws."

Twice a day, rain or shine, Jackson slipped away from camp and found a secluded field. He perched on the edge of a fence and prayed for an hour, hands clasped, face turned upward, tears spilling, mouth forming noiseless words—a ritual that may or may not have had something to do with a recurring fear that he was possessed. He was reluctant even to read a letter from his wife (whom he called "my little dove") on Sundays. He forbade profanity, alcohol, and mingling with the camp followers, although he considered himself a genuine and ardent admirer of true womanhood, and was said to never pass a lady—of high or low degree—without tipping his filthy cap.

Despite his piety he was utterly unfazed by the prospect of murder or death. "He would have a man shot at the drop of a hat," one Confederate soldier declared, "and he'd drop it himself." He ordered the execution by firing squad of a soldier, a father of four, for assaulting a man of higher rank. The general prayed over the incident and found, as he always did, that God's will matched up with his own. During one battle, he inquired sharply about a missing courier and was told the young man had been killed. "Very commendable, very commendable," Jackson muttered soberly, and put the matter out of his mind.

Belle would never forget the first time she saw Stonewall Jackson, just before the Battle of Manassas, when he was still a colonel. He was not called "Stonewall" then nor nationally known, but he was already worshipped by secessionists in the Valley. She tracked him down at Ramer's Hotel in Martinsburg, finding him in the parlor, surrounded by people of all ages and both genders desperately vying for his attention. Men crushed toward him with outstretched hands. Children pressed close to gawk and clutch the hem of his coat. Belle stood back while a trio of women cut the Confederate buttons off his clothes. "Really, ladies," Jackson joked, "this is the first time I was ever surrounded by the enemy!"

At the right moment Belle approached, holding a bouquet she had gathered just for him.

"Colonel," she asked, "won't you give me a button?"

He whirled around, held out his arms, and said, "Look at me," indicating that there wasn't one button left to hold either coat or vest together. Belle laughed and then stepped forward, close enough to smell his skin, to see every pore and whisker: "He did not look old, but he was a man with heavy lines in his face. There was a great deal of character in them. He was not your ideal of a beautiful soldier, but he was a very courteous man, very good, and there was something attractive about him."

While acting as a courier over the past few months Belle mostly interacted with Jackson's aides, only rarely scoring a peek at the general himself. But this was Jackson's time, the year he would triumph over the North and cement the Confederacy, and she vowed to make him notice her.

SHE WILL FOOL YOU OUT OF YOUR EYES

WASHINGTON, DC

Any day Rose expected to receive her "special" fruitcake, in which apples and raisins were replaced by a wad of Confederate bills and a letter detailing her escape route to Richmond. When a guard finally came to her room, holding a cake aloft, he said that he had investigated her original confection with the aid of a penknife and had found it unsuitable: perhaps she would instead accept this version, with all of the usual ingredients? His lips tugged up into a smirk, waiting for her response, for her fury to rise like cream, and she couldn't stop herself from yielding. She yanked the cake from the platter and hurled it down the stairs, watching it bounce once and split against the wood, listening to the guard laugh behind her.

Two weeks later, on Saturday, January 18, Rose was reading in the library while Little Rose sat at her feet, playing with her dolls. A guard appeared in the doorway and told her that, after five months of house imprisonment, she was being transferred to a military prison on Capitol Hill. She had two hours to pack, including the time it would take to examine each article before it was placed in a bag. To her surprise, she was permitted to bring her pistol but no ammunition, a situation she hoped to remedy from behind bars. Little Rose would be accompanying her, staying in the same cell, and she also hoped that together they could continue spying from the inside.

Pinkerton ordered that the windows of Rose's house be boarded and all paper removed from the premises. He wanted to transport her in a covered wagon with military escort, but one of her more sympathetic guards, Lieutenant Sheldon, insisted that she ride in a private carriage. "Good-bye, Sir," Rose said, and paid him the highest compliment she could muster for a Yankee: "I trust that in the future you may have a nobler employment than that of guarding defenseless women." Little Rose looped her arms around his neck and hugged him.

Rose seized one more opportunity to bedevil her captors, leaving two bottles of synthetic ink, the kind used to encode messages, next to her sewing machine—a way to distract attention from her real means of communication. She took one last look at the framed portrait of Gertrude—her daughter had inherited her "same strange fancy of the eye"—and left her home, she feared, for the very last time.

Old Capitol Prison.

The Old Capitol Prison stood on the east side of First Street, a dreary jumble of structures that had temporarily housed Congress when the British burned the unfinished Capitol during the War of 1812. Its heart was a three-story pile of brick dominated by a large arched window, barred with wood latticing to deter escapes. For Rose, the building embodied a piece of personal history. In a previous incarnation it had been Hill's Boarding House, her aunt's establishment and her very first home in Washington, the place where she and her sister came to live after their father was killed and their mother could no longer care for them.

She was only three or four years old, living in Montgomery County, Maryland, on a farm called Conclusion, where fifteen slaves planted wheat and tobacco and obeyed her father's every order. John O'Neale (the final *e* was dropped from the family name in Rose's early childhood) was a libertine, with a particular taste for horse racing, cockfighting, and drinking. He often brought his favorite slave to the local tavern. Jacob was his most prized possession, twenty years old and freakishly double jointed, able to execute astounding acrobatic feats for the amusement of John's friends. One night in April 1817, John and Jacob went out drinking, but only Jacob came home.

Early the next morning Jacob set out to look for his master and saw him on the road, about 150 yards from the house, thrown from his horse and lying on the ground, bleeding from the head. He rushed home and sought help from another slave, who advised him go back and make sure that John was dead—or else he might blame Jacob for his injuries. Jacob allegedly complied, hoisting a rock and crushing his master's skull. A doctor's examination discovered three head wounds, one from the rock and two from the fall, but it was impossible to conclude which was the fatal blow. Jacob asserted his innocence, was tried and found guilty, and was hanged six months later. For Rose, her father's murder became a hazy, shapeless memory, but one that colored every moment of her life.

Rose's mother, Eliza, found herself a widow at twenty-three, with four daughters and another on the way, and only $2,600 to her name. The price of wheat and tobacco was plummeting and she would soon be forced to sell the farm. In 1828, eleven years after John's death, she sent her second and third daughters, the "extra" ones, Ellen and Rose, to their aunt's boardinghouse in Washington, hoping the city would be kind to them.

Behind the walls of this building, as a teenager, Rose lived among Whig and Democratic congressmen from North Carolina, Kentucky, and Tennessee, who stayed there when Congress was in session. Rose and Ellen, two years older, always joined the men for dinner, passing heaping platters of crab cakes and baked perch from the Chesapeake Bay, listening to political gossip, learning the nuances of cronyism and the etiquette of the backroom deal. These conversations, more than her occasional lessons in a class-room, constituted Rose's true education. When weather permitted, she donned her best bonnet, took a carriage to the Capitol, and watched debates in the Senate chamber, where hickory logs smoked in two rusty stoves and streams of tobacco juice trickled across the floor. She scrutinized the politicians' body language as closely as their speech, noting how the slightest gesture—a twitch of the jaw, a dip of the brow—could betray doubt or weakness. She delighted in the hidden meanings skimming below the surface of their sentences, the art and power of the unspoken. She grew acutely aware of her own physical idiosyncrasies and quirks, of how her voice sounded inside her own ears, and she practiced con-trolling its resonance and timbre; no one would ever have cause to misconstrue her words.

The men noticed Rose, too, with her black eyes and sleek skein of hair, her figure that curved like a vase. She had learned enough about politics to compare rumor and truth, and culti-vated an air both regal and flirtatious. "She was a celebrated belle and beauty," one acquaintance said, "the admiration of all who knew her," and numerous men vied to escort her about town.

Tennessee congressman Cave Johnson was an early suitor and, at twenty years her senior, thirty-six to her sixteen, old enough to be the father she'd barely known. He introduced her to his powerful friends, including Pennsylvania congressman James Buchanan, who, nearly thirty years hence, would seek Rose's advice during his successful presidential campaign. Every night, Rose especially looked forward to sitting in the parlor with South Carolina statesman and future vice president John C. Calhoun, who became her "kindest and best friend." She was at Calhoun's bedside when he died in 1850, and now, entering the prison, she recognized the room in which he had drawn his last breath.

A guard led her and Little Rose to the office of Superintendent William P. Wood. He was thirty-six years old and a solid block of a man, seemingly as wide as he was tall. One colleague described him as "short, ugly, and slovenly in his dress; in manner affecting stupidity and humility, but at bottom the craftiest of men."

Little Rose stepped forward and looked up at him. "You have one of the hardest little rebels here you ever saw," she said. "But if you get along with me as well as Lieutenant Sheldon"—the one guard both she and her mother liked—"you will have no trouble."

For once, Rose objected to her daughter's bravado: "Rose, you must be careful what you say here."

The Old Capitol Prison had originally been intended only for Confederate prisoners of war but soon housed a motley assortment of inmates. The two northerly buildings were occupied by blockade runners and bounty jumpers—men who enlisted in a regiment, collected the on-the-spot cash bounties given to volunteers, and then deserted, only to magically turn up a few weeks later in a different state, under a different name, to join another regiment. Confederate soldiers captured in battle were confined to the lower floor and basement, and a four-story structure in the middle section held suspected spies.

Rose and her daughter settled into their quarters, a room in the back with a view of the prison yard; Superintendent Wood feared Rose might connect with Southern sympathizers strolling along the sidewalk out front. The space was ten feet by twelve and furnished with a straw bed, soiled cotton sheets, a small feather pillow ("dirty enough," Rose observed, "to have formed part of the furniture of the Mayflower"), a few wooden chairs, a wooden table, a fireplace, and a cracked mirror. Every breath carried the overwhelming stench of the latrines, located across the prison yard in a wooden building that doubled as the hospital and apothecary.

A carpenter arrived to install bars on the windows. He told Rose that the prison's commanding general had sketched the drawings himself, placing the bars to block maximum air and sunlight. When Superintendent Wood protested that this measure wasn't necessary, Andrew Porter, Washington's provost marshal, scoffed. "Oh Wood," he said, "she will fool you out of your eyes. She can talk with her fingers."

The following morning she opened her barred window, hoping for a sip of fresh air.

"Get away from that window!" a guard called from the yard below. He raised his musket and she responded in kind, leveling her unloaded pistol at his head. She'd heard he had orders "not to shoot the damned Secesh woman, who was not afraid of the devil himself," and kept her aim steady long after he had given up.

The guards knew exactly how to retaliate, Rose had to admit, filling the cells around her with escaped slaves, who streamed into Washington from Virginia and were admitted to the prison as an act of charity. She did not appreciate being taunted by these Negroes, who told her, "Massa Lincoln made me as good as you," and insisted that they be called "gem'men of color." Whenever a Negro servant entered to clean Rose's room, the guards bolted and locked the door. She screamed and rapped her knuckles raw against the iron, begging to be let out.

"The tramping and screaming of Negro children overhead was most dreadful," she wrote. "The air was rank and pestiferous with the exhalations from their bodies; and the language which fell upon the ear, and sights which met the eye, were too revolting to be depicted. . . . Emancipated from all control, and suddenly endowed with constitutional rights, they considered the exercise of their unbridled will as the only means of manifesting their equality."

As the news spread about her incarceration, Rose became a curiosity, the oddest and most exotic creature in the jungle, expected to perform on command. Large parties of Yankees paid guards $10 for just a brief glimpse of the "indomitable rebel," staring into her cell through the barred door. An editor from a Rochester journal tried flattery, telling Rose she had been detained because of her writing talent, which was on a par with that of Madame de Sévigné, a French aristocrat known for the wit and brilliance of her letters. Next descended a group of Boston society women, one of whom challenged: "Confess that it was love of notoriety which caused you to adopt your course, and you have been certainly gratified." They called out to Little Rose, asking her to twirl around and say a few words.

Rose and Little Rose in the courtyard of Old Capitol Prison, 1862, taken by Mathew Brady.

"My little darling," Rose whispered to her daughter, "you must show yourself superior to these Yankees, and not pine."

"Oh Mamma, never fear—I hate them too much," Little Rose said. "I intend to dance and sing 'Jeff Davis is coming' just to scare them!"

The days slogged into weeks. At night Rose spent hours sifting through her daughter's hair for bedbugs and burning them off the walls. They lay in squeamish silence, listening to rats scuttling across the floor. The straw bed prickled against their skin, and Little Rose wept against her neck: "Oh Mamma, the bed hurts me so much." When she was able to sleep, she screamed herself awake from dreadful dreams. She was occasionally allowed outside to play in the prison yard, only to return in tears after the guards insulted her mother. The girl cried constantly from hunger. "To-day the dinner for myself and child," Rose wrote, "consists of a bowl of beans swimming in grease, two slices of fat junk, and two slices of bread. Still, my consolation is, 'Every dog has its day.'"

Little Rose awoke one morning in mid-February unable to rise from the bed, seized by a slow and nervous fever. Her tongue turned red, with matching spots sprouting across her chest and down her limbs, and her eyes were all pupils. Rose pressed her hand to her daughter's forehead, noting that her "round chubby face, radiant with health, had become pale as marble." She had camp measles, one of the fastest-spreading diseases; during the previous summer, one of every seven Confederates serving in northern Virginia had contracted measles, with more than eight thousand cases reported in three months. It could be fatal, leading to a collapse of the immune system and engorgement of the lungs, and Rose feared that her daughter was "fading away." She thought of the Lincolns' son, eleven-year-old Willie, on his deathbed from typhoid fever.

On February 18 she sent a letter to the provost marshal, begging for a visit from the family doctor, "unless it be the intention of your Government to murder my child."

A few hours later the prison physician, Dr. Stewart, let himself into her room. "Madam," he said, "I come to see you on official business."

Rose positioned herself between Little Rose and the doctor, whom she silently nicknamed "Cyclops." She did not trust this Yankee doctor to touch her daughter; she would hold out for her personal physician, even if it meant delaying Little Rose's care. "Sir," she replied, "I command you to go out. If you do not, I will summon the officer of the guard and the superintendent to put you out."

The doctor stepped around Rose and started for the bed, where Little Rose lay shivering. Rose thrust out an arm. "At your peril but touch my child," she warned. "You are a coward and no gentleman, thus to insult a woman."

Stewart refused, his voice shaking as he spoke: "I will not go out of your room, madam."

Rose pushed around him and pounded at her door. "Call the officer of the guard!" she shouted, and a lieutenant appeared. "Sir," Rose said, "I order you to put this man out of my room, for conduct unworthy of an officer and a gentleman, and I will report you for having allowed him to enter here."

The lieutenant glanced at the doctor, who was his superior officer, and back at Rose. He tried diplomacy: "I am sure Dr. Stewart will come out if you wish it."

She shook her head. "Do your duty; order your guard to put him out."

The doctor gave up. As soon as the door shut behind him she began to laugh, and the very sound of it—so strange and unexpected, and so long since she'd heard it—set her off again, pounding at the walls in bleak hysteria. "It was farcical in the extreme," she wrote, "this display of valour against a sick child and careworn woman prisoner," and she slid to the floor, giggling into her knees, giddy with the notion that the careworn woman had won.

Rose's personal doctor visited and tended to Little Rose. The girl became herself again, eyes brightening and cheeks fattening, but Rose now sensed herself fading away. Her mind felt dim, slower to latch on to thoughts, and she developed a nervous tic, an involuntary jerking of the head. It was March, two months since she'd been taken from her home, with no resolution or even word of her case. She heard the guards boast about the Union's recent capture of Fort Henry and Fort Donelson in Tennessee, major victories for Ulysses S. Grant; Kentucky would remain in the Union, and now Tennessee was vulnerable to an invasion. Curious visitors still came to gawk and prod and taunt. A *New York Times* correspondent observed that her "vivacity is considerably reduced since her rebel plumage has been clipped with Yankee shears."

As much as Rose loathed these intruders, she realized their potential value; they might divulge further news of the outside world if she asked sly enough questions. When her attempts failed, she summoned up all of her strength and pried loose a plank in the floor of her cell, creating a gap for Little Rose to slip through. Clutching her daughter under her arms, she lowered her to the cell below, which held Confederate prisoners of war. The men caught the girl by her legs, soon sending her back up with reports about the Union army.

Rebel Vixens of the Slave States

WASHINGTON, DC, AND ON THE MARCH

In March, Colonel Orlando Poe, the commander of the 2nd Michigan, relieved Emma of her nursing duties and appointed her regimental mail carrier; soon she would be promoted to postmaster for the entire brigade. Poe would later say he chose her for the job because, "as a soldier, Frank Thompson was effeminate looking," and he wanted to "avoid taking an efficient soldier from the ranks." She got a new horse and privately named him Frank, as if to reinforce her own secret identity, an extension of herself over which she had complete control.

Emma relished her new role, riding Frank from camp to camp to Washington and back, knowing hers was the face everyone most wished to see. "The mail was even more heartily received than other things," she wrote. "It was nothing short of a calamity for a heavy mail to be captured by the enemy." Occasionally she encountered Jerome Robbins along her route but they maintained a cool distance, avoiding all talk of the past, and he receded further from her mind when word came that they were finally to advance on the Confederates.

On March 14, General McClellan issued an announcement to be read to all troops at roll call. "I will bring you now face to face with the rebels," he told them. "Ever bear in mind that my fate is linked with yours. . . . I am to watch over you as a parent over his children; and you know that your General loves you from the depths of his heart." Instead of marching through northern Virginia, where he feared a massive Confederate force lay waiting, the Union commander planned to ship his Army of the Potomac to the southern tip of the Virginia Peninsula by sea, and then fight his way west to Richmond.

McClellan's strategy for his Peninsula Campaign required one of the most ambitious maritime movements in American military history. It would take three weeks and four hundred vessels—steamers, barges, sloops, and canal boats—to ferry the Army of the Potomac to Fort Monroe: 121,500 men, 14,594 horses and mules, 1,150 wagons, 44 batteries of artillery, 74 ambulances, pontoon bridges, tons of provisions, tents, telegraph wire. One British observer likened the feat to the "stride of a giant." Emma's regiment boarded a steamer called the *Vanderbilt*, a process that took several hours. Company after company clustered together on board until every foot of space was occupied. They disembarked three days later during a ferocious rainstorm, marching twenty-three miles in what Emma called "a fair specimen of Virginia mud," sinking, in some spots, all the way to their knees.

They set up camp, bivouacking on the marshy ground, short on provisions, horses starving. The rebels found them soon enough, lobbing shells and thirty-two-pound cannonballs that burst over their heads or fell within feet of their tents. A party of fugitive slaves—"contrabands," they were called—also made their way to camp, having escaped from rebel territory where they worked on fortifications at the James River. They arrived late one night in the midst of a relentless rain, falling to their knees and shouting, "Glory! Glory to God!"

The commotion awakened Emma, and she emerged from her tent to investigate. "There they were," she wrote, "black as midnight, all huddled together in a little group, some praying, some singing." She watched as Jerome rushed to help the regimental surgeon tend to one man, who had been shot by the rebels as he fled. Her heart clenched at the sight of her old friend, lost in his work, winding a bandage with delicate precision.

Other soldiers built a fire and served hot coffee and bread. Emma sat down with the contrabands, and was struck by how intelligently they could converse about Christ. "Why should blue eyes and golden hair," she wrote, "be the distinction between bond and free?" The issue of slavery became real to her in a way it hadn't been before; she understood, to a lesser degree, what it meant to feel vulnerable inside your own skin.

Emma was often sent to procure supplies for the hospitals and food, seeking local citizens willing to exchange butter, eggs, milk, and chicken for Federal greenbacks. The area of the Peninsula around Fort Monroe remained under Federal control, but between the fort and Richmond sprawled sixty miles of strange and hostile territory, the land a swampy patchwork of gullies, creeks, and ravines. Emma kept watch for both rebel pickets and "rebel vixens of the slave states," whom she compared to Parisian women during the Reign of Terror. She carried her Moore seven-shooter revolver at all times.

One morning she set out for a five-mile ride to an isolated farmhouse and was surprised to find it in good condition—a rarity for any home in the Peninsula—with the fences still standing and cornfields thriving as if there were no war at all. She rode up to the house and dismounted, hitching her horse and ringing the bell. A tall woman, about thirty years old, invited Emma inside with a gracious sweep of her arm. Within the past three weeks she had lost her father, husband, and two brothers in the rebel army and was now in "deep mourning," which lasted up to two and a half years and required an all-black ensemble: dress, veil, bonnet, cape, and jewelry made of polished coal, containing small photos of the deceased and locks of their hair.

The woman spoke in a languorous murmur, unfurling each vowel: "To what fortunate circumstance am I to attribute the pleasure of this unexpected call?"

Emma said she was a Union soldier and willing to pay for food. She noticed agitation creep across the woman's face, a nervous shift of expression. The woman directed her to sit in another room while she prepared the items, but Emma declined, wanting to remain in a position where she could watch every movement. Her hostess seemed to be stalling, walking around in her stately way without clear purpose, and Emma feared "her ladyship" was contemplating the best mode of attack.

Emma stood and asked if her things were ready.

The woman smiled. "Oh," she replied, "I did not know that you were in a hurry. I was waiting for the boys to come and catch some chickens for you."

"And pray, madam, where are the boys?"

She paused, glanced toward her door. "Not far from here."

Emma took a step backward. "Well," she said, "I have decided not to wait; you will please not detain me any longer."

The woman nodded and began packing eggs and butter into a basket. She was "trembling violently," Emma saw, "and pale as death." With shaking hands she gave Emma the basket and refused to accept a greenback, insisting it was "no consequence about the pay."

Emma thanked her and started for the door. The woman followed a pace behind. Emma unhitched Frank, climbed into the saddle, and set off. She had ridden just a few yards when she heard the snap of a gunshot behind her, the ball hurtling just above her head.

She whipped back around as the woman fired again. The bullet veered wide, missing her by inches. She found her seven-shooter and considered where to aim: "I didn't want to kill the wretch, but I did intend to wound her." The woman dropped her gun, raising her arms in a tentative surrender, and Emma sent a bullet through the meat of her left palm.

The rebel woman dropped to the ground, keening, transfixed by the hole in her hand. Emma took the rebel's pistol, cinched a halter strap tight around her wrist, and tied her to her saddle. Remounting her horse, she dragged the woman fifty feet along the road, her black dress grinding against the dirt, her legs flailing under a froth of crinoline. Over the pounding of hooves Emma heard her begging for release, and worried that she might draw the attention of any rebels nearby.

Emma halted, turned around in the saddle, and raised her revolver.

"If you utter another word or scream," she said, "you are a dead woman."

Emma hoisted her captive into the saddle, fashioned a tourniquet from her handkerchief, and took her to the hospital. She was unnerved at the ease with which she'd pulled the trigger, especially since her target was a woman.

In order to advance to Richmond via the York River, McClellan had to drive the Confederates from Yorktown—no easy feat, as the rebels had built earthworks on top of positions abandoned eighty years before, when the British surrendered to George Washington and ended the American Revolution. He also had no idea how many enemy troops had dug in at Yorktown, a lapse in intelligence that Confederate general John Magruder rushed to exploit. A flamboyant, dramatic Virginian (during the Mexican War he had staged a performance of *Othello* in which a young Ulysses S. Grant, dressed in crinolines, tried out for Desdemona), Magruder executed a masterly display of special effects, frantically shifting his eight thousand soldiers around from one part of the line to the next, parading them in an endless circle. As an added touch, he kept up a sporadic barrage of artillery and ordered his bandsmen to play well into the night.

The charade worked. McClellan believed that Magruder's force numbered one hundred thousand, an overestimation supported by both Allan Pinkerton and Dr. Thaddeus Lowe, an aeronautical expert who pioneered the use of hydrogen balloons to gather intelligence. Emma watched the professor in action, "making balloon reconnaissances, and transmitting the results of his observations to General McClellan by telegraph from his castle in the air, which seemed suspended from the clouds." McClellan in turn telegraphed President Lincoln: "It seems clear I have the whole force of the enemy on my hands." He asked for reinforcements and proposed to embark on "the more tedious but sure operations of siege."

Lincoln implored him to advance: "I think you had better break the enemy's line. . . . The country will not fail to note—is now noting—that the present hesitation to move upon an entrenched enemy is but the story of Manassas repeated . . . *you must act*."

Once again McClellan ignored him. "The President very coolly telegraphed me yesterday that he thought I had better break the enemy's lines at once," he wrote to his wife. "I was much tempted to reply that he had better come and do it himself." Instead the general oversaw the construction of ever more imposing earthworks and sent Thaddeus Lowe back up in his balloon. Confederate general Joseph Johnston took advantage of the lull, moving his army from Richmond down to the Peninsula. "*No one* but McClellan," he said, "could have hesitated to attack."

Pinkerton began reassessing the Confederate forces when he received tragic news: three of his top detectives in Richmond, including Pryce Lewis, who had helped apprehend Rose Greenhow, had been captured and sentenced to death. Another, Timothy Webster, would be hanged imminently, the first American to be executed as a spy since Nathan Hale in 1776. Still, Pinkerton inexplicably failed to recruit Union sympathizers already living in Richmond, people who were intimately familiar with its politics and personalities, people who might have connections in the nascent Confederate government—or the wherewithal to recruit and place such connections.

As word of the spies' fate spread throughout the camp, Emma received a visit from the 2nd Michigan's chaplain.

"I know of a situation I could get for you," he said, "if you have sufficient moral courage to undertake its duties. It is a situation of great danger and of vast responsibility."

She understood that she was being asked to take Webster's place, and contemplated the reasons why she, above all other men, had been chosen. Aside from Jerome, the chaplain knew her better than anyone else in her regiment. He had stood with her on the field at Bull Run, prayed with her at hospital beds, accompanied her to watch the skirmishing pickets, helped her dig graves for the dead. He knew she had the admiration and respect of her regiment, that her fellow soldiers considered her "brave, willing and cheerful" and a person of good moral character. He knew, too, that Colonel Poe wished to keep his most efficient soldiers in the ranks to prepare for battle, and that her particular brand of valor could be used in other ways. Her lack of experience wasn't a disadvantage; Pinkerton had a history of hiring detectives and spies who had never worked in law enforcement, including the men who had guarded Rose.

Emma figured that Poe—who had done secret service work for McClellan earlier in the war, outside of Pinkerton's auspices—had approved the idea and recommended Frank Thompson for the job. She told the chaplain she would have to give it some thought. "The subject of life and death was not weighted in the balance," she wrote. "I left that in the hands of my Creator, feeling assured that I was just as safe in passing the picket lines of the enemy, if it was God's will that I should go there, as I would be in the Federal camp. And if not, then His will be done."

She accepted, and reported to Washington for an examination.

WISE AS SERPENTS AND HARMLESS AS DOVES

RICHMOND

The Confederates executed Timothy Webster at Camp Lee, three miles from Elizabeth's home, using a defective cotton rope that slid up his neck. He fell on his back to the ground, half hanged, and spoke his final words after they hoisted him again: "I suffer a double death. Oh, you are going to choke me this time." They did.

That afternoon Elizabeth took her carriage to Castle Godwin, a former "Negro Jail" that now housed anyone suspected of disloyalty. The number of inmates had surged since early March, when Jefferson Davis, acting on a tip that there were "designing men in this City plotting treason against our Government," declared martial law in Richmond and ordered the arrest of thirty suspected Northern sympathizers, including Franklin Stearns, her friend and fellow Unionist. Stearns owned a distillery on Fifteenth Street that was doing such robust business with Confederate soldiers—$5,000 per day, by one account—that Richmond newspapers called for his execution. "It is the universal conviction," ranted the *Examiner*, "that Franklin Stearns, by means of his whiskey, has killed more of our men and done more to disorganize our army than all the balance of the Yankee nation put together."

As Stearns was led to his cell, Elizabeth heard, he'd thrown his guards a contemptuous look and said, "If you are going to imprison all the Union men you will have to provide a much larger jail than this." In any event, the authorities, lacking evidence, released him after several weeks.

She was equally unnerved by the imprisonment of several women, all charged with disloyalty and giving aid to the United States government. Elizabeth didn't know any of them personally, but she took their arrests as a sign that the Confederates now intended to make examples of female traitors.

She decided, nevertheless, to keep to her plans.

Her carriage stopped before the prison, a new brick building along Lumpkin's Alley. She told the guard she wished to visit Timothy Webster's widow, who had been found guilty of complicity, a lesser offense, and was housed on the second floor. If only Confederate authorities knew the truth: the presumed widow was actually Hattie Lawton, also a Pinkerton operative, one of the female detectives who had guarded Rose Greenhow and subsequently worked undercover in Richmond. Disguised as "Mrs. Webster," she insisted she was a Southern woman who wished only to return to her native Maryland. Elizabeth, also unaware of Lawton's true identity, hoped to convince prison officials to permit her to serve her yearlong sentence at the Van Lew mansion.

Lawton wept real tears over the execution of her partner, whom she had exhorted to "die like a man." Elizabeth sat with the "poor agonized creature," offering her prayers and a handkerchief, and was about to look for the prison's commander, Captain George W. Alexander, when he appeared at the cell door.

Elizabeth forced herself to smile at the "desperate brigand looking villain." He was dressed all in black: black trousers buckling at the knees, loose black shirt relieved only by a white collar, black whiskers bending into a frown atop his lips. His dog, a black Bavarian boarhound named Nero, snarled by his side. Nero had been imported as a pup and trained to fight, winning three matches against full-grown bears. Rumor had it that he'd been trained to attack anyone wearing blue.

"The body of Webster has been brought back to the prison," Captain Alexander said. His obsidian eyes shifted to Elizabeth, and he asked, pleasantly, if Miss Van Lew would like to view it.

Hattie Lawton sobbed. Nero let a thick thread of spit fall from his jowls. Captain Alexander watched her, waiting.

The question disturbed Elizabeth, as much for Alexander's oddly polite tone as for the implication beneath his words: there was a reason, he seemed to be saying, that she should witness what happens to traitors.

Just as politely, she declined, and asked if his widow might be permitted to stay with her.

Alexander refused, and suggested that Elizabeth had visited long enough.

It was probably just as well, she thought on the ride home, what with the ongoing strain between her brother and his wife, exacerbated of late by John's interest in her work with Union prisoners. While aiding Elizabeth, he, too, maintained a facade of loyalty to the Confederacy, and was even picked as a juror in the murder trial of one rebel guard accused of killing another. And Mary was still abusing the family servants to such a degree that they had begun quitting the household; the Van Lews' cook left just after the New Year, with others—a washer and ironer and a seamstress—soon to follow.

She thought of Mary Jane Bowser, still adjusting to life without her husband and to her role as a servant for the Confederate first family, her eidetic memory cataloging everything the president said. It was not yet time for Elizabeth to take the next step . . . but it would be soon. She did not want to risk her beloved servant's life until she had no other choice.

Elizabeth turned her attention to Libby Prison, the city's newest, converted from a tobacco warehouse on the waterfront. Since the prison occupied its own city block and was separated from the other buildings, Confederate authorities believed it would be relatively easy to guard. In an attempt to secure it further, they'd whitewashed the dark exterior brick wall so that any prisoner trying to escape would make a clear target for the armed sentries roaming below.

Elizabeth heard daily about the horrid conditions within its walls. Inmates were forbidden to go outside for fresh air or exercise and so spent every hour of every day in the prison's six upper rooms, each one 105 feet long, 44 feet wide, and 8 feet high, and holding more than a hundred men at a time. Every floor had one water closet—literally a closet with a trough used as a toilet. The place was overrun with vermin, which scurried over prisoners' feet and faces and hands. One man awakened during the night to find a large rat perched on his head. In a desperate attempt at extermination, the men began trapping and cooking the rodents for dinner. Guards would take any opportunity to shoot prisoners, a pastime they called "sporting for Yankees," killing them for such minor infractions as leaning too close to the windows. "To 'lose prisoners' was an expression very much in vogue," Elizabeth wrote, "and we all understood that it meant cold blooded murder."

Libby Prison.

She found out what she could about the men in charge of the prison: commander Thomas Pratt Turner, twenty-one, whose "utter depravity," according to one prisoner, "gained a full and complete expression in every lineament of his countenance"; Dick Turner (no relation), twenty-three years old and second in command, infamous for kicking dying prisoners just for lying on the floor; George Emack, twenty-five, who'd once held a gun to the head of a sick prisoner and threatened to shoot if he didn't get up; and Erasmus Ross, the twenty-one-year-old clerk, well known for swinging his bowie knife and terrorizing his charges, and known not at all as the Northern-sympathizing nephew of Elizabeth's friend Franklin Stearns.

The inmates called the clerk "Little Ross," a nod to his diminutive stature. His main responsibility was keeping track of the prisoners through daily roll calls. "He never called the rolls without swearing at us and abusing us and calling us Yankees," said Captain William Lounsbury of the 74th New York. "We all hated him." One evening at roll call Ross struck Lounsbury in the stomach and hissed, "You blue-bellied Yankee, come down to my office. I have a matter to settle with you."

Lounsbury took note of Ross's two revolvers, the gleam of his omnipresent bowie knife. He could hear his comrades whisper, "Don't go—you don't have to," and they reminded him that others whom Ross had called out had never returned. But Lounsbury followed the clerk down to his office in the corner of the prison. Ross held the door open for his prisoner and checked up and down the hallway before pulling it shut.

"See here," Ross said, "I have concluded to try you and see if you can do cooking." He pointed behind a counter, lifting his eyebrows and speaking his next words in italics: "Go in there and look around. See what you can find."

Lounsbury backed up to the counter, keeping his eyes on the clerk, and when he looked down he was shocked to see a Confederate uniform. It was a size too small, but he tugged it on, hopping from one leg to another and pulling on the jacket, re-creating himself as the enemy. He walked out of the prison, looking to his right and left and back, wondering if Ross had set him up and was preparing to shoot him for sport. The clerk followed a few steps behind, his rifle down by his side. Only a slim orange peel of sun remained in the sky.

Lounsbury quickened his pace, half skipping. Ross was still behind him, now accompanied by a sentry, both men keeping close watch. He broke into a full run, cutting through a vacant lot clotted with weeds. Out of nowhere a Negro appeared, stepping into his path.

"Come wit me, sah, I know who you is," the man said.

Lounsbury followed him into the dusk, walking a half dozen blocks and stopping beneath the soaring columns of the Van Lew mansion. The Negro pointed to the front door and without another word scurried around to the back, leaving Lounsbury alone.

Elizabeth answered on the first knock. She had been expecting someone—not Lounsbury in particular, but any prisoner randomly picked by Ross. Franklin Stearns had assured her that his nephew was trustworthy and loyal to the Union, but the operation remained rife with risks: Ross could defect to the Confederacy or, more likely, be discovered and tortured until he named his accomplices. She pushed such fears from her mind and studied the prisoner, out of breath, doubled over in his butternut shell jacket and gray trousers. If sister-in-law Mary happened to rise from bed, the uniform would provide a perfect cover; Elizabeth could say she was merely offering a weary rebel soldier something to eat and a place to rest.

She pulled Lounsbury inside and escorted him upstairs to the secret room, hoping to evade her nieces. In the morning she brought him corn bread and gave him detailed instructions, mapping out the safest route to the James River, the precise path through the woods that would circumvent the rebel pickets; by nightfall he should be safely with Union troops. From her parlor window she watched him leave, dressed in that shabby Confederate coat that stretched tight across his shoulders and ended high above his wrists, swiveling his head like a periscope, making sure no one guessed what he really was. Small, smaller, gone.

She stationed two more of her bravest, brightest servants on the perimeter of Libby Prison. They sent word to employees on the inside, Negro men and women, to take notice of the older white lady who often strolled by, and tell the prisoners that a safe refuge awaited them should they escape. It was dangerous to trust that such knowledge wouldn't fall on the wrong ears, but her servants were discreet and knew which prison employees stood on the Union's side. Her next goal was to secure more allies in Confederate uniforms, men with a tenuous devotion to the cause, men who could be paid to ignore her activities or bribed if they threatened to reveal her.

Despite several visits to General Winder, during which Elizabeth rolled out her usual roster of compliments—his generosity, his intelligence, his glorious mane of hair—he refused to allow her into Libby Prison, explaining that such privileges were, at the moment, reserved for ministers and others whose motives seemed unimpeachable. He stressed that last word, letting it rise into a question.

Elizabeth smiled, and replied that there were inmates at several other institutions in need of charity; surely the general, as a fellow Christian, would understand her urgency.

She had ideas about how to circumvent Winder's rule and gain access to Libby, but in the meantime she increased her visits to Castle Godwin, always cradling an antique French plate warmer in her arms. The device had a double bottom in which she had crammed wads of Federal greenbacks, hundreds of dollars with which soldiers might buy decent food and clothing on the prison black market, or bribe guards to look the other way. Most escapees had cut their blankets into thirds and fashioned them into ropes, lowering them from windows in the middle of the night, the sentry below keeping lackadaisical watch. Others tracked down a fellow soldier who was up for exchange and bribed him to swap identities. During one visit she heard a guard comment to another, "I think I'll have a look at that the next time she comes in."

So the next time, before leaving home, Elizabeth filled the double bottom with boiling water and slipped it inside a cloth holder. When he demanded to see the warmer, she handed it over, removing it from its covering. As soon as his skin touched the scalding metal the guard yelped in pain, sucking on his fingers while Elizabeth apologized— how thoughtless and clumsy of her; he should see a nurse at once. Her plate warmer was not inspected again, but the guards kept a hard eye on her, following her as she wandered from cell to cell. Once, as she prepared to leave, she heard two sets of footsteps behind her, one heavy and one light. She turned to see Captain George Alexander and Nero the boarhound, a low growl rumbling in his throat.

The captain bared his teeth when he spoke: "You have been reported several times."

She looked at him blankly and left without a word.

A few days later she was reading the Richmond papers in her library when a servant led a young man into the room. He wore torn trousers and muddy boots, and nervously kneaded his wool cap.

"I wish to tell you something that will interest you greatly," he began, "and the government also."

Her mouth went dry. She felt a twist of dread. Surely this was one of General Winder's "plug-ugly" detectives sent to entrap her. Maybe her sister-in-law knew this visitor; maybe Mary was the one who had reported her to Captain Alexander.

Elizabeth kept her voice even: "Nothing of that sort would be of any interest to me."

"Let me board here," he said—no, *insisted*. "Let me sleep any-where. In the library, on the floor . . ."

She demurred, but asked a servant to make tea for the visitor. She was calm and polite, chatting about everything but the war. Afterward she watched him shuffle down Grace Street, hands still working his cap, not once looking back.

"We have to be watchful and circumspect," she wrote that night in her journal, "wise as serpents and harmless as doves, for truly the lions are seeking to devour us."

A few days later she saw the same man, marching in a Confederate uniform.

A Woman Usually Tells All She Knows

WASHINGTON, DC

The door of the black carriage shut, locking Rose inside, and it set off westward along Pennsylvania Avenue, bouncing over the rough, sinking cobblestones, wheels churning up patches of mud. It was unseasonably cold for springtime. Fat snowflakes dotted the sky, and the sun, she observed, "was obscured by clouds as dark as Yankee deeds." She pressed her forehead against the window and watched the city blur past: carriages carrying officers in full uniform, laughing with their gaudy courtesans; newspaper correspondents dashing into and out of Western Union; the office of Dr. Von Moschzisker, whose "ethereal ear inhalator" promised to cure deafness caused by cannon fire. It was her first trip outside the prison in more than three months. The carriage stopped at a once-elegant mansion, where the two-man US Commission Relating to State Prisoners waited to try her case.

As Rose approached the building, she realized it represented another piece of her past; it was the former home of William Gwin, erstwhile Democratic senator from California and gold rush millionaire, who had been arrested a few months earlier on suspicion of trying to flee south to join the Confederacy. At the height of his power the senator and his wife had spent $75,000 annually to run their household; a significant portion of this amount had funded lavish costume balls, described by one witness as "the most magnificent entertainments of the kind ever given in this country." Rose had attended the final soiree, held before the outbreak of the war, uncharacteristically disguised as an old plantation housekeeper, complete with a bishop-sleeved housedress made of spotted dark chintz, its white linen collar stiffly cinching her neck. She entered the now dingy hallway where Mrs. Mary Gwin, dressed as a "Marquise of the Court of Louis XIV," had once greeted her with a curtsy and a glass of champagne. With a guard close behind, Rose pushed past swarms of soldiers and spectators—all of them there to see her— and climbed the stairs to the third floor, where she was left to wait alone in a freezing, fireless room. At least the solitude would give her time to strategize.

The commission was charged with resolving the cases of thousands of civilian prisoners being held on suspicion of disloyalty. Lincoln hoped to alleviate overcrowding by freeing prisoners who were willing to either take an oath of allegiance to the Union or sign a "parole of honor" pledging to refrain from aiding the rebels. Rose had one distinct advantage: the head commissioner, John Adams Dix, was an old family friend, both a former colleague of her husband's and a cabinet officer in the Buchanan administration when she'd been close with the president. Major General Dix was, in Rose's view, "one of the few Northern politicians in whose integrity I entertained any confidence, or for whom I felt any respect," and she was grateful he hadn't recused himself. She was not required to answer any questions, and she would not be speaking under oath. She would obfuscate, deny, put her interrogators on the defensive, shame them for torturing a poor woman and her child. She would make them doubt the validity of the case even as they proved her guilt.

An hour had passed, the room seeming to grow smaller and colder. She might have stood in this same spot years earlier, listening to the clear, sad notes of violins, watching liveried servants bustle and bow. The guard outside the door rattled his musket as a reminder of his presence. She watched her breath appear and vanish, appear and vanish. Her fingers fought her attempts to bend them; she couldn't feel her toes inside her shoes. Without warning the door swung open, admitting a gust of warm air. Prison superintendent William Wood appeared and cleared a path for her, Union soldiers parting like blue walls of water on either side. As she came to the hearing room, she heard a voice boom her name. She had been announced; she was on.

Dix and his co-commissioner, former New York superior court judge Edwards Pierrepont, stood, appearing awkward and ill at ease. She could not have scripted their mood better herself.

"Gentlemen," she ordered, "resume your seats."

They obeyed.

What had once been a chamber in the Gwin mansion now resembled a makeshift courtroom. A large table occupied the center, with Dix at one end and Pierrepont at the other. A reporter hunched over a small desk in the rear, his pen nib poised in the air, prepared to transcribe the session verbatim. Rose's own seat was midway between the commissioners, in full view of everyone. She glanced at Dix, now sifting through a towering pile of papers and looking uncertain what to do.

She pounced: "I recognize the embarrassment of your positions; it was a mistake for your government to have selected gentlemen for this mission. You have, however, shown me but scant courtesy in having kept me waiting your pleasure for nearly an hour in the cold."

Both men apologized, insisting they hadn't been notified of her arrival. She craned her neck, aiming for a better look at Dix's pile of papers, and recognized several half-charred scraps Pinkerton and his men had retrieved from her stove. Her mind mulled over potential responses to questions not yet asked.

Judge Pierrepont took the lead. He was a few years younger than she, Rose surmised, with a sparse sprig of hair on top of his head balanced by a full dark beard. "The charge that is brought against you," he began, "is for aiding the enemy to military information."

"I am a Southern woman, with Southern sentiments," she replied. "I have a right to aid my cause in any manner that lies in my power." She stood abruptly. "You look upon me as having committed treason, and all because of the view I entertain. Is it not so?"

"You can say whatever you think best," Pierrepont said evenly. "You need not answer any questions which you do not want to."

"Go on with your questions. Is this a kind of mimic court?" She gave a short bark of a laugh. "Of course I am not obliged to commit myself. In fact, I have been cheated so much by this Government that I have no respect for it."

She watched Pierrepont struggle to remain calm. "You can make any reply you please to the charges."

"Charges! How many have you? Now isn't this a farce! Isn't it solemn! It's a perfect farce."

Pierrepont's mouth twisted beneath his beard. He tried again, stating that she was accused of corresponding with the enemy and providing information about Union troops.

"Where is your proof?" Rose demanded. "This is a charge without any proof. I have corresponded on many occasions with my friends. This much is true. Cannot I write what I please? Must my private letters be torn open and laughed over, and my sentiments restricted? Besides, I am not in a position to get important information. None of those connected with the Government will give me any. All that has been good and glorious in this Government, I believe to have come from those who are now in the Confederate Government." She paused to sit down again, exhaling theatrically. "I have no doubt you will charge me with counseling Mr. Davis how to lead his armies."

So far, so good, Rose thought. A careless listener would have imagined that the judge was endeavoring to defend the Government rather than incriminate her. She glanced at Dix, still fingering his stack of papers.

"You are charged with corresponding with the rebels," Pierrepont reminded her, "and particularly with giving them information of our army in relation to the battle of Bull Run."

"I am not aware of that. It is certain that if I had the information, I should have given it. I should consider that I was performing a holy duty to my friends."

Exasperated, Pierrepont got to the point. "How would you like to go to the other side of the lines?" he asked.

Rose was cagey. "That would be a question."

"Suppose the government should conclude to let you go. Would you consider that you owed allegiance to it; and would you be willing to be bound by the rules of war?"

Rose stiffened; she would never pledge allegiance to the North. Her preference, at the moment, was to stay where she was, agitating the Yankee government and cementing her status as a Confederate martyr and heroine. She would leave Washington only on her own terms, and Little Rose would have to stay strong until Federal officials acquiesced. "This is my home," she said. "I have been taken from my home and carried to a prison, to be insulted and subjected to a treatment of the most outrageous kind. Every association of my home has been broken up and destroyed. If the government deigns to send me across the lines as an exile, I have no alternative but to go as such."

Pierrepont intimated that he was speaking on behalf of President Lincoln: "It has been proposed that we make this suggestion to you and to see if you would like to accept it."

Rose let the judge wait for her answer. The quick scratch of the hearing reporter's pen was the only sound in the room.

"With the privilege of aiding and abetting my cause?" she finally asked.

Pierrepont didn't hesitate: "Yes, if that is your desire."

"In other words, you mean to tell me that if I do not accept it, I will be forcibly exiled."

"Would it be exiling you to send you South among your friends?"

"It is exiling me to use any force, to send me South from my home."

"I merely made it as a suggestion."

"Is there any other charge of treason against me?" Rose taunted him. "It may be that I am charged with having spies in the government. It is really absurd to suppose that I could get important information of the Government's designs when I am not in the Government; or that I could get such information about the Government when no one in power would give it to me; or that if such information that you say I have, I must have got it from sources that were in the confidence of the Government. I don't intend to say any more. I merely throw this out as a *suggestion*."

The judge sighed and turned to Dix. "General," he pleaded, "I think you had better talk to Mrs. Greenhow. You are an old friend of hers."

"I don't know that I have anything to say," Dix muttered.

Rose smiled and said sweetly, "You both see how pleasant it is."

"You might ask Mrs. Greenhow about the correspondence in cipher," Pierrepont suggested.

Rose interrupted him. "I am not obliged to answer that question."

"You are not obliged to answer any question that you do not please to," Dix conceded. "Those that you do answer will be recorded, so you see you have perfect freedom."

Rose couldn't quell her sarcasm. "Liberty of speech!"

Her friend's response was tight and controlled: "You have the privilege of putting the answer in your own form."

"Of course, General, I decline to answer questions of that sort," she replied, returning to the query about her cipher and throwing it back at him: "Although I believe your chief detective"—meaning Allan Pinkerton—"found some cipher at my house while I was sick. It was a very good cipher, and it might have been found useful, but I did not get an opportunity to use it."

Dix rummaged through his pile and held a paper aloft. "I want to know in regard to this letter which was found in your house," he said, waving it in the air. "It reads: 'There are 45,000 on the Virginia side; 15,000 around the city, to wit, up the river above Chain

Bridge. . . . If McClellan can be permitted to prepare, he expects to surprise you . . .' Did you write this?"

"I have no recollection of it," Rose replied.

Dix took a step closer to her. "There are other letters of the same sort."

"Are those letters said to be found in my house?"

"Yes."

"I think that is false. I think that is unequivocally false, I don't swear it is. I have been in the habit of entertaining guests at my house. So far as I am myself concerned, I pronounce it unequivocally false."

"You know, Mrs. Greenhow," Dix scolded, "that in a context like this, while the very existence of the Government is in danger, the communication of such information as this, which tends to subvert the interests of the Government, should certainly be considered a very serious offense."

Rose seized the opening. "I don't know about that," she said. "When the President, with his 100,000 men, has to hold in confinement a poor woman and children, I think he had better give up the ghost. Since his Black Republican party came into power—and I detest it in my very soul—I have kept entirely out of the world. I lost my child a short time before. I have not been in the world during that time; therefore any information I may have got must have been brought to my house, and brought to me. Brought to me by traitors, as you call them, in that party. Therefore I am to be held responsible for that? If Mr. Lincoln's friends will pour into my ear such important information, am I to be held responsible for all that? Could it be presumed that I could not use that which was given to me by others? If I did not, I would be unjust to myself and my friends. It is said that a woman cannot keep a secret. I am a woman, and a woman usually tells all she knows." She let those words settle and then took a gamble, asking for her papers to be returned to her. "I don't see any treason," she insisted, "in all that you have told me."

The naïveté of Pierrepont's response surprised her. "I don't think you are bent so much on treason," he admitted, "as mischief."

Mischief? Rose thought, offended by the downgrade of her contributions. She again went on the offensive. "Let me tell you," she said, "I have studied the Constitution and law of this country. I have informed myself about the Constitution as well as I have about my Bible. Your Government ought to be ashamed of itself for allowing me to be scandalously treated; in allowing such scoundrels as it did to get to my letters, and read them and laugh over them. I have been made the victim of every kind of villainy."

Pierrepont took a turn, rattling a piece of paper.

"So you know J. S. Sheldon?" he asked, referring to the one guard who had been kind to her and Little Rose.

Rose hesitated, wondering what the judge knew. "Yes, very well."

Now Pierrepont sounded smug. "There seem to be some letters here that passed through his hands."

Pierrepont was bluffing when he said "letters"—there had been only one—but Rose understood: that single instance when she'd asked Lieutenant Sheldon to mail a letter, he instead delivered it straight to Union authorities. Rage crept into her voice. "Call him up and let him answer. Call him up and we will see what he has to say about that."

Pierrepont and Dix both kept quiet, waiting for her next words to convict her.

Rose spoke haltingly. "I have no intention . . . I have no recollection of it at all. During the first month of my imprisonment I had perfect freedom to write what I pleased, and the letters were carried through the provost marshal's office."

"The question is not as to whether you have written the letters," Pierrepont pointed out, "but, also, whether you have a right, knowingly, to try and subvert the purposes of the Government."

"I don't consider that I was," she replied, regaining her footing. "I look at it in an entirely different light. I have a right to write what I please. I always did so."

Pierrepont cut her off. "It is wartime now."

Dix picked another piece of paper from his pile. Rose recognized her pink stationery and wild scrawl. It was a letter to her daughter Florence in which she ranted against "black Republican dogs" and gossiped that Secretary of War Stanton was planning to replace McClellan.

"This letter is equal to declaring determined hostility to the Government," Dix charged. "I believe there are a great many others of the same purport."

"The Government knew what my feelings have been, and always were," Rose said. "I have not changed them. I have no other feelings than those now."

She felt a shift in the room, an invisible tilt of a lever, spilling momentum in her direction. Dix seemed to give up; he knew they had gotten all they would from Rose. He returned her letters to his pile and turned to his co-commissioner: "Judge, I don't know as we wish to ask Mrs. Greenhow any more questions."

Rose left them with some patronizing advice. "In these war times," she said, "you ought to be in some more important business than holding an inquisition for the examination of women. I look upon this as nothing more than an inquisition."

The court reporter transcribed the final words: "Further consideration of the case was postponed, and the prisoner remanded."

She was escorted back to her cell at the Old Capitol Prison, feeling triumphant even as the guards locked the door behind her. The noise jarred Little Rose awake from a nap. "Mamma," she said, "tell me a story." Rose sat the girl in her lap and obliged, making it up as she went along and letting the good guys win.

A few days later Superintendent Wood appeared at Rose's cell to convey an offer from the commissioners. They would exile her to the South on one condition: she must stay there until the war was over. She wrote a letter in response, informing them she accepted her "banishment" but did so under protest, and would not promise to avoid the North. "I ask of your clemency time and freedom to make the necessary arrangements for clothes for myself and little child," she wrote, and couldn't resist a parting shot: "Of course, if this is granted me, I shall bind myself for the period allotted not to blow up the President's house, equip a fleet, break open the treasury, or do any other small act which you may suppose comes within my limited powers to perform."

Two weeks passed with no further word from the commission. Little Rose once again grew wan and listless. An ex–US Army officer being held on Rose's floor on suspicion of disloyalty defied a sentry's order to back away from his cell window. The sentry fired a shot through the bars, killing the prisoner inside his cell—an act Rose called murder "in cold blood." She stopped sleeping, aware of every shadow and sound, intent on coaxing Little Rose out of her nightmares.

A small measure of hope came via an unexpected visitor: Alfred Ely, the New York congressman who was captured at Bull Run and befriended by Elizabeth Van Lew while imprisoned in the Confederate capital.

"I had been treated well and kindly at Richmond," he told Rose, "and I've come to see what I can do for you."

She had nothing to lose, so she took him up on the offer, asking if he might ascertain why his government was delaying her release. Ely returned four days later, on April 22, and said, according to Rose, that General McClellan himself had objected to her release on the grounds that she "knew his plans better than Lincoln, and that he did not wish me sent South at this time." Another possible reason was that Lincoln, furious over the execution of Union spy Timothy Webster, was holding Rose in retaliation.

Ely asked for her carte de visite and told her he would try again.

"No," Rose said, "you will be refused a pass. They are afraid lest my fearless denunciations of their infamies may open the eyes of their followers, and make them question the orthodoxy of Aboli-tionism."

Ely suppressed a smile, but in a sense Rose was right: he was never permitted to visit her again.

Partly out of boredom and partly for revenge, she promised two rebel prisoners she would aid in their escape, confirming plans during a walk through the prison yard, sneaking one bribe money and the other her gun. She considered how her complicity might affect her and Little Rose, and decided it was worth the risk.

A Slave Called "Ned"

WASHINGTON, DC, AND THE
VIRGINIA PENINSULA

Emma's carriage pulled up to Secret Service headquarters, a two-story, lead-colored building at 217 Pennsylvania Avenue, widely said to be "the nation's most feared address." A guard ushered her into a room where three Union generals, including George McClellan, waited to interview Frank Thompson. Emma, accustomed to seeing her military hero from afar, galloping through camp on his horse, Dan Webster, was both thrilled and terrified to stand so close to him, as in awe of the general as he clearly was of himself. McClellan was shorter than she'd expected and wonderfully square, thick-throated and broad-chested, with a soft, boyish face and auburn mustache curtaining his mouth. She pulled herself up to her full height and emptied her face of expression, wanting to appear as soldierly (and manly) as possible, and wondered what these men saw when they looked at her.

McClellan and the other generals, Samuel P. Heintzelman and Thomas F. Meagher, questioned and cross-questioned Emma about her views of the rebellion and her motive for wishing to "engage in so perilous an undertaking." She would be going undercover, wearing clothing other than her military uniform, so if the Confederates caught her, they would treat her not as a prisoner of war but as a spy, and hang her as they did Timothy Webster.

Emma cited her strong Christian faith and her belief that slavery was against God's will. Next came the easy part, a test of her skill as a marksman. She was most nervous about the final phase, a physical examination. She recalled her first medical exam with the army, when the doctor merely measured her height and shook her hand, and worried that it might be more thorough this time around.

This doctor, fortunately, focused mainly on her head, both internally and externally, a strangely intimate blend of psychoanalysis and scalp massage. With the doctor's fingers kneading her skull, she answered dozens of queries about Frank Thompson, even venturing into the years before she became him. The doctor stretched a measuring tape from ear to ear and sketched a rough diagram of her head, estimating the development of various regions of her brain. She silently prayed that her head did not betray her sex; phrenological studies on women often concluded that their organs of "adhesiveness," cautiousness, and procreation were so prominent as to elongate, and even deform, the middle of the back of the head. The doctor poked and prodded with his caliper and scratched notes on a pad. Emma felt stifled inside her frock coat, drops of sweat sliding down between her breasts. He determined, finally, that Frank Thompson indeed had the head of a man, with "largely developed" organs of secretiveness and combativeness. Emma acted as though she'd expected to hear as much, and took the oath of allegiance.

She had three days to prepare for her first mission: slip into Yorktown disguised as a slave and determine the number of Confederate troops stationed there, and the strength of the fortifications. The request was not unusual; spies on both sides used costumes and props and even handicaps, real or feigned, to deceive the enemy. There was a sudden proliferation of newsboys, actors, peddlers, doctors, and itinerant photographers, all of whom had a natural pretext for passing through the lines. Lafayette C. Baker, the Federal government's chief detective and a professional rival of Pinkerton's, wandered around rebel camps with an empty camera box. A Confederate spy named Benjamin Franklin Stringfellow capitalized on his slight figure and delicate features by dressing as a woman and frequenting Northern balls, gleaning information about Ulysses S. Grant from his dance partners. One of Pinkerton's men, Dave Graham, feigned stuttering and epileptic fits as he wandered in Confederate territory, pretending to be a peddler. John Burke, a rebel scout for Generals Beauregard and J. E. B. Stuart, went undercover by removing his glass eye. Another Confederate spy, Wat Bowie, donned a homespun dress and scoured his body with burned cork, posing as a slave girl so convincingly that he fooled the Union detectives on his trail. Emma recognized the irony of her mission, layering a temporary disguise over the one she could never remove.

In the morning she set out for Fort Monroe to gather the elements of her disguise: a plantation-style suit, purchased from the contraband camp; a black wool wig from the postmaster (she explained that it was for "reconnaissance business"); and a vial of silver nitrate from the hospital medical supplies. A barber sheared her curls. She darkened her head, face, neck, hands, and arms; donned her suit and wig; and returned to camp near Yorktown, where everyone believed she was a contraband in search of work: "I found myself without friends—a striking illustration of the frailty of human friendship." Nonetheless she was relieved; if the disguise fooled her comrades, she would certainly go undetected by the rebels. Staring at the mirror, she barely knew herself. She spent an hour practicing Negro dialect and picked a new and temporary name, "Ned."

That evening, after the bugle call signaling lights-out, Emma filled her pocket with hard crackers, loaded her revolver, and hid a pencil and folded squares of paper under the inner sole of her shoe. She passed the Union guards outside Yorktown and walked as quietly as she could, at one point coming within yards of a Confederate sentry without being observed. She thanked heaven for her good luck, stretched out on the cold wet ground, and waited for morning.

At sunrise a group of slaves passed by, returning from bringing breakfast to the rebel picket lines. She followed them to camp, where they were ordered to work on fortifications, but she had neither tools nor any idea what to do. A young Confederate officer noticed her confusion and approached.

"Who do you belong to, and why are you not at work?"

"I dusn't belong to nobody, Massa," Emma replied, adding that she had always been free and was heading to Richmond to find work. The officer nodded and called for the overseer. "If he don't work," the officer said, "we'll tie him up and give him twenty lashes just to impress upon his mind that there's no free niggers here while there's a damned Yankee left in Virginia."

Emma thought of the fugitive slaves back at the Union camp, falling to their knees when they realized they were free.

The overseer furnished her with a pickax, shovel, and wheelbarrow. Emma watched the other slaves and followed their lead, pushing a load of gravel—the smallest she could get away with—up a narrow plank to the parapet. It was an arduous task for even the strongest man, each hoist of the shovel and stab of the ax grating against her skin, and by dusk she was raw from wrists to fingertips. If she wasn't able to work at all the next day, the rebel officers would discover her deception. An idea struck: she paid a fellow slave to switch places; she would carry water for the troops while he built the fortification.

She took advantage of her new position, roaming freely about the camp, occasionally ducking behind a tree to sketch fortifications and jot down the number of mounted guns, 151 in all: three- and four-inch rifled cannon, thirty-two-pounders, forty-two-pounders, eight- and ten-inch Columbiads, nine-inch Dahlgrens, ten-inch mortars, and eight-inch siege howitzers. Lingering with one brigade, she overheard snatches of talk about the arriving Confederate reinforcements and was thrilled to glimpse Robert E. Lee. The men whispered that the general had come to inspect the Yankee fortifications, and that he had pronounced it impossible to hold Yorktown after McClellan opened his siege guns upon it. Another rumor claimed that the Confederates planned to evacuate Yorktown, the final piece of intelligence Emma needed. She had to return to her own camp before someone realized she was not what she seemed to be.

For the next two days Emma carried water, listened, and waited for a chance to escape. She was losing her disguise. Her scalp itched, her wool wig shifted askew, the silver nitrate peeled away to reveal patches of pale flesh. One slave studied her quizzically and joked to another: "I'll be darned if that feller ain't turnin' white."

She panicked and fumbled for a response. "Well," she said, "gem'in I'se allers 'spected to come white some time; my mudder's a white woman." While they laughed she backed away, one foot behind the other, until she scuttled out of sight.

Early that evening she was among a group of slaves sent to deliver dinner to the rebel pickets. She walked down the line, dropping salted beef on tin plates, ducking minié balls shot by Yankee pickets half a mile away. The other slaves returned to camp but she hung back as long as she could, taking a seat on the ground. No one paid her any mind until two boots stomped into view and a faint shadow arched over her, accompanied by a booming order: "You come along with me."

Emma obeyed, following the Confederate sergeant to a gap along the line where a picket lay wounded. Another rebel officer squatted over him, hoisting a canteen to his lips. The sergeant pointed at Emma and said, "Put this fellow on the post where that man was shot until I return."

The officer nodded, and the sergeant turned back to Emma, thrusting a rifle against her chest and advising her to use it "freely." Without warning his hands rose and encircled her neck. He gave her a hard shake, thumbs pressing at her throat.

"Now, you black rascal," he said, "if you sleep on your post I'll shoot you like a dog."

"Oh no, Massa," she croaked, "I'se too feerd to sleep."

There was a new moon that night and the sky was lightless. The air turned dense and gave way to rain, drizzle becoming a downpour. Her color washed away, feature by feature, layer by layer. She heard a rustling noise, the rebel pickets taking cover behind the trees. Now.

Gripping her rifle—she was *not* going to lose that prize—she sprinted toward the thick forest dividing the two lines. As she neared Union territory, she pulled back, recognizing the absurdity of her situation: she was now in more danger of being shot by her own pickets than by the enemy. She waited for hours until a speck of sun appeared, flashed the Union sign, and stumbled back to where she belonged.

She gave McClellan her sketches and told him that the Confederate troops numbered 150,000, a third more than even the general had believed, and five times as many as there actually were. Such an inflated estimate was common among novice spies; even Pinkerton, for all of his success in foiling assassination plots and arresting traitors, knew little of war or warfare or how to gauge enemy strength, and sent out agents as untrained as he, even relying on observations from escaped slaves. The cautious McClellan was always eager to accept estimates of Confederate strength that confirmed or exceeded his own calculations.

Emma also relayed the most vital piece of intelligence: the rumor that the Confederates planned to evacuate Yorktown to defend the entrenchments around Richmond. Although others (including a slave who escaped at the same time as Emma) reiterated this claim, both the general and Pinkerton were dismissive, believing that the enemy intended to hold Yorktown at any cost.

But on this point she was right. Three days after she returned to camp, at two thirty in the morning, General Magruder and his rebel force fled, heading northwest toward Williamsburg. They had spent a week preparing for their departure, dismounting their guns, filling wagons with ammunition and provisions, and transporting 2,500 of their sick and wounded to hospitals in Richmond. When Union troops moved through, several men were killed by sub-terra explosive shells—an early version of land mines—hidden in the ground. Tents were left standing, with caricatures of Yankee soldiers scribbled on the canvas. They discovered several letters lying unfolded on a table. One was addressed to Abraham Lincoln, another to "The First Yankee Who Comes," and the last to General McClellan:

> *You will be surprised to hear of our departure at this stage of the game, leaving you in possession of this worthless town. But the fact is, McClellan, we have other engagements to attend to, and we can't wait any longer. Our boys are getting sick of this damned place . . . so goodbye for a little while.*

The Union army was exhilarated by the news, which, Emma wrote, "spread throughout the Federal army like lightning; from right to left and from center to circumference the entire encampment was one wild scene of joy." Jerome Robbins also noted the "unequalled excitement." In a letter to Washington, McClellan characterized the retreat as a Union victory, declaring that "our success is brilliant & you may rest assured that its effects will be of the greatest importance. There shall be no delay in following up the rebels." Despite McClellan's bravado, he had no strategy for an organized advance. He improvised a pursuit with his cavalry and five divisions of infantry, including the 2nd Michigan, ordering them to "leave, not to return."

Emma and her comrades were at the end of the force chasing Magruder's rebels, marching through endless rain, the roads a perfect sea of sludge. They were encased in mud to the waist, tripping and tumbling headlong. With each step the battle roared louder, the crash of musketry reverberating through the woods, horses rearing, trees plunging. Rebels sent spray after spray of bullets into the advancing ranks, but still they charged forward through ditch and mire, loading and firing as they went, bodies dropping at Emma's feet: "There was plenty of work for me to do here, as the ghastly faces of the wounded and dying testified."

Orders came at her from all directions: Go to the front with a musket in your hands. Mount your horse and take an order to this general. Grab a stretcher and help carry the wounded from the field. At one point she saw a colonel fall and rushed over to him. Another "poor little stripling of a soldier" followed to assist. To-gether they carried him through a hail of bullets and set him down by the surgeon's feet, lingering long enough to see if the wound was fatal. The surgeon opened the colonel's shirt and found no holes, no blood. He examined him piece by piece—not even a scratch to be seen, and yet the patient seemed in too much pain to speak.

The surgeon stood and said, "Colonel, you are not wounded at all. You had better let these boys carry you back again."

The colonel sprang to his feet, indignant. "Doctor, if I live to get out of this battle I'll call you to account for those words."

The surgeon leaned in close. "Sir," he said, "if you are not with your regiment in fifteen minutes I shall report you."

Emma backed away in disgust, "mentally regretting that the lead or steel of the enemy had not entered the breast of one who seemed so ambitious of the honor without the effect." In the future she would determine whether a man was wounded before she moved to help him.

The hours passed and the rain came harder, drenching the living and the dead. Some lay on the ground, fully alive but helpless, their legs and arms too chilled and cramped to move. Emma made countless rounds from the front of the lines to the surgeon's tent and back again, sinking under the weight of her stretcher, the mud suc-tioning her boots. Watching a young surgeon perform an amputation reminded her of unskilled hands preparing a turkey: "It was his first attempt at carving and the way in which he disjointed those limbs I shall never forget." It had been a bloody day for the 2nd Michigan, with 17 killed, 38 wounded, and 4 missing, and even bloodier for the 5th Michigan, which lost 170 men. Both sides claimed victory in the Battle of Williamsburg, General McClellan calling it "brilliant" and the Confederates believing they had delayed the Federals, allowing their own army to retreat toward Richmond.

All told, more than 2,200 Union troops and 1,600 rebels lay scattered in heaps, the wounded and dead from both sides entangled in ravines. The dead lay in all postures, but mostly on their backs, heads tilted, mouths slightly open, one hand placed over the wound. One man remained on his hands and knees, with his head shot off. Two men lay face-to-face, each with his bayonet through the other's body. An endless chorus of moans drifted across the field. During the darkness of the night soldiers fetched water from the ditch, plunging canteens amid the piles of corpses. A captain from the 5th Michigan emptied out the balance of his canteen to discover that it was "quite red."

The enemies called for a temporary truce to collect their soldiers. "It was indescribably sad," Emma thought, "to see our weary, exhausted men, with torches, wading through mud to their knees piloting ambulances over the field, lest they should trample upon the bodies of their fallen comrades." A friend from Michigan was among the seriously wounded, shot clear through the thigh. She spent the next two weeks in a makeshift hospital in Williamsburg, tending to Union and Confederate soldiers alike and contemplating her next visit to rebel territory—this time with yet another layer of disguise, a woman impersonating a man impersonating a woman.

Perfectly Insane on the Subject of Men

THE SHENANDOAH VALLEY, VIRGINIA

Belle's mother worried about her safety in Martinsburg—the village was once again in Union hands—and in mid-May sent her, along with Belle's servant, Eliza, farther south to stay with relatives in Front Royal. On the way Belle was detained briefly on suspicion of being a spy, but John Adams Dix, the commissioner who had presided over Rose's hearing, determined that there was no "clear evidence of guilt" and ordered her release. Belle didn't let the unpleasant incident ruin her trip. She attended every party, whether or not she'd been invited, and boasted indiscriminately of her exploits to acquaintances new and old.

Women, Belle soon learned, made for a surly audience, their lack of interest in her stories directly proportionate to her insistence on telling them. One Front Royal neighbor, Lucy Buck, found Belle "all surface, vain, and hollow" and complained of being carried "captive into the parlor." Another, a teenager named Kate Sperry, was dismayed when Belle knocked on her door. "Of all fools I ever saw of the womankind she certainly beats all," Kate wrote. "Perfectly insane on the subject of men . . . she is entirely crazy." Even when Belle told them they could "write the boys by me" and offered to deliver the letters beyond the lines, the girls were not impressed ("Poor boys!" quipped Lucy Buck).

Front Royal, at the northern reaches of the Shenandoah Valley, spanned 9.5 square miles and had only five hundred residents, but its strategic importance belied its size. This town, too, changed hands often, and it fell under Federal control around the time that Belle and Eliza arrived. Belle's aunt, Mrs. Fanny Stewart, was proprietor of the Fishback Hotel, a tidy, three-story structure with balconies jutting irregularly from windows, its cheerful yellow facade out of place amid clusters of homes wrecked by warfare. In the rear, tucked away from the bawdy noise of High Street, a winding brick path lined with violets and quaker-ladies led to a private cottage, where Belle stayed during her visits. "It is here," warned the *Philadelphia Inquirer*, "that some of the most accomplished women in the valley assemble, with purpose and design to *pump* from our young lieutenants, who know little of the stern realities of war, the name and number of their regiments— how many effective men their particular regiment can muster, and what their state of efficiency is."

At that moment Belle was focused on Union general James Shields, who had claimed the Fishback Hotel as his headquarters. One night, after dining with her family in the cottage, she left her calling card for the fifty-one-year-old general, and he came promptly to pay his respects. Unsavory gossip seemed to follow Shields; his men reportedly thought him "disloyal or insane." He defied the urgent protest of the medical director of his division by attending the examination of a young rape victim, and then insisted upon examining her himself. Nevertheless Belle found him charming: "He was an Irishman, and endowed with all those graces of manner for which the better class of his countrymen are justly famous; nor was he devoid of the humor for which they are no less notorious."

She chatted casually, asking him how he found the town and his accommodations at the hotel, and inquired if he might be so kind as to grant her a pass to Richmond. Shields laughed and replied that Stonewall Jackson's army was so demoralized that he dared not entrust her to their "tender mercies." But, he confided, the rebels would be annihilated in a few days, after which she could wander as she pleased.

Belle smiled and said nothing, giddy because the general had inadvertently tipped his hand. His joking tone betrayed his confidence about a swift and decisive Union victory. "He was completely off his guard," she thought, "and forgot that a woman can sometimes listen and remember."

During the next few days Belle socialized with the general and his staff, sharing her stash of cigars, memorizing all their names, and showing off her latest pet, a crow with a split tongue, which she was training to talk; so far its vocabulary included "Miss Belle," "Stonewall," and "General Lee." *Harper's* author and illustrator David Hunter Strother, working as a civilian topographer for Shields, found her "looking well and deporting herself in a lady-like manner. I daresay she has been much slandered by reports. She sported a bunch of buttons despoiled from General Shields and our officers and seemed ready to increase her trophies"—tokens she had been given or had taken from Union soldiers she seduced into providing her with information. She managed all of this, one admirer mused, "without being beautiful."

Another Yankee visitor, a correspondent for the *New York Tribune*, observed Belle in a different mood, seducing General Shields so thoroughly that she remained "closeted four hours" with him and subsequently wrapped a rebel flag around his head. Next she moved on to Shields's aide-de-camp, Captain Daniel Keily, who courted her with flowers and love poems. To "Captain K," as Belle called him, she was "indebted for some very remarkable effusions, some withered flowers, and last, not least, for a great deal of very important information"—most notably the time and location of a Union council of war.

On the specified night Belle crossed the brick path to the Fishback Hotel and crept upstairs. A certain chamber was positioned directly above the drawing room, where the officers planned to gather, and she shut herself in a closet to lie prone on the floor. She knew that someone, sometime, had bored a hole in the wood (for purposes of espionage, she liked to think), and now she pressed her ear tightly against it, pleased to discover she could hear everything unfolding below: the scrape of moving chairs, papers rattling, a closed fist pounding the table.

The men discussed their army's positions and plans. General McClellan was advancing toward Richmond, and General Irvin McDowell would support his drive. Shields and General John Geary would move to reinforce McDowell, leaving General Nathaniel Banks, currently fifteen miles west of Front Royal, stripped of much of his force. McDowell had wired Shields to say that Stonewall Jackson was on the line toward Richmond, "so in coming east you will be following him." The men detailed the route they'd take to join McDowell and how, exactly, they could trap Jackson.

Belle had what she wanted. More important, she had what Jackson wanted; the information from this war council was the most important intelligence she'd gleaned thus far. The men finally disbanded at one in the morning. She heard the general lumber upstairs and down the hall to his chamber, yawning and clearing his throat. At the click of his door she tiptoed outside, the flicker of gaslight guiding her back to the cottage. She needed to deliver the information at once to Turner Ashby, Stonewall's cavalry commander and a friend to her many relatives in the Confederate army.

She transcribed the conversation and gathered a couple of passes, acquired through "various circumstances" from paroled rebel officers returning south. Passes customarily required a specific name, but some Union staff officers used vague wording, making them out to "bearer" or leaving them blank altogether. At this hour of the night, Belle hoped the Union sentinels would be too drunk or too tired to question her. If she carried a pass from divisional headquarters or higher, they would assume she was a Federal agent. For added insurance, she unlaced her corset, kicked off her crinoline, and dressed in the garb of a boy: trousers, shirt, a worn cotton kepi.

She slipped out to the stables, saddled a horse, and headed for the mountains, estimating the route to be about fifteen miles. Twice Federal guards stopped her, and twice they let her through after a quick glimpse at her pass. The path was rough, with hard climbs up the stony beds of brooks and leaps over deep gorges and ravines, and it was around 3:00 a.m. when she arrived at Ashby's temporary residence, the home of someone she knew as "Mr. M."

She sprang from her horse and sprinted up the steps. The house was still and dark, and she rapped at the door with tight fists.

"Who is there?" a voice called from a second-story window.

Belle took a step back, lifted her head, and shouted, "It is I!"

"But who are you? What is your name?"

She removed the kepi from her head and called, "Belle Boyd. I have important intelligence to communicate to Colonel Ashby. Is he here?"

"No, but wait a minute. I will come down."

Mr. M opened the door and pulled her inside. "My dear, where did you come from? And how on earth did you get here?"

"Oh, I forced the sentries, and here I am. But I have no time to tell you the how, and the why, and the wherefore. I must see Colonel Ashby without the loss of a minute. Tell me where he is to be found."

Ashby himself then appeared at the door.

"Good God!" he exclaimed. "Miss Belle, is this you? Where did you come from? Have you been dropped from the clouds? Or am I dreaming?"

Belle rested a hand on his arm and insisted that, yes, he was wide awake, and that her "presence was substantial and of the earth—not a visionary emanation from the world of spirits." Quickly she explained the contents of her note, detailing how Union forces intended to trap Stonewall Jackson, and then handed it over.

As she went back to the cottage it began to rain, ragged hot wires of lightning crackling all around her. Belle kept on, soaked to the skin, her horse kicking up divots of mud. As she approached Union lines, a flash of lightning illuminated the figure of a guard, rifle poised.

"Who comes there?" he challenged.

She thrust a hand into her pocket, seeking her pass. Nothing. Somehow she'd lost it during the ride. Her mind cast around for the Union countersign, but for once she couldn't remember. She felt her throat stitch tight in panic. Another streak of lightning lit the sky, and she saw a corporal standing behind the guard.

"Let the boy pass," the corporal said. "I know him."

Belle sped off before he could change his mind. By dawn she was back in bed at the cottage, dreaming of Stonewall reading her words.

THE VIRGINIA PENINSULA

Having won control of the York River, McClellan prepared to move his base of operations to White House Landing, twenty-three miles east of Richmond, and begged Lincoln to send reinforcements from the Shenandoah Valley, where Stonewall Jackson and the Confederates were on the offensive. While he waited for troops to arrive he prepared to lay siege to Richmond, mending the lines of the railroad tracks so he could get his guns into place, and bringing up dozens of cars and locomotives by water from Fort Monroe. His engineers soon had the tracks open as far as the Chickahominy River, a marshy ribbon of water that bisected the northern tip of the Peninsula, separating the troops from the Confederate capital. The river was generally fordable by infantry, but the artillery required bridges, which the rebels had destroyed on their retreat from York-town. The engineers reconstructed those as well, progressing seven miles all the way to the depot at Fair Oaks, where the church spires of Richmond could be seen poking the sky. McClellan wrote to his wife, sharing his conviction that the final, critical battle of the war would take place imminently on the outskirts of the city, "which must in that event suffer terribly, and perhaps be destroyed."

The general was desperate to know what waited for him on the other side of the river. A contraband who left Richmond reported that the city was full of sick soldiers and that citizens were flocking in from the surrounding country. The Richmond newspapers quoted Jefferson Davis as saying he did not anticipate the fall of the capital, and believed that the war could be successfully carried on in Virginia for twenty years. Allan Pinkerton delivered estimates by local farmers that the Confederate force numbered anywhere from 100,000 to 150,000 men, nearly three times its actual strength. McClellan believed that "reconnaissances, frequently under fire, proved the only trustworthy sources of information," and Emma readied herself for another mission behind rebel lines.

She'd decided the slave disguise was both too risky (she could be recognized as the cowardly picket who deserted his post, a crime worthy of death) and too punishing to her skin. Instead she would assume an identity that would come more naturally: an Irish peddler woman. Emma thought of her Irish-born mother, with that voice like clotted cream; she could easily mimic Betsy's brogue and borrow her expressions. She bought supplies from an Irish peddler who'd been following the Union army: a hood, a basket, green spectacles, several peasant dresses. She packed the disguise in her basket and rode her "noble steed," Frank, to the edge of the Chickahominy River. The bridges were not yet complete, so she swam Frank across the water, gave him a farewell pat, and sent him back to the other side, where a soldier awaited his return.

She pulled on several peasant dresses, layering them to create girth, and yanked the hood six inches down her face. The clothing was soaked from the river and she roamed the strange territory around the Chickahominy swamps, breathing in the stench of rotting vegetation, starting at every roar of cannon and scream of shell. The sky dimmed and the air cooled and she was plagued by a fever and shaking chills. She had seen dozens of men die from malaria and she recognized it encroaching on her own body. She settled on the dank ground and pillowed her head against the basket. Her brain dipped into delirium, "tortured by fiends of every conceivable shape and magnitude," tracing each link of her life that brought her to the spot where she now lay. A line from John Greenleaf Whittier's

"Maud Muller" looped through her mind: "For of all sad words of tongue or pen, the saddest are these: 'It might have been!'" a wistful earworm that lulled her to fidgety sleep.

For two days Emma remained in the swamp, fighting to lift her body and calm her mind, telling herself she'd rather die upon the scaffold at Richmond than in this inglorious manner. On the third morning she was roused by the snap of gunfire and peeled herself off the ground. Her costume was rancid and stiff and her arms hung like wilted flowers. She walked toward the enemy lines, guiding herself by the sounds of battle, and invented yet another new name, "Bridget."

She came upon a small white house and circled it. Peering into a window, she spotted a rebel soldier resting upon a straw tick on the floor and let herself in. He'd been ill with typhoid fever for several weeks when he got separated from his company during a skirmish, he told her. Unable to find his way back, he took refuge in this abandoned farmhouse, hoping the Yankees wouldn't discover him.

Emma could tell he was nearing death. She kindled a fire, rummaged through the kitchen for cornmeal, and made a hoecake, the edges of the batter sizzling into lace. The soldier thanked her "with as much politeness as if I had been Mrs. Jeff Davis" and pleaded for her to stay and talk with him. "Rebel though he was," she did, asking if he professed to be a Christian.

"Yes, thank God!" the man rasped. "I have fought longer under the Captain of My Salvation than I have yet done under Jeff Davis."

This was his last conversation, Emma thought, a chance for her to lead his thinking to a different and righteous path. She realized she'd forgotten her brogue only after the words left her mouth: "Can you, as a disciple of Christ, conscientiously and consistently uphold the institution of slavery?"

The sudden shift of her voice seemed to wake him up. His eyes flipped open, darting like fish, skittering across her face and beyond, as if he were seeing two of her. After a moment he asked her to pray with him, and when she said "Amen," he grasped her hand.

"Please tell me who you are," he whispered. "I cannot, if I would, betray you, for I shall very soon be standing before that God whom you have just addressed."

She promised that as soon as he grew stronger she would tell him her story, and forgave herself that lie.

Instead he began telling his own story. His name was Allen Hall, and he had one last request: If she should ever pass through the Confederate camp between here and Richmond, would she give this gold watch to Major McKee, of General Ewell's staff?

Emma agreed. Her patient relaxed then, settling into himself.

"Am I really dying?" he asked.

This time she told the truth: "Yes, you are dying, my friend. Is your peace made with God?"

"My trust is in Christ," he replied. "He was mine in life, and in death He will not forsake me." They were the same words she once heard a Union soldier say on his deathbed. Allen Hall died at midnight, his hand still in hers. She covered him with a blanket, clipped a lock of his hair, and fell asleep on the floor next to him.

At daybreak she searched the house for anything that might aid her disguise: a new basket, ochre, a bottle of red ink, black pepper, and court plaster, an adhesive cloth used to cover open sores; she affixed a patch the size of a dollar to her cheek. The ochre ruddied her complexion, and the ink etched shadows under her eyes. One final prayer for her rebel friend and she set off down Richmond Road, stopping to bury her pistol for fear she might be searched. She walked five miles before she saw a rebel picket in the distance.

She found the black pepper in her basket, sprinkled some in her handkerchief, and pressed it against her face. Her nose watered and her eyes reddened, dropping fat tears. She hobbled forward, poking her eyes with the handkerchief. In a thick English accent, the guard asked what business she had in the Confederate camp.

A message for Major McKee, she responded in her brogue. It's from a brave, fallen Confederate soldier, have mercy on his dear soul.

The man seemed less moved by the death of his comrade than by encountering another foreigner. Go anywhere you please, he said, adding, "I wish I was at 'ome with my family . . . Englishmen 'ave no business 'ere."

Good for you, Emma thought, you are one after my own heart; but she replied, "Och, indade I wish yez was all at home wid yer families."

She wandered beyond the lines, weaving among the white canvas tents. A banjo twanged and a group of soldiers played baseball, whacking a yarn-wrapped walnut with a strip of fence rail, filched from a nearby farm. Feeling a tap on her shoulder, she turned to find another picket.

"One of our spies has just come in and reported that the Yankees have finished the bridges across the Chickahominy," he told her, "and intend to attack us either today or tonight." He warned that it wouldn't be safe to stay at camp too long, although the artillerymen had prepared a number of masked batteries—pieces of artillery concealed by terrain or trees. "There's one," he said, pointing to a brush heap, "that will give them the fits if they come this way."

She felt a flutter of panic; there wasn't much time. She hurried to headquarters and asked for Major McKee. "He's gone to set a trap for the damned Yankees," an aide told her. For hours she roamed about the camp, assessing troop placements and artillery strength, listening for gossip about plans for the coming battle. Crouching over her basket, she jotted down the number of guns and every plausible rumor. The major returned in the late afternoon. She took another long whiff of pepper before approaching his tent, willing herself to cry.

With a deep curtsy, Emma told McKee that she was the bearer of tragic news. One of his brave soldiers, Allen Hall, had gone home to the Lord, and his last wish was for the major to have his watch. She lowered it into the major's hands.

He began to sob. Emma waited, patting his arm, and after a moment he lifted his head, looking her up and down. She kept absolutely still.

"You are a faithful woman," he said, "and you shall be rewarded. Can you go direct to that house and show my men where Allen's body is?"

Emma said she could.

He pulled a $10 Federal bill from his pocket and pressed it into her hand. "If you succeed in finding the house, I will give you as much more."

She uncurled his fingers. Thank you, she said, but she couldn't take the money.

His brows pressed together, flattering his eyes into slits. At once she realized the gravity of her mistake: a poor Irish peddler would never refuse money. Bursting into tears—real ones, this time—she explained, "Oh, Gineral, forgive me! But me conshins wud niver give me pace in this world nor in the nixt, if I wud take money for carrying the dyin missage for that swate boy that's dead and gone— God rest his soul. Och, indade, indade I nivir cud do sich a mane thing, if I im a poor woman."

The major nodded, his brows sliding back into place; he believed her. She blew her nose and said she would be glad to show him where Allen Hall had died, so long as he brought her a horse. The farm was a bit of a distance away, and she had not been feeling well.

McKee called for a horse and a number of rebels to escort her. She mounted and turned to say good-bye to the major, who was struggling not to cry. She thought of the intelligence she'd gathered, crammed deep into her basket, and couldn't separate it from the major's grief. She felt shamed by her own duplicity, as if reporting the rebels' plans would kill Allen Hall all over again.

Her guilt dissolved with Major McKee's very next words: "Now, boys, bring back the body of Captain Hall, if you have to walk through Yankee blood to the knees!"

They covered the five miles in silence, Emma in front, reaching the little white house at sundown. One of the soldiers asked Emma to go on down the road to see if there were any Yankees in sight. She agreed, riding all the way to the Chickahominy River and back to her own lines, keeping her horse, "Reb," as a souvenir.

While the Union command appreciated the intelligence Emma delivered, some officers began to wonder how "Frank" so convincingly impersonated a woman.

My Love to All the Dear Boys

THE SHENANDOAH VALLEY, VIRGINIA

After her midnight ride to Turner Ashby, Stonewall Jackson's cavalry commander, Belle grew increasingly confident about her value to the Confederate cause, seeking any opportunity to expand her already distinguished record. When she heard of a courier assignment in Winchester, twenty miles north of Front Royal and also in Union hands, she immediately offered her services. Failing to secure a pass from the provost marshal, she sought out Lieutenant Abram Hasbrouck, an officer with the 5th New York Cavalry who always succumbed to her charms, even though her pet crow wasn't able to pronounce his name.

"Now, Lieutenant Hasbrouck," Belle said, "I know you have permission to go to Winchester, and you profess to be a great friend of mine." Her gloved hand perched on his forearm, her fingertips pressing into his sleeve. "Prove it by assisting me out of this dilemma, and pass us through the pickets."

The lieutenant relented.

He dismounted from the carriage before they arrived at Winchester, explaining that he had some matters to attend to at the Federal camp on the outskirts and would rejoin them the following day. Belle and her servant, Eliza, continued, heartened by the sight of so many secessionist women, identifiable by their "Jeff Davis" bonnets, featuring side panels that concealed the face from prying Yankee eyes. "The women are almost universally bitter Secesh and spit out with venom," observed one Union officer, while "the men are more discreet and keep their sentiments to themselves." Winchester changed hands continually (doing so, on at least one occasion, thirteen times in a single day), and the female population never failed to harass the enemy, shooting at Union officers from windows, tossing hand grenades, dumping pails of scalding water and the contents of chamber pots. At the home of a family friend, where Belle had arranged to meet her contact, the women were gathered in the parlor, sewing uniforms for the rebels and invoking the Lord to send Stonewall Jackson soon.

The next morning, May 22, Colonel William Denny of the 31st Virginia Militia Regiment came to the house, a leather bag slung over his shoulder. He knew Belle's father and her other relatives in the rebel army, many of whom hailed from the area.

"Miss Boyd," he said, reaching into the bag, "will you take these letters and send them through the lines to the Confederate army? This package is of great importance; the other is trifling in comparison."

Belle stacked the two packages in the crook of her arm. Then Denny, explaining that it was the most valuable of them all, pressed a piece of paper into her palm. "Try to send it carefully and safely to Jackson, or some other responsible Confederate officer. Do you understand?"

Belle nodded, her mouth a solemn slash. "I do, and will obey your orders promptly and implicitly."

She hid the more important package on Eliza, tucking it under the waistband of her homespun dress, knowing the pickets would never search a slave. As a precaution, she scribbled "Kindness of Lieutenant Hasbrouck" on the "trifling" package so it appeared to be a harmless gift from a Yankee admirer, and placed it inside an open basket. She held the most vital item, the note, casually in her hand, occasionally using it to fan her face.

She took no notice of a servant, dusting in the parlor, watching her and Eliza conduct their business and exchange smug smiles. As soon as Belle and her maid left, this servant told Union authorities exactly what he'd seen.

A t a local florist shop, Belle ordered a handsome bouquet sent to the Winchester provost marshal, who was, she knew, "never displeased by a little flattery and a few delicate attentions." She included a card inquiring if he might be so kind as to grant her a return pass to Front Royal, and within an hour his aide delivered one. Having finished his own business at camp, Lieutenant Hasbrouck ordered a carriage and together they began the journey back.

At the picket lines two Union detectives approached on horseback, stopping on either side of the carriage. One peered into the window and rested his gaze on Belle. She found him "repulsive-looking."

"We have orders to arrest you," he said.

"For what?" Belle asked.

"Upon suspicion of having letters." Turning to the coachman, he ordered him to drive to the headquarters of Colonel George Beal, commander of the 10th Maine Infantry. Belle could feel Eliza trembling beside her. The motion set her off and she too began shaking, their bodies meeting in quick and nearly imperceptible collisions. She contracted every muscle, willing herself to be still.

The group spilled out and filed into the office, Eliza still close by her side.

"Do you have any letters?" the detective asked.

If Belle said no, she would be searched. She reached into her basket and pulled out the package her contact had deemed of minor consequence.

Colonel Beal read the lettering on the package, mouthing the words "Kindness of Lieutenant Hasbrouck" and then saying them aloud: "What! What is this? 'Kindness of Lieutenant Hasbrouck!' What does this mean?"

Belle glanced at the lieutenant. He stood motionless, his upper lip shiny with sweat. She might need him in the future and wanted to keep him on her side.

"It means nothing," she insisted. "It was a thoughtless act of mine. I assure you Lieutenant Hasbrouck knew nothing about the letter, or that it was in my possession."

They all watched as Colonel Beal ripped open the package and studied its contents: a copy of the *Maryland News Sheet*, a staunchly rebellious journal.

Again Belle insisted that the lieutenant was innocent, protesting with as much vigor as she could muster. Colonel Beal turned to her and demanded, "What is that you have in your hand?"

She unfolded her fingers to expose the note, the most important piece of intelligence. "What?—this little scrap of paper?" she asked. "You can have it if you wish; it is nothing. Here it is."

She held it out, waiting for him to come to her, but she had no intention of relinquishing it. If necessary, she planned to follow the example set by Harvey Birch in James Fenimore Cooper's *The Spy*, or, more recently, Rose Greenhow, and swallow it whole. But the colonel, still clutching the *Maryland New Sheet*, redirected his wrath back to the seemingly traitorous Union lieutenant "from the guilty to the guiltless," Belle thought—and she and Eliza were dismissed. (Another Union officer, after hearing of the incident, remarked incredulously that "with her usual adroitness and assumed innocence she got clear of any charge of treachery.")

Back in Front Royal, behind the closed doors of her chamber in the cottage, Belle pulled the note from her corset and memorized its contents, an update on the information she'd overheard during the Union war council: General John Kenly had only a thousand men at Front Royal. Irvin McDowell's troops had not yet arrived, and General Banks, with four thousand men, was northwest at Strasburg. Generals Shields and Geary were a short distance southeast of Front Royal, with John Frémont farther west, just beyond the Valley.

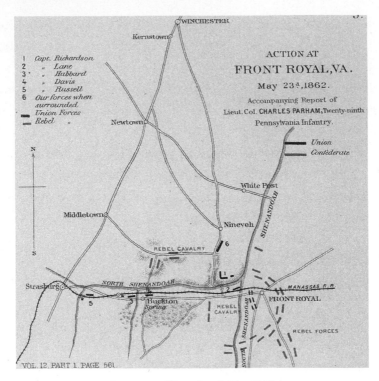

Battle map of Front Royal, May 1862.

Belle understood that while Front Royal itself was not well protected, all of these separate Union forces could unite against Stonewall Jackson, who was driving his sixteen thousand troops—her father among them—rapidly down the Valley.

She didn't have much time to warn them.

Something was imminent; everyone in Front Royal could sense it. The Union troops had enforced martial law and no citizen was allowed to leave, not even to fetch fuel or provisions. The morning of May 23 was oppressively warm, but everyone gathered outside to gossip and search for news. A Yankee soldier wandered through the crowd, asking if he could "buy some pies and pigs," the word "pigs" followed by sudden, sharp reports of a rifle in rapid succession.

At that moment Belle heard Eliza's familiar voice: "Oh, Miss Belle, I t'inks de rebels am a-coming, for de Yankees are a-makin orful fuss in de street."

Belle jumped from her seat. The streets in every direction were filled with a roving crush of blue. She rushed outside and clutched the sleeve of a passing Federal officer, pressing him for details. He replied that the Confederates were approaching the town in force under Jackson and Ewell. They had surprised and captured the outside pickets and had managed to advance within a mile of the town without the attack even being suspected. "Now," he added, "we are endeavoring to get the ordnance and the quartermaster's stores out of their reach."

"But what will you do with the stores in the large depot?" Belle asked.

"Burn them, of course!"

"But suppose the rebels come upon you too quickly?"

"Then we will fight as long as we can by any possibility show a front," the officer said, "and in the event of defeat make good our retreat upon Winchester, burning the bridges as soon as we cross them, and finally effect a junction with General Banks's force."

Belle thanked him and returned to the Fishback Hotel, bumping into A. W. Clarke, a correspondent for the *New York Herald*, on the stairs. Since he began boarding at the Fishback, Clarke had, in Belle's words, made several attempts to "intrude his society" upon her, and she'd found his attentions "extremely distasteful"; once she was even forced to bolt her bedroom door after slamming it in his face. He seemed to think, Belle wrote, that to "insult an innocent young girl was to prove [his] manhood."

"Great heavens! What is the matter?" Clarke asked as she pushed past him.

"Nothing to speak of," she replied. "Only the rebels are coming, and you had best prepare yourself for a visit to Libby Prison."

He turned around and hurried to his room. In her own chamber Belle found her opera glasses and the dispatch she'd retrieved in Winchester the previous day. As she passed Clarke's room, she spotted him packing his clippings and files. Poking her head in, she asked, "Where are you going, Mr. Clarke?"

Without looking up, he replied, "I'm going to skedaddle."

His room key was on the outside of the door. She couldn't resist: quietly she closed the door, turned the key in the lock, and dropped it into her pocket.

From the balcony she could see the Southern advance guard about three quarters of a mile away. Back downstairs, she rushed to the street and found a group of men she knew to be devoted to the cause.

Was there one among them, she asked, who would be willing to carry her information to General Jackson? She gave a quick summary: the Confederates were advancing, but Stonewall might be uncertain as to enemy strength and holding back on the main attack. He could strike against Front Royal, and perhaps other isolated Federal units, provided he knew where all the Union forces were and could avoid being trapped if they converged.

"No, no. You go!" one said, and with those words she understood that her next action would be the most consequential of her life, that the sum of those few moments would forever exceed the whole of all her years.

Belle began to run. She wore Yankee colors, ironically, a white bonnet and a royal blue dress, her heart lurching behind its buttons, her perfect size-two-and-half feet hot in her lace-up boots. Through the streets of Front Royal, bodies falling out of her way, and then the fields opening up to her, unblemished and immaculate, the quick slap of gunfire coaxing her faster. The Federal pickets took aim. Minié balls struck the ground at her feet, kicked dust up into her eyes. She was acutely aware of her own mechanics, the hoarse foghorn croak of each breath, the dig of nails inside tightly fisted hands. Her message looped through her mind, knowledge she carried with a keen and lovely pride.

The fire came from both sides now, bullets whispering past her ears, each one bearing a message meant only for her. She felt strange pulls of pressure at her skirt and glanced down long enough to see that shots had sheared the hem. "I shall never run again as I ran on that, to me memorable day," Belle wrote. "Hope, fear, the love of life, and the determination to serve my country to the last, conspired to fill my heart with more than feminine courage, and to lend preternatural strength and swiftness to my limbs. I often marvel, and even shudder, when I reflect how I cleared the fields, and bounded over the fences with the agility of a deer." She began waving her bonnet in grandiose loops, a signal for the rebels to continue to press forward. Many of the men recognized her and cheered as they dashed by: "Their shouts of approbation and triumph rang in my ears for many a day afterwards, and I still hear them not infrequently in my dreams." There she was: Belle Boyd, the famous rebel spy, her most fantastic act of espionage performed for all the world to see.

A private with the 2nd Virginia Infantry named Henry Kyd Douglas was shocked to see the figure of a girl gliding up a ravine, heeding neither weeds nor fences, whipping the air with a bonnet. He motioned to Stonewall Jackson, and the general sent him to see what she wanted. Douglas readily agreed: "That was just to my taste and it took only a few minutes for my horse to carry me to meet the romantic maiden whose tall, supple, and graceful figure struck me as soon as I came in sight of her." He was twenty-one, just a few years older than Belle, and had known her since childhood. She called his name and he ran to her, marveling that "she was just the girl to dare to do this thing."

Her breath was coming in raspy gusts, one hand pushing against her heart.

"Good God, Belle, you here!" Douglas said. "What is it?"

She produced the dispatch and spoke in gasps: "I knew it must be Stonewall when I heard the first gun. Go back quick and tell him that the Yankee force is very small—one regiment of Maryland infantry, several pieces of artillery and several companies of cavalry. Tell him I know, for I went through the camps and got it out of an officer. Tell him to charge right down and he will catch them all." She took another breath, and urged that the cavalry must rush to seize the bridges before the Union could destroy them. "I must hurry back. Goodbye. My love to all the dear boys—and remember if you meet me in town you haven't seen me today."

Douglas raised his cap and Belle blew him a kiss. He watched her sprint back through the weeds and ravines, waving her bonnet all the way, finally disappearing in a dip in the land. Belle would have been gratified to know what happened next: Stonewall Jackson rode up on Little Sorrel to ask who that "young lady" was. With a smile, he suggested that Douglas rush to the front of the line and into town to see if he could find her again.

The conflict raged. Turner Ashby's Confederate cavalry made a brave charge in which two of his best captains were killed. The Confederate 1st Maryland met the Federal 1st Maryland, with disastrous results for the latter. In the end about 85 Union soldiers were killed or injured and 700 captured, with the Federals retreating so quickly that they left all of their casualties behind.

Henry Kyd Douglas found Belle in front of the Fishback Hotel, holding court with a group of Union soldiers (now prisoners), her cheeks "rosy with excitement and recent exercise and her eyes all aflame." He stooped from his saddle to let Belle pin a crimson rose to his uniform.

"Remember," she said, "it is *blood red*, and it is my color."

The finale of Belle's big day came when a courier arrived at the hotel and dropped a crumpled slip of paper into her hand:

Miss Belle Boyd,
    I thank you, for myself and for the army, for the immense service that you have rendered your country to-day.

                    Hastily, I am your friend,
                    T.J. Jackson, C.S.A.

Seventy miles east, at the Old Capitol Prison in Washington, DC, Rose Greenhow was walking in the yard with Little Rose when an inmate from Virginia tossed a piece of paper at her feet. Its contents made her heart "leap with joy": Stonewall Jackson had soundly defeated Union forces at the Battle of Front Royal in the Shenandoah Valley. "All honor to the brave Jackson," she thought, "who is now the special terror of the Yankees!"

The news was especially encouraging in the wake of recent events. Her plan to help two inmates escape had gone badly awry. One, a suspected Confederate spy, was shot as he rappelled down the prison wall and later died from his wounds; the other backed out of the plan at the last minute. When guards visited Rose's cell to interrogate her, she admitted she would consider it a "point of honor" to aid the escape of any Confederate prisoner. As punishment they reduced the size of her and Little Rose's meals. "My child is so nervous from a repetition of these dreadful scenes that she starts and cries out in her sleep," Rose wrote. "I am nearly starved. I had a fowl served up to me to-day (or rather a small piece of one), which must have been the cock which crowed thrice to wake Peter; we could not get our teeth through it."

She'd also been distressed by the Confederate defeat at Shiloh, Tennessee, which made the Mississippi River Valley vulnerable to a Union advance, and she wept "tears of blood" over the fall of New Orleans. The final blow had been a visit from her old friend and alleged lover, Senator Henry Wilson, who cheerfully boasted that the rebellion was crumbling and Richmond about to fall. Despite her involvement in the escape plan, and despite one Union general's warning that she was a "dangerous, skillful spy," she and Little Rose would soon be freed and exiled to the South, a decision she attributed to the Yankee government's desire to finally be rid of her. She expected to be in the Confederate capital if and when the end came.

One Grain of Manhood

OUTSIDE OF RICHMOND

Six days after the Battle of Front Royal, on May 29, McClellan sent General Heintzelman's corps, which included Emma's regiment, across the Chickahominy River to join the other troops on the outskirts of Richmond. Emma dared to hope that the Confederate capital would fall imminently, but the following evening a violent thunderstorm swept the Peninsula. For hours the rain fell, heaved in all directions by great gusts of wind, accompanied by thunder one rebel soldier described as "hell's artillery." Lightning bleached the sky, one strike instantly killing four men of the 4th Alabama. "Had it not been for McClellan's faith in the Bible and in God's covenant with Noah," Emma wrote, "he would no doubt have seriously contemplated building an ark in order to save himself and his army from destruction. The Rebels seemed to think this flood was sent as a judgment from the Almighty upon their hated enemies."

Confederate general Joe Johnston did view the storm as a godsend. The Army of the Potomac was now divided by the Chickahominy, and the rising floodwaters would make it nearly impossible for McClellan to reinforce his troops across the river during battle. Two days later, around noon, Johnston ordered his men to attack, hoping to surprise the most advanced Union regiments and drive them back toward the river. The general then moved his headquarters to a small house nearby and waited for the sounds of battle. The fighting raged around him—cannon thundering, shells shrieking, rifle balls stuttering *t-h-t! t-h-t*—but he heard none of it. The newspapers called such strange mirages of noise "silent battles." The sounds were most often absorbed by woods and hills or deflected by wind currents, bounced away to a distant location, even hundreds of miles away. Johnston had no idea his men were fighting until several hours later.

Emma, dressed in Frank Thompson's full uniform and mounted on "Reb," the horse she'd acquired during her last undercover assignment, acted as orderly for General Philip Kearny. The general had lost his left arm in the Mexican War and compensated by gripping his reins in his teeth and brandishing a sword in his right hand. She rode alongside him on the battlefield, delivering messages and dodging shots and shells, watching her comrades march across the one remaining rickety bridge, held together by a single stringer. She was about to ride off with another message when a ball struck and shattered the arm of General Oliver Howard, standing nearby. She hitched Reb and rushed to the general, pouring water on the wound and down his mouth. His arm was limp, sandbag-heavy and awash in blood. As she rummaged for bandages in her saddlebag Reb sank his teeth into her arm, stripping flesh from her bone, then turned and kicked her in the gut with his hind feet, hurling her through the air.

All of the breath rushed out of her and she felt turned inside out, whipped like laundry on the line. She gasped violently, greedily gobbling at the air, ordering herself to her knees, to her feet, wobbling back over to General Howard. Her own arm was so swollen she couldn't lift it above her head, but she wrapped the general's wound and reported to an old sawmill that had been converted to a field hospital. It was crowded with injured men who had crawled in from the battlefield, dragging themselves forward inch by inch using nothing but elbows and chins.

At the moment she was the only nurse there. She bound her arm into a sling and went to work removing the soldiers' clothing. She had neither scissors nor a knife and so bit through the soaked, stiffened garments, blood seeping into her mouth and staining her teeth. When she ran out of bandages she started off in the direction of two houses, about a mile away. She hoped the rebel folk inside would notice her condition and show some mercy.

The occupants of the first house refused to let her in, and she limped along to the next. A man came to the door, holding it open just a crack, and told her he had nothing she could use for bandages. Emma snapped, her patience and strength both exhausted. She drew two pistols from her belt and aimed them at the man, the hand of her injured arm shaking under the weight of the metal.

The man looked into the guns' shiny gaze and that was all the convincing he needed. He called to his wife, who brought an old sheet, a pair of pillowcases, and three yards of cotton cloth, for which she demanded five dollars. Emma gave her three.

On the way back to the sawmill she felt a strange whirling in her head. She removed her sling and saw that she'd grazed her arm with her pistol, reopening the wound, fresh blood blooming across her skin. Her breath thinned, and she stopped to rest by the side of the road.

An hour passed, and she heard the distant sound of hooves drawing closer. She looked up to see a Union chaplain. His eyes flickered down, taking her in, but he continued without offering her a ride. She assumed he was hurrying to help the wounded at the sawmill, but after she staggered back she found him wrapped in a blanket, sleeping on a pile of hay.

She considered asking the chaplain if he would be so kind as to bring her horse—just for the sake of having the vicious "Reb" give him "a little shaking up"—but instead she stood over him in quiet disgust. She muttered that he was "not the possessor of one grain of manhood," hoping he was awake enough to hear the words. Slumping down beside him, she began to cry, for once not caring if she looked like a girl.

The night brought respite, Yankees and rebels lying on their arms within speaking distance of each other. Fighting resumed at seven thirty in the morning. Emma spotted McClellan riding along the battlefront, the men cheering as he passed, all of them comforted by the sight of their commander. The rebel army was falling back toward Richmond and dealing with an unexpected emergency: General Joe Johnston had been struck in the chest by a heavy fragment of shell, knocked off his horse, and carried off the field on a stretcher. President Davis asked General Robert E. Lee, his chief military adviser, to assume command of the Army of Northern Virginia, the Confederacy's main fighting force in the eastern theater of the war. Recognizing the precarious state of his troops, Lee continued the withdrawal toward Richmond.

By noon the Battle of Fair Oaks (known as Seven Pines to the South) was over, a Union victory at immense cost: 790 killed, 3,594 wounded, and 647 missing; the Confederate casualties numbered 1,000 more. It was the worst battle yet in the eastern theater, and Emma, standing by the surgeon's tree, could do little more than survey the horrific scene. "The ground around that tree for several acres in extent was literally drenched with human blood," she observed, "and the men were laid so close together that there was no such thing as passing between them." For the next two days she helped load the wounded onto railroad cars bound for White House Landing, where they were transferred to hospital ships for the trip north. Men died along the way, their bodies crammed in close boxcars along with the living, maggots burrowing into their wounds.

Publicly McClellan swaggered, boasting about the "great victory" and issuing an optimistic address to his troops: "Soldiers of the Army of the Potomac! I have fulfilled at least a part of my promise to you. You are now face to face with the rebels, who are held at bay in front of their Capital. The final and decisive battle is at hand. . . . The enemy has staked his all on the issue of the coming battle. Let us meet him, crush him here, in the very center of the rebellion." But in letters to his wife the general confessed a distaste for the realities of warfare.

"I am tired of the sickening sight of the battlefield," he wrote, "with its mangled corpses & poor suffering wounded! Victory has no charms for me when purchased at such cost." In perfect position to advance on the Confederate capital and possibly end the rebellion, McClellan instead blamed the poor weather for his inaction and badgered Lincoln for more men.

While the general waited for his bridges to be rebuilt and for reinforcements to arrive, an odd lull ensued. Union and Confederate pickets lined the swampy lowlands on the south side of the Chickahominy River, in some spots separated by a mere hundred yards, trading newspapers and telling stories and hoping the enemy would honor the informal agreement not to shoot. Emma temporarily resumed her duties as postmaster and mail carrier, riding Frank from the front lines to Fort Monroe and back again, a nearly sixty-mile route that took her through areas heavily populated with rebel sympathizers; she had heard reports of another mail carrier being robbed and murdered by bushwhackers on the very same path. It was always late when she passed over that "most lonely spot" of the road, so dark she knew it only by the rustle of paper under her horse's feet, and she prayed that God would keep her safe until that dreadful noise was gone.

THE MADAM LOOKS MUCH CHANGED

RICHMOND

Rose arrived in Richmond—the capital of *her* country—on June 5, just after the Battle of Seven Pines, along with railcars, wagons, and hacks carrying thousands of Confederate casualties bound for the city's forty-two hospitals or Oakwood Cemetery. Coffins piled up faster than gravediggers could bury them, and the heat caused dozens of bodies to swell and burst the wood. The finest stores on Main Street had been transformed into hospitals, with cots of wounded men pressed up to the entrances, a macabre stretch of window displays. "The weather was very warm, the doors were open and no curtain or screen shielded them from the gaze of passersby," Elizabeth wrote. "So sickeningly fetid was the atmosphere that we could not sit in our grounds."

Union officials released Rose under two conditions: she had to sign an oath vowing not to "return north of the Potomac River during the present hostilities," and also promise to use her influence to secure the release of two Pinkerton detectives, Pryce Lewis and John Scully, both of whom had guarded her at her home and who were now languishing in Castle Godwin. The Richmond newspapers celebrated her arrival, declaring her name had been "rendered historic" by her work for the Confederacy, and by her defiance of "the monsters" at Washington.

A carriage dropped Rose and Little Rose off at the Exchange and Ballard Hotel at Franklin and Fourteenth Streets, two buildings—

one a four-story Greek Revival, the other a five-story Italianate—
joined by a cast-iron pedestrian bridge. The hotel was the city's most
prestigious, with a guest book signed by Charles Dickens in 1842,
when he sat perspiring in his room and longing to return to cooler
climes. Edgar Allan Poe, during his last visit to Richmond in 1849,
lectured on "The Poetic Principle" and "Philosophy of Composi-
tion" in the hotel's parlor and spent his nights at the nearby Swan
Tavern. A few months earlier, in January, John Tyler, the former
president and Rose's cousin by marriage, had died of a stroke in his
room. A bellboy took Rose's valise up to her quarters and delivered
letters from the Pinkerton detectives, both of them imploring her to
intervene. Rose ignored these and other pleas to come.

Instead she began working on her memoir, focusing on her imprisonment and the horrors of the Lincoln administration, and took
Little Rose for walks, shielding the girl's eyes and nose from the
sights and smells of the wounded soldiers. She noted approvingly
that "all was warlike preparation and stern defiance and resistance
to the invader." McClellan's campaign had stalled but the city was
still on edge, expecting an attack at any moment. In anticipation,
Jefferson Davis sent his wife and children temporarily to Raleigh,
North Carolina, a day's journey by train and still in Confederate
hands. "I belong to the country but my heart is ever with you," he
told Varina, and taught her how to load and shoot a pistol. Elizabeth, too, was so certain the Union would take Richmond that she
prepared a "charming chamber" for McClellan in her home, fixing
it with "new matting and pretty curtains" and calling it "General
McClellan's room."

Despite her efforts on behalf of the Confederacy, Rose knew she
wouldn't be accepted by the proper ladies of Richmond, all of
whom had heard the gossip about her late-night callers; "She must
not come handicapped with her old life," Mary Chesnut admon-
ished. But the city's leading men had no such judgment. General

John Winder, the provost marshal, and Jefferson Davis both came to her suite at the Ballard to pay their respects. Rose gave the Confederate president a gift of jelly and three oranges, one of which he promised to give to the wounded general Joe Johnston. Little Rose inquired about Davis's seven-year-old daughter, Maggie, and wondered "if there were any Yankees where she was."

The Confederate president was struck by Rose's appearance, the lovely face that was now deeply lined and sallow. "The Madam looks much changed," he confided to his wife, "and has the air of one whose nerves were shaken by mental torture." He avoided talk of her time in prison and thanked her for her service, telling her, "But for you, there would have been no Battle of Bull Run." For Rose, hearing such praise from the Confederate president was "the proudest moment of my whole life." Davis would soon have another assignment for her, one that would prove vital to the Confederacy's attempts to gain legitimacy, and one that would cost her more than she could have ever imagined.

The Secesh Cleopatra

THE SHENANDOAH VALLEY, VIRGINIA

In the hope of disrupting McClellan's forces around Richmond, Robert E. Lee ordered Stonewall Jackson to vacate the Shenandoah Valley and march down to the Virginia Peninsula. With the Confederates gone, Belle watched the Union troops spill across High Street in Front Royal, once again taking control of the town. They captured 160 prisoners, including Belle, escorting her to the cottage behind the Fishback Hotel and stationing a squadron of guards outside her door. She knew her recent dash across the battlefield had made her known throughout the country, with Southern papers anointing her the "Secesh Cleopatra" and Northern papers denouncing her as a "camp cyprian" and "accomplished prostitute" who has "passed the first freshness of youth." She remained so imprisoned for several days until her old Union contact and rumored paramour, General Shields, arrived in town and released her.

General Nathaniel Banks followed, establishing headquarters at the Fishback, and Belle called on him right away, hoping to make another high-ranking friend among the enemy. For the occasion she donned a dark green riding dress, her gold palmetto breast pin, and a hat festooned with Confederate brass buttons and topped with a piece of palmetto frond, rising like an antenna from her head. The finishing touch: a pair of shoulder straps bearing what she claimed as her new insignia of rank: "Lieutenant Colonel, 5th Virginia Regiment, Confederate Army."

She welcomed General Banks to Front Royal, and said she hoped he was enjoying the accommodations at her aunt's hotel. She had one minor favor to ask: Would the general mind granting her a pass to travel farther south?

"Where do you wish to go?" Banks asked.

"To Louisiana, where my aunt resides."

"But what will Virginia do without you?" he said, and laughed. Belle was surprised, and leaned closer.

"What do you mean, General?" she asked.

"We always miss our bravest and most illustrious," he replied. "And how can your native state do without you?" The general proceeded to speak "with the utmost good-nature and pleasantry" about her daring dash across the battlefield and her role in his own army's defeat, but refused to let her leave the state.

Belle headed back to her cottage, the palmetto frond on her hat listing to starboard, and mentally replayed the exchange. She sensed an edge lurking beneath the general's light tone, a stiffness in his demeanor, as if the enemy at last understood the nature of her power and the force of her intent, the very things she'd known all along.

It was about time.

Belle filled her days with flirting and spying, the two being interdependent in her mind, focusing on a Major Dick Long of the 73rd Ohio, who possessed "a killing set of whiskers" and boasted of his conquests with the "rosy-cheeked rogue." Another Yankee private said he admired her courage in defending her home and presented her with a pistol, which Belle planned to regift to Stonewall Jackson. She paid a visit to General Shields's aide-de-camp, Captain Keily, the officer who had written her love poems and given her information about the war council. He had been shot in the face recently but still Belle managed to question him, interspersing her interrogation with comforting words about his recovery. She was "the sensation of the village," acknowledged one Union man, and had a "large following of Federal officers who were ready to do her homage."

She was vigilant about Union spies—they still roamed everywhere in the Valley—and began noticing a man with a brusque accent trailing her wherever she went. She confronted him one day when both were loitering in the office of the Union provost marshal.

"I suppose you came to report me again," she said, and spun back around, denouncing "Southern Union men" so loudly the entire office could hear.

The informant, a German immigrant named Eugene Blockley, reported to Allan Pinkerton, who started a file on Belle. "She gets around considerably, is very shrewd, and is probably acting as a spy," the detective wrote. "She is an open, earnest, and undisguised secessionist, and talks secession on all practicable occasions. . . . Informant considers her more efficient in carrying news to the rebels of our operations than any three men in the valley."

Belle didn't let the incident interfere with her next mission, the seduction of Dr. George Rex, medical director of the First Army Corps. She courted him with tactical precision, assessing his strengths and exploiting his weaknesses, calculating when to advance and when to retreat. In her more fanciful moments, she imagined that this wealthy, influential doctor would one day propose marriage and sail off with her to Europe.

Her efforts caught the attention of a brigade surgeon, Dr. Washington Duffee, who sent a wry letter to Secretary of War Stanton in Washington:

"I communicate to you a fact," he wrote, "that the celebrated Belle Boyd the 'Rebel Spy' now at Front Royal has apparently fallen in love or is anscious [*sic*] to make a victim of the Medical Director of the 1 Army Corps. Where [*sic*] that used by higher authority at the War Department Jackson and all the Rebel officers with whom she is in direct communication might be trapped. Where a skillful 'Detective' placed by the Government easily could these rascals be led into captivity."

This was the second notice Stanton had received about Miss Belle Boyd, after Allan Pinkerton's report about her activities in the Valley, and he was already on the case. It seemed the girl was willing to engage with any man in uniform, so long as he told her what she wanted to hear.

One June morning, sitting in the drawing room of her cottage, Belle noticed a Confederate soldier standing by a flag-tent in the courtyard, where the Union provost marshal granted or refused all passes. She had never seen him before, and wanted to do him the favor of introducing herself. Tying on her lucky white bonnet, she wandered into the courtyard and asked for his name.

"C. W. D. Smitley," came the reply. He looked to Belle like a younger version of General George McClellan, whom she'd seen in cartes de visite, sold in packets of eight for a dollar. Smitley was taller and leaner than the general, but had the same sharply parted brown hair and tidy beard. He told her he was a parolee hoping to head south to Richmond.

She wished him luck, joking about how one couldn't "travel the slightest distance without a pass signed by some official." In fact, she added, there was once a picket stationed between her family farm and the dairy, and the dairymaid wasn't allowed to milk the cows without a pass signed by the officer of the day. This protocol became such a nuisance that Belle devised a way around it, filling out a pass to read: "These cows have permission to pass to and from the yard and dairy for the purpose of being milked twice a day, until further orders." She secured the proper signature and pasted the pass between the horns of one of the cows.

The boy laughed, and Belle asked if he might dine with her that evening at the home of a good Confederate neighbor.

"With pleasure," he said, and Belle excused herself to get ready.

To Smitley's surprise, he found her to be "a lady of culture" and "a brilliant conversationalist." During the course of the evening a number of young ladies called, all accompanied by Federal officers, and they hovered around Belle like moths, coming and going with their demands: Would she talk about her work as a courier, about smuggling quinine, about her mad dash across the battlefield at Front Royal, her only protection a bonnet and crinoline? She had perfect recall, each anecdote as bright and polished as one of her souvenir buttons, and when they no longer wished to hear her talk, she offered to perform. Settling on the piano bench, nestled in her swirl of a dress—no frock coat or shoulder straps tonight—she plucked out the notes of "Bonnie Blue Flag" and began to sing:

And here's to old Virginia
The Old Dominion State
Who with the young Confederacy
At length has linked her fate . . .

She commanded Smitley to join her and he obeyed, his voice cracking on the words "Davis, our loved president." In between verses Belle peeked at him and smiled.

The following day Smitley told her he had to leave Front Royal shortly—he'd finally secured a pass—and would be gone before supper. Belle let the corners of her lips sink, but said, "I am glad that you will soon be free." She studied him a moment, his face half-shadowed beneath a musty slouch hat, and decided he was who he claimed to be.

"Won't you take a letter from me to General Jackson?" she asked.

He agreed, and Belle left him in the drawing room to write a note to Stonewall, beginning with pleasantries and compliments and concluding with information about Union general John Pope's forces, totaling about thirty-five thousand men and heading toward the Shenandoah Valley. When she returned, Smitley stood and held out his hand.

"Will you promise me faithfully, upon the honor of a soldier, to take the utmost care of this, and deliver it safe to General Jackson?" she asked.

He vowed "by all the host of heaven" that he would, and that was the last she saw of him.

That evening she went to another party, and was about to take her place at the piano when a Union officer approached. Belle knew him. He had proved useful in the past but turned cool once he realized she wanted nothing from him but information. He slapped his hand against the wall, trapping her, and remarked that she'd been seeing quite a bit of Officer Smitley lately, hadn't she?

Belle didn't see how it was any of his business.

It was his business, the officer countered, because Smitley was actually a Federal spy.

She felt a hot brick of dread in her gut. She denied it, and said she was insulted on the good Confederate's behalf.

It was true, the officer insisted. He himself had sent Smitley to entrap her. He smiled again, and Belle began to cry, as distressed by the romantic betrayal as by what might happen next.

The Bright Rush of Life,
the Hurry of Death

RICHMOND

Just as McClellan was about to order the first "decisive step" of the final advance on Richmond, he received another ominous report from Pinkerton. Under interrogation, a deserter from Stonewall Jackson's command had revealed that Jackson's entire force, still en route from the Shenandoah Valley, would attack Union general Fitz John Porter's corps on the north side of the Chickahominy River. McClellan nevertheless kept to his plan, ordering two divisions to advance just twelve hundred yards, taking a dense oak grove that sprawled between the Union and Confederate front lines.

Emma's regiment was held in reserve, but she could feel the heat of the gunfire and see the men in the rifle pits, muddy water up to their knees, looking like "fit subjects for the hospital or lunatic asylum." By late afternoon she heard voices rise high and clear, the ringing cheers of the Union laid over the sharp, wild yip of the rebels, a malevolent refrain on endless repeat. When it finally stopped, Union troops had given McClellan half of what he wanted—six hundred yards, gained at a cost of one casualty per yard.

The general declared that his men had "behaved splendidly" at Oak Grove, the first of what would become known as the Seven Days Battles, but two hours later found himself in a panic. A fugitive slave, newly escaped from Richmond, reported that rebel officers boasted of two hundred thousand troops ready for the "big fight," prompting McClellan to telegraph Washington: "I regret my great inferiority in numbers, but feel that I am in no way responsible for it. . . . I will do all that a general can do with the splendid army I have the honor to command, and, if it is destroyed by overwhelming numbers, can at least die with it and share its fate."

In reality Lee had just eighty thousand men and was taking an enormous risk throwing fifty thousand of them at McClellan's right flank, leaving a mere thirty thousand to defend Richmond— less than half of the Union force now surrounding the Confederate capital. His plan depended on precise timing, flawless coordination between his generals, and a belief that McClellan would fail to capitalize on his numerical advantage and instead cower under the rebel assault. Lee would strike the following morning, at first light.

But Stonewall Jackson was late, delayed by exhausted troops and confusing, vaguely mapped roads through the thick lowlands, and Lee had to move without him. At last, at 4:00 p.m. on June 26, a dozen guns sounded by the Chickahominy, the booms melding into thunder, as many as forty shots per minute. Emma's division was near the front line and she heard the rebels before she saw them— again that feral, primal yell that evoked the cry of hounds on the scent.

Elizabeth heard it, too, riding in her carriage to Henrico County, just northeast of the city, to visit her Unionist friend John Minor Botts. "The excitement on the Mechanicsville Turnpike was more thrilling than I could conceive," she wrote that night in her diary. "Men riding and leading horses at full speed; the rattling of their gear, their canteens and arms; the rush of the poor beasts into and out of the pond at which they were watered. The dust, the cannons on the crop roads and fields, the ambulances, the long line of infantry awaiting orders . . . I realized the bright rush of life, the hurry of death on the battlefield." The windows of Botts's home rattled without pause, offering a view of bursting shells and dingy coils of smoke. She still kept McClellan's room—the charming spare chamber with the pretty curtains—ready for him, complete to the water drawn for his bath.

The battle of Beaver Dam Creek at Mechanicsville lasted six hours and culminated in a Union victory, 361 casualties to the rebels' 1,484, and McClellan sent a triumphant wire to Washington: "I almost begin to think we are invincible." He changed his mind after learning that Stonewall Jackson had finally arrived, but not in time to fight—which meant that his right flank had faced only a portion of Lee's force, and that the rebels were sure to attack again the next day. Although the Confederates were disorganized and outnumbered, and Richmond was more vulnerable than ever, McClellan decided that his most imperative goal was to move his base of operations from White House Landing on the York River to Harrison's Landing on the James River, which snaked along the downtown neighborhoods of Richmond. Once he was settled there, he reasoned, he could refocus his attention on the Confederate capital. He would call it not a retreat but a change of base—"one of the most difficult undertakings in war."

But a retreat it was. Lee began chasing the 120,000-man Army of the Potomac, with its wagons and supply trains, its guns and provisions, even its 2,500 head of cattle, as it fled south to the safety of the James. On the third day, as Lee battered Union forces north of the Chickahominy, Emma volunteered to warn surgeons, nurses, and all ambulatory patients to flee as soon as possible or else risk falling "into the hands of the enemy." All morning she rode from one hospital to another, repeating these words, shouting over the percussion of gunfire, desperate to reach the one person who mattered.

She found Jerome Robbins just after noon at his hospital in Talleysville. He looked older than he was, overtired and dusty. She was conscious of her own flushed face and dampened hair, the wild, spinning panic in her eyes. He had to leave, she told him. They were coming, closer and closer. They were out there now—listen. He refused to leave his patients. She tried again, screaming now. Still he refused. She left him there, standing by the beds, guarding the neat rows of mutilated men, and found her way back to the river.

The rebels chased the Union army through the marshy muck of White Oak Swamp, two days of marching with no food or sleep, until they reached Malvern Hill, eighteen miles southeast of Richmond. There they got busy building defense works and positioning more than 250 guns until the men finally collapsed—all except the guards, who paced back and forward in front of the line, ready to arouse the sleepers at any moment. Emma awoke to the sound of cannon and the battle raged all day with terrible fury: "hour after hour, the enemy advancing in massive column, often without order but with perfect recklessness; and the concentrated fire of our gunboats, batteries, and infantry mowing down the advancing host in a most fearful manner, until the slain lay in heaps upon the field." General McClellan, meanwhile, was safely ensconced on a navy gunboat on his way to Harrison's Landing, just six miles downstream from Elizabeth's family farm.

Elizabeth was quietly furious about the Union retreat. Richmond should have fallen; the war should be over. She should be celebrating Independence Day by unpacking her Union flag and raising it over her home. Instead her neighbors wore supercilious smiles and spoke in pious tones, asking for donations for the Confederate bazaar at Ninth and Broad Streets, or for the "Ladies Aid and Defense Association," which raised funds for the construction of a gunboat to defend the James River. Elizabeth found the whole enterprise "supremely ridiculous," and consoled herself by working harder on behalf of the Union.

Increasing numbers of escaped officers were finding their way to her home, or to one of many Union safe houses she'd established, through her money and connections, in and around the city. She supplied all of the men with food, clothing, and forged exemption tickets, which identified them as employees on Confederate government contract works, and her servants guided them toward the Northern lines, keeping watch for pursuing cavalry. Once the escapees made it to Union territory, they headed straight to McClellan's headquarters, reporting prison gossip about the rebels and their plans. Now, with McClellan stationed at Harrison's Landing, so close to her family's vegetable farm, Elizabeth devised another way to pass on information about the enemy. Her connection—Mary Jane Bowser's husband, Wilson—was already set up at the farm on the other end, but any number of things could go wrong before her dispatches made it that far.

One afternoon in the first week of July she sequestered herself in her library, a straight pin between her fingers and a book on her lap, chosen not for its text but for the quality of its pages: clean and stiff, with no dog-ears, tea stains, or rips. She had to be careful to hide her actions from her nieces, now eight and five years old. Their mother had recently told John she did not want her children being raised by any "damned Yankees," an intended slight against both her husband and Elizabeth. Mary kidnapped the girls; fled to her family's estate, Bellefield, just outside the city; and went on a celebratory bender—what Elizabeth called a night of "awful sin." John rescued the kids, bringing them and their "mammy" to live permanently at the Church Hill mansion. It was a terrifying, traumatic event, and Elizabeth wanted to return some peace and routine to the girls' lives, insomuch as peace and routine were possible during a war.

They were smart, curious girls, always asking questions about what was happening to their city, counting the charges of musketry as if they were raindrops tapping the roof, noticing the people who crossed to the other side of the street rather than acknowledge their aunt. Once, unbeknownst to Elizabeth, Annie followed her upstairs on tiptoe, watching her disappear into the secret room at the end of the hall. When her aunt crept back downstairs, Annie went to investigate. She found the door to the room slightly ajar. Peeking behind it she saw two men, foul-smelling, faces streaked with dirt. One of them pressed a grimy fingertip to his lips.

"Keep it all a secret," he whispered, "or else we will die."

Annie backed away slowly and pushed the door closed. When she checked the room the next day, the men were gone.

Now the girls sat by Elizabeth's feet, wondering why their aunt was sticking a pin through the pages of her book. Elizabeth asked Annie if she would take her sister to her room and play quietly for a while; if they needed anything, they could find Grandma. Her niece obeyed, and she was alone again.

She began on page 1, poking precise holes into a sequence of letters, the letters forming words, the words forming requests: *Give me the names and ranks of imprisoned officers. Are you hearing anything about Confederate troop numbers and movements? About Lee's plans? About Confederate forces in Manassas?* Satisfied, she asked a servant to load several items into her carriage: a table, an armchair, a looking glass, a dozen new towels, pen and ink, paper. She carried her book and a bundle of clothing, making sure to leave pins pushed through the collars, and asked her driver to take her to Libby Prison, Twentieth and Cary Streets. Officials still forbade her to enter the prison proper but gave her access to its hospital, the first room on the east side.

She alighted from her carriage, carrying the books and clothing, being careful not to disturb the pins in the collars. Her servant stayed behind, unloading the rest of the items to drop off at the main prison. A man stood by the doorway. As far as Elizabeth could tell, he was a civilian—he wore no rebel uniform—but he held up a hand in a way that made her stop.

"Are any of the prisoners related to you?" he asked.

"No," she answered, and pulled her package closer.

"Are you acquainted with any of them?"

She glanced skyward toward the prison's upper floors, where the windows offered a clear view of the bridges across the James River to Manchester, from which trains departed south and west. The men had nothing but time, which they spent playing endless games of cards, covering the walls with graffiti ("The Union Must Stand and Shall Be Preserved" was a favorite slogan), and whittling a variety of items from soup bones: pipes, spoons, crosses, chess pieces, intricate earrings and brooches for their sisters or wives. They could easily keep watch out those windows, looking for things of interest to Washington.

"I do not know any of them," Elizabeth said. "I visit them on the grounds of humanity."

The man shook his head once, back and forth. "You could show your humanity better by visiting *our* sick soldiers around the city."

She lifted her chin and spoke with calm certainty: "The Bible states we must visit our enemies."

"But it does not say we must visit them before visiting *our* sick friends."

She assured the man that she had charitable intentions to spare, slid past him, and headed left to the hospital.

Scanning the room, she memorized the layout, noting the proximity of the guards. In one far corner a small partition concealed medicine and supplies. Four rows of cots extended the entire length of the space, all of them occupied by sick and wounded men. She found an officer who seemed awake and alert, his eyes flitting in her direction as she sat next to him. She thanked him for his service and said how honored she was to bring him and his fellow inmates some new clothing; he must excuse her if she inadvertently left some pins in the fabric. Then she placed the book in his hands, floated her mouth by his ear, and whispered, "Read the pinpricks."

A few days later Elizabeth retrieved the book and transcribed the answers—information about imprisoned soldiers and scuttlebutt from the rebel guards—connecting the dots letter by letter, word by word, sentence by sentence. She crept out to the stables and slipped the letter beneath the driver's seat of a wagon. One of her servants would make the trip to the farm. He would be unaware that he was carrying anything but manure; she wanted to involve as few people as possible for as long as possible. At the other end Mary Jane's husband would be waiting, and knew exactly what had to be done.

General McClellan established headquarters at Berkeley Plantation on Harrison's Landing, six miles downstream from Elizabeth's family farm.

That night, lying in bed, Annie and Eliza sound asleep, Elizabeth pictured it as it happened: Wilson Bowser, message hidden in his shoe or inside his drawers, lowered himself into a rowboat and pushed off, drifting in midriver under a pale and lusterless moon. He coasted to the Union gunboats patrolling near the Appomattox-James confluence, holding up his hands to show he was unarmed. He explained to the officers that he had a note for the Union high command from a knowledgeable source in Richmond, information that needed to be delivered to Harrison's Landing right away. The dispatch was sent up the chain, landing, hopefully, in the hands of a Union colonel Elizabeth never named, an old family connection from the North.

Over the next few weeks Elizabeth learned that a portion of Lee's army had begun to move away from Richmond and toward northern Virginia, threatening Union troops at Manassas. If McClellan vacated Harrison's Landing to reinforce them, she would have to establish another route for her dispatches. And considering the Union's faltering performance on the Peninsula, those dispatches should contain more reliable and vital information than what she'd gleaned from prisoners. She needed to connect with Mary Jane Bowser and finally turn her servant into a spy—a much riskier prospect now than it was last year: one mole in the Confederate White House had already come to light.

Jefferson Davis's enslaved coachman, William Jackson, had recently escaped from his employers and fled to Union lines at Fredericksburg, Virginia. During a meeting with Secretary of War Stanton, he provided information about the rebel army that could have come only from inside the Confederate White House and gave interviews to the Northern press, calling his former boss sickly and irritable and Varina a devil. Elizabeth knew that the Davises would from now on keep a wary eye on other servants. One small indiscretion could lead to Mary Jane, and that would implicate Elizabeth.

Nevertheless she strolled across the street to Eliza Carrington's home to meet with her friend's seamstress, whose name no one ever divulged. Elizabeth inquired about Mary Jane's schedule, wanting to know which days and times she dropped off the First Lady's dresses. She told the seamstress she'd return soon, on a day her servant would be there.

A few days later, while she was walking through Church Hill, a man dressed in civilian clothing fell into step beside her. She lifted her skirts an inch, picking up her pace.

He kept up with her. His voice was a gruff whisper: "I'm going through the lines tonight. Do you have anything for me?"

She stopped and turned, facing him straight on. She had never seen him before but she knew what he was, and what he was trying to do. She shrugged and walked on, muttering to herself, as if she hadn't understood a word he said.

SHE BREATHES, SHE BURNS

THE SHENANDOAH VALLEY, VIRGINIA

On the morning of Tuesday, July 29, Belle sat on the balcony of her cottage behind the Fishback Hotel, watching the first sliver of sun crest above the hills, the only sound the steady pounding of hooves. It had been a few weeks since Union spy C. W. D. Smitley had betrayed her, and now came the repercussions: a large body of Union cavalry approaching, the men in twos and threes. Shading her eyes with a hand, she stood and scanned High Street, spotting an empty carriage parked in front of the hotel. She knew it was meant for her.

Eliza called up the stairs: "Miss Belle, de Provo' wishes to see you in de drawing room, and dere's two oder men wid him."

After her initial trepidation Belle had resigned herself to the idea of prison; a part of her even welcomed it. She had read the Southern reports about Rose Greenhow's imprisonment—the inadequate food, the lack of fresh air and exercise, the generally "outrageous persecutions," as one newspaper seethed, "which have been inflicted upon this true and noble woman"—and Belle expected the same treatment. She would be elevated to Rose's status, a true Confederate martyr honored as much for her sacrifice as her daring.

Belle descended the steps slowly, as if her legs had invented walking—legs one rebel soldier declared "the best-looking in the Confederacy." The assembled men watched and waited: provost marshal Arthur Maginnis, Major Francis Sherman of the 12th Illinois Cavalry, and a civilian with an unruly beard and restless, darting eyes—one of Secretary of War Stanton's henchman, she assumed. He reminded her of Edgar Allan Poe's raven.

"Miss Boyd," the provost marshal intoned, "Major Sherman has come to arrest you."

Belle lifted an index finger to her chin. "Impossible," she said. "For what?"

Major Sherman stepped forward with a document:

Sir: You will proceed immediately to Front Royal, Virginia, and arrest, if found there, Miss Belle Boyd, and bring her at once to Washington.

> *I am, respectfully,*
> *Your obedient servant,*
> *E. M. Stanton*

Stanton's detective, the Raven, followed her to her room, pawing through her dresses and petticoats and undergarments, examining every scrap of paper in her desk. To Belle's dismay he confiscated the pistol she'd saved as a special present for Stonewall Jackson. Eliza fell to her knees, locking her arms around Belle's calves until the Raven pried her away.

Citizens and soldiers clogged the streets, waiting to witness Belle's departure. She saw mostly "sorrow and sympathy" on their faces, but also some looks of "exultation and malignant triumph." Her neighbor Lucy Buck reflected the latter category, and recorded the scene in her diary: "Belle Boyd was taken prisoner and sent off in a carriage with an escort of fifty cavalrymen today. I hope she has succeeded in making herself proficiently notorious now." The carriage rattled through Winchester, escorted by a regiment of two hundred soldiers, and stopped at a Union camp just north of Martinsburg. Another carriage pulled up and Belle saw her mother, dressed in deep mourning and wearing a heavy veil. Belle rushed to her, arms outstretched. "My poor, dear child!" Mary Boyd cried. She begged the guards to let Belle await her departure at home, where her siblings wished to say good-bye, but her request was denied. In the near distance Belle spotted a maple tree, its branches stretching long and curling upward, as if beckoning her, and she pictured herself hanging from the highest one.

By nine the next morning Belle's train had reached Washington, and a carriage waited to convey her to the Old Capitol Prison. She was delighted to learn that Rose's old cell was directly above her own. The door moaned open to reveal Superintendent William Wood, and she straightened herself to match his modest height. He bowed vaguely in her direction.

"And so this is the celebrated rebel spy," he said. "I am very glad to see you, and will endeavor to make you as comfortable as possible. So whatever you wish for, ask and you shall have it. I am glad I have so distinguished a personage for my guest." He promised to send her a servant, and that she would not be there for long if she was a model prisoner.

He shut the door gently behind him.

Belle stopped herself from calling him back and scolding him for his courtesy. She didn't understand. Could it be that Union authorities considered Rose Greenhow a true and dangerous threat, and her a mere nuisance? She contemplated how to remedy the situation. If she behaved badly enough, if she compounded every demand with another, more absurd demand, perhaps Superintendent Wood would be inspired to do his job and torment her properly.

She paced around her cell, taking inventory of her effects: a washstand, a looking glass, an iron bedstead, a table, a couple of chairs. Two windows stretched the entire length of one wall, giving her a partial view of Pennsylvania Avenue and, in the distance, the home of John Floyd, secretary of war during the Buchanan administration. There she had passed "many happy hours" after her soci-

ety debut, her hair tied up in combs and ribbons, dancing to "Old
Zip Coon" and telling the loudest stories in the room. She pressed
her face between the bars and called for a sentry. She needed a rocking chair and a fire immediately, she said; the cell was too gloomy
for her to bear.

Wood returned at 8:00 p.m., accompanied by a detective sent
by Secretary of War Stanton. Belle ordered them both to sit. Behind
them the fire hissed like a perturbed cat.

"Ain't you tired of your prison a'ready?" the detective asked.
"I've come to get you to make a free confession now of what you've
did agin our cause. And, as we've got plenty of proof agin you, you
might as well acknowledge at once."

"Sir," Belle replied, "I do not understand you; and, further-
more, I have nothing to say. When you have informed me on what
grounds I have been arrested, and given me a copy of the charges
preferred against me, I will make my statement, but I shall not now
commit myself."

He launched into a stuttering monologue about the enormity of
her offense and the futility of her cause, and suggested she sign the
oath of allegiance.

Belle stepped closer and said, "Tell Mr. Stanton from me, I hope
that when I commence the oath of allegiance to the United States
Government, my tongue may cleave to the roof of my mouth; and
that if ever I sign one line that will show to the world that I owe the
United States Government the slightest allegiance, I hope my arm
may fall paralyzed by my side."

The detective scratched her speech across the pages of his notebook. "Well, if this is your resolution," he said, "you'll have to lay
here and die, and serve you right."

Once again she thought of Rose, and how the older spy managed to get her bold condemnation of the Yankee government
printed in the newspapers. Her next words aligned quickly in her
mind, and she recited them at full volume: "Sir, if it is a crime to
love the South, its cause, and its President, then I am a criminal. I

am in your power; do with me as you please. But I fear you not. I would rather lie down in this prison and die, than leave it owing allegiance to such a government as yours. Now leave the room; for so thoroughly am I disgusted with your conduct towards me that I cannot endure your presence longer."

Cheers and cries of "Bravo!" rang up and down the floor, and she wished Stonewall Jackson could see her now.

Save for the want of exercise, Belle suffered no privations. Washington secessionists supplied her with meals of soup, beefsteak, chicken, boiled corn, tomatoes, potatoes, Irish stew, bread and butter, and a variety of fruits, all served to her in her room by a contraband whom Wood had assigned as her servant. One of the sentinels brought her every newspaper that mentioned her name and allowed visits from curious journalists, which Belle welcomed. "She was dressed today," reported a correspondent for the *Washington Star*, "in a plain frock, low in the neck, and her arms were bare. Jackson, it appears, is her idol. . . . She takes her arrest as a matter of course, and is smart, plucky, and absurd as ever. A lunatic asylum might be recommended for her." When the new provost marshal, twenty-five-year-old Major William Doster, came to call on her, he found her relaxing by her fire, reading *Harper's* and eating peaches.

"I can afford to remain here," Belle told him, "if Stanton could afford to keep me. There is so much company and so little to do." Besides, she added, it was an excellent chance to brush up on her literature and get her wedding outfit ready; she was certain one of her fellow inmates was destined to be her husband. The provost marshal laughed, but Belle was serious. If she weren't going to be tortured, she wanted to at least be pursued.

At first she did the pursuing, slipping her mouth between the bars of her cell and singing:

She is not dead, nor deaf, nor dumb
Huzza! She spurns the Northern scum!
She breathes! She burns! She'll come! She'll come!
Maryland! My Maryland!

Every prisoner stopped to listen, entranced. Those somehow unfamiliar with Belle's name sought to rectify this omission, asking who she was, where she had come from, what she had done. "The pathos of her voice," wrote one, "her apparently forlorn condition, and at these times when her soul seemed absorbed in the thoughts she was uttering in song, her melancholy manner affected all who heard her, not only with compassion for her but with an interest in her which came near on several occasions [to] bringing about a conflict between the prisoners and the guards." Another prisoner admitted that Belle's voice brought "a lump up in my throat every time I heard it. It seemed like my heart was ready to jump out as if I could put my finger down and touch it. I've seen men, when she was singing, walk off to one side and pull out their handkerchiefs and wipe their eyes, for fear someone would see them doing the baby act."

Her intentional emphasis of the words "Northern scum" further enflamed emotions, once prompting a sentry to yell, "Hush up!"

"I shan't do it!" Belle retorted. Picking up a broom, she repeated the offending line and swept at the guard's feet, shooing him away.

Sunday morning proved to be the optimal time to assess her romantic prospects. Superintendent Wood (an avowed atheist, it was said) stopped at every floor to announce services: all who wished to hear the Lord God according to Abe Lincoln should convene in Room 16, and all who preferred the Gospel according to Jeff Davis would be accommodated in the yard—the only way, Belle reasoned, to "separate the goats from the sheep." Wearing a small Confederate flag, the stem tucked in her bodice, she sat near the preacher, close enough for him to reach out and touch her head. She read intently from her prayer book, which she'd inscribed "Belle

Boyd, Old Capitol Prison," and smiled at the rebel prisoners as they passed—a gesture, admitted one, that "did them more good than the preaching."

She began corresponding with a group of prisoners from Fredericksburg, confined in Rose's old cell, who'd found a loosened plank in the floor—the one through which Rose had lowered Little Rose. They tied strings around their letters and dropped them through the slat to Belle, and she always made them wait before sending up her reply.

She found herself drawn to one rebel soldier in particular, Lieutenant Clifford McVay, whose acquaintance she'd first made years earlier, during her time as a socialite in Washington. He'd been wounded during the ongoing Peninsula Campaign and left for dead on the battlefield, where Federal forces captured him. His cell conveniently faced hers, and at night, if her favorite guard came on duty, she winked at McVay and signaled for him to open his door. He sang her a love song, and she responded in kind, trilling the lyrics to "My Southern Soldier Boy." As she hit the final note she peeled off her glove, pausing at her wrist, letting it dangle from the tips of her fingers. She pitched it to him in a graceful arc. He caught it and, keeping his eyes on her, reached into the silk to withdraw a billet-doux. Scribbling a response, he tucked it inside the glove and aimed it at her heart.

One night her servant brought her a sugar loaf with her dinner, a treat she wished to pass to her suitor. She asked the sentry on duty for permission to deliver it across the passage.

He shrugged. "I have no objection."

As her outstretched hand made contact with McVay's, the sentry swung his musket, connecting with the points of her knuckles, smashing her thumb. He had set her up. Without meaning to, she began to cry. She faced the sentry and demanded to see the corporal of the guard. The sentry refused, and Belle stepped forward.

"Go back," the sentry ordered, "or I'll break every bone in your body."

He raised and thrust his bayonet, a shiny streak coming straight at her, carving a scoop of flesh from her arm, pinning her to the wall by her dress. He held her there for a moment, like a prize catch on a line, while she screamed and thrashed. She had never been physically assaulted in her (now) eighteen years. Beneath the pain she felt confused, as if the guard had made a mistake, and she wanted him to take it back. When he finally let her go, her mind refocused, shifting back to defiance, its proper and natural setting. The scar from that wound became one of her favorite souvenirs from the war. She rolled up her sleeve for anyone who asked, fingering the scar like a talisman.

THE STILL, SMALL VOICE

MANASSAS, VIRGINIA, AND WASHINGTON, DC

On August 14 General McClellan, per Lincoln's orders, began to withdraw his army from Harrison's Landing and head for northern Virginia, where Union forces were being threatened at Manassas. Emma and her comrades were demoralized at the thought of retreating from Richmond. The entire campaign had been a waste: all of those entrenchments built, all those tracks and bridges repaired, and all those Union boys lost, fifteen thousand graves scattered along the length of the Peninsula, remains sinking in the marshy ground, waiting to be yanked up and eaten by hogs or desecrated by rebel hands. They had been so close, and now all of the Army of the Potomac swore at the prospect of giving up. In a letter to his wife McClellan detected the hand of God in his failure: "I think I begin to see his wise purpose in all this. . . . If I had succeeded in taking Richmond now the fanatics of the North might have been too powerful & reunion impossible."

Emma was cheered only by learning, finally, what had happened to Jerome Robbins after she left him in the hospital at Talleysville. The following evening a group of Confederate cavalry, led by Robert E. Lee's nephew, Colonel Fitzhugh Lee, clattered into the clearing. To Jerome's surprise the rebels were courteous, and instead of being taken prisoner he was offered parole, a policy mandated by Stonewall Jackson, who believed treating medical personnel as enemy combatants was immoral. A steamer collected Jerome and 105 others and brought them to Camp Parole on the Chesapeake Bay, a facility set up to house parolees until they could be exchanged. Emma sent him a letter expressing her relief and a $5 note. He wrote to thank her, his tone and his words somehow reaching back to the place he had broken—that had been broken for as long as she could remember, if she were honest about it—and both Frank Thompson and Emma Edmondson forgave him, this time for good.

On the march north the men bivouacked in the street, many of them drunk, and several others mildly scandalized by the fact that Colonel Orlando Poe, the married commander of the 2nd Michigan and Emma's friend, invited a woman to stay with them. "She slept in the tent with us & he lay next to her," one soldier confided to his diary. "I believe she said she belonged to some Soldiers' Aid Society though the exact nature of the aid she did not state to me." The 2nd Michigan continued to Manassas, planning to reinforce General John Pope, whose troops had been in the vicinity of the Shenandoah Valley for days with no rations but the rotting fruit and corn in the fields. Emma was delayed in joining them, ordered instead by General Samuel P. Heintzelman to go on a scouting mission across the rebel lines to determine Confederate strength and positions.

She boarded a train to Washington at Warrenton Junction, gathered items for a disguise as a female slave—calico dress, headscarf, and silver nitrate, her skin be damned—and returned the same night. Since connecting with the group of contrabands who came to camp during the Peninsula Campaign, Emma interacted with slaves whenever she had the chance, listening to their stories and hoping she might one day teach them. Her choice to disguise herself again as a slave was, in her current circumstances, the best way she knew to show empathy.

Union forces under General Pope skirmished with Lee's advancing forces all along the Rappahannock River, the lines between the two armies fluid and evolving, and Emma fell in with a group of nine contrabands in the neighborhood of Warrenton, nine miles from the train depot. The contrabands, Emma deduced, "preferred to live in bondage with their friends rather than to be free without them."

At headquarters they began cooking rations, the air filling with the scent of "coosh"—bacon grease (and salt pork, if they were lucky), cornmeal, and water fried up into a flimsy pancake, a Confederate staple similar to a common meal of plantation slaves. Rebel officers lingered nearby, murmuring among themselves. Emma strained to listen, and within a few hours she learned the troop numbers at several important points and the number expected to arrive during the night. The lines were thick with pickets and she waited until dawn to break away.

On the journey back she found herself in the cross fire of a skirmish and sought refuge in the cellar of an old house. The firing grew hotter, shot and shell shaking the foundation, the floor above her opening up in patches and raining down in crumbling stone. She recalled the story of Elijah remaining in the cave during the tempest, the earthquake, and the fire, and afterward came the still, small voice. She closed her eyes, hoping to hear that same small voice speak to her.

Back at camp Emma learned that Stonewall Jackson's division had torn through the Union supply depot at Manassas Junction, burning a hundred railroad cars and seizing all of the ammunition and subsistence stores his men could carry; only a half dozen barrels of hard bread and as many hams lay scattered around the tracks. The cars were pierced with rifle shots and still burning. Bridges were torched and tracks torn up. Smoke drifted from the scorched remains of a bakery. From Manassas Junction the general slashed his way to the railroad bridge over Bull Run, destroying it after he crossed, the reflection of the flames lighting the sky for miles. The Second Battle of Bull Run (Second Manassas) was about to begin, Yankees and rebels meeting on the same ground as in the previous year, but this time the Federals were on the defensive, and Emma's regiment was on the front lines.

At 5:00 a.m. the 2nd Michigan moved down the road to the old battlefield at Manassas and deployed as skirmishers, taking position in the woods about a mile from Jackson's headquarters. The rebels caught sight of them and opened with a section of artillery, firing high into the oak trees, sending splinters the size of stove wood plunging to the ground. One piece swiped the cap clear off a lieutenant's head but left him otherwise unharmed.

Emma was sent to the front with messages and mail, her mount a mule instead of a horse, and along the way she detoured to a side road and came upon a wide ditch. She tried to cross, but instead of leaping the mule reared and fell headfirst into the gap, heaving her against the side; everything turned black.

The dull thump of cannon pounded above, rousing her. She didn't know how long she'd been unconscious, or how her mule had managed to extricate his hooves from the mud, but there he stood at the edge of the ditch, looking down at her, quizzical and waiting. *The mail!* she thought, and began crawling out of the ditch on elbows and knees, panicked at the idea of it going undelivered.

Each motion forward revealed another injury. Her left thumb was limp and a rope of pain stretched through her left side, from foot to breast, but she crept inch by inch toward the mule. The mailbags, spattered with mud, still hung beneath the animal's stomach. With broken leg, foot, and finger she remounted and made her delivery, every bump and jostle bringing a sharp, hot pain, substantial enough that she took a risk and visited the hospital corps. The surgeon wanted to examine her but she demurred, fearing what he might discover, and convinced him she needed only some morphine sulfate and chloroform.

The Union defeat at Second Bull Run cost sixteen thousand casualties, five times as many as were lost in the first conflict there. Inside Pope's headquarters at Nalle House, an elegant brick mansion belonging to a captain in the US Navy, the entrance hall and parlor looked "more like a butcher's shambles than a gentleman's dwelling," said one witness, with the dying and wounded lying head to foot. Silk settees and curtains were polka-dotted with blood. Beside the piano stood the amputating table, its instruments scattered across mantelpieces once dedicated to leather-bound books and vases of flowers.

The 2nd Michigan spent the next several days near Fairfax Station, where Emma worked in a field hospital set up by the train depot, and where she watched thousands more of the wounded wait in agony for trains to Washington. The city's hospitals were overflowing, and two thousand cots were filled and laid out in the Capitol; lawmakers had to navigate carefully to avoid tripping over bandaged and bloody limbs. Belle leaned out the window of her cell in the Old Capitol Prison and taunted the returning soldiers as they passed by: "How long did it take you to come back from Bull Run? Are you going on to Richmond? Where's General Pope's headquarters?"

"Hush up, you damn bitch," one retorted, "or I'll shoot you!"

"Shoot me?" Belle laughed. "Go meet men, you cowards! What are you doing here in Washington? Stonewall Jackson is waiting for you on the other side of the Potomac. Aye, you could fight defenseless, imprisoned women like me, but you were driven out of the Shenandoah Valley by the men of Virginia."

While McClellan let his exhausted troops recuperate, Lee followed up on his victory at Bull Run, sending his ragged, barefoot Confederates into western Maryland, the first invasion of Union territory. Counting on McClellan's tentativeness, Lee hoped to capture Harrisburg, Pennsylvania; Philadelphia; Baltimore; or even Washington. The Federal capital braced for an attack just as it had after the first debacle at Bull Run. A steam warship was anchored on the Potomac, ready to carry Lincoln and his cabinet members to safety, and government employees were to be armed with muskets to assist in the capital's defense. All was going according to the Confederates' plan until two Union soldiers, resting in a field in Frederick, Maryland, found a piece of paper wrapped around three cigars: a copy of Lee's Special Order No. 191, outlining his strategy. When the news reached McClellan, he turned to his officers and exclaimed, "Here is a paper with which, if I cannot whip Bobby Lee, I will be willing to go home!"

He launched a series of assaults against the outnumbered Confederates along Antietam Creek, southwest of Frederick, in what would be the deadliest single-day engagement in the history of American warfare, with nearly five thousand men killed and twenty thousand wounded. Two days later the Union army's staff photographer, Alexander Gardner, visited the scene, marking the first time an American battlefield was photographed before the dead were buried. He employed a new technique, stereographing, in which two lenses captured two simultaneous pictures that, when seen through a viewer, created a three-dimensional image. Although Gardner intended to repulse the public with his unflinching depiction of the war—"here are the dreadful details," he wrote, "let them aid in preventing another such calamity falling upon the nation"—he only summoned them ever closer. Americans, throughout the North and South, decorated their parlors with his grotesque images, featuring such titles as "Federal buried, Confederate unburied" and "Bloody Lane, Confederate Dead, Antietam," so authentic and visceral when compared with the stylized, patriotic renderings in *Harper's Weekly*. Gardner's success inspired other war photographers, a primitive gaggle of paparazzi who rushed to capture the carnage when it was still raw.

Antietam was a tactical draw; McClellan succeeded in driving the Confederates out of Maryland but allowed them to retreat to Virginia unmolested. Nevertheless he boasted about his "complete" victory and confided to his wife, "I feel some little pride in having with a beaten and demoralized army defeated Lee so utterly & saved the North so completely . . . my military reputation is cleared—I

Bloody Lane, Confederate dead, Antietam.

have shown that I can fight battles & *win* them!" While Lincoln remained concerned about McClellan's timid performance, he took Antietam as a sign that God had "decided this question in favor of the slaves." On September 22, five days after the battle, he issued a preliminary emancipation proclamation, warning Confederate states that unless they returned to the Union by January 1, their slaves "shall be then, thenceforward, forever free." Elizabeth celebrated by arranging for the private purchase of a slave, Louisa Roane, so she could be reunited with her husband, a Van Lew servant and member of the Richmond Underground.

Emma claimed to have been on the field at Antietam, helping carry the injured and dead from the field. She would remember coming upon one young wounded soldier, a thin rope of dark blood encircling his neck.

"I stooped down," Emma wrote, "and asked him if there was anything he'd like to have done for him. The soldier turned a pair of beautiful, clear, intelligent eyes upon me for a moment in an earnest gaze, and then, as if satisfied with the scrutiny, said faintly: 'Yes, yes; there is something to be done, and that quickly, for I am dying.'

"Something in the tone and voice" made Emma look closely at the soldier, studying the lines and dips of the face, the way the bones fitted together. Her suspicion was confirmed. "I administered a little brandy and water to strengthen the wounded boy," she recalled, "for he evidently wished to tell me something that was on his mind before he died. The little trembling hand beckoned me closer, and I knelt down beside him and bent my head until it touched the golden locks on the pale brow before me."

"I can trust you," the soldier said, "and will tell you a secret. I am not what I seem, but am a female. I enlisted from the purest motives, and have remained undiscovered and unsuspected. . . . My trust is in God, and I die in peace. I wish you to bury me with your own hands, that none may know after my death that I am other than my appearance indicates." Again, Emma said, the soldier looked at her with that same earnest scrutiny, and whispered, "I know I can trust you—you will do as I have requested."

She claimed to have honored the soldier's wish, making a grave for her under the shadow of a mulberry tree, apart from all the others. But Emma had not been on the field at Antietam at all; she, along with the rest of the 2nd Michigan, had stayed in Alexandria until shortly after the battle. The story let her believe, in some small way, that people might extend the same kindness to her, keeping her secrets and letting her know that she was not alone.

RICHMOND UNDERGROUND

RICHMOND

Belle had no desire to leave the Old Capitol Prison, as she was still enjoying the company of Lieutenant Clifford McVay, who had taken to writing amorous notes on tissue paper, wrapping them around marbles, and rolling them across the floor. She let the notes disappear beneath her skirts, keeping her eyes on his as she retrieved them, and mouthed each word as she read it. He seemed close to proposing marriage, and she was confident enough that if she refused him twice, as per Southern custom, he would persist for the third and final time. She wished to stay in the capital long enough to assemble her wedding trousseau: a cloak in dark stamped velvet; an underskirt cut in deep scallops and finished with rows of pale beads; a morning dress trimmed in white lace and pink satin ribbons, the sleeves fixed to reveal her brilliant jewel of a scar.

So she was not at all prepared for the September afternoon, about a month into her imprisonment, when Superintendent Wood stood outside her cell and boomed, "All you rebels get ready! You are going to Dixie tomorrow and Miss Belle is going with you!" The Federal and Confederate governments had arranged for a formal exchange of inmates, two hundred Yankees for two hundred rebels, in an attempt to alleviate overcrowding in their respective prisons, and Belle was specifically mentioned in the order: "I forward likewise Miss Belle Boyd," wrote Brigadier General James Wadsworth, "a young lady arrested on suspicion of having communicated with the enemy. I have agreed that she shall be placed over the lines by the first flag of truce." She was not to return to the North for the duration of the war.

Belle suspected that officials were eager to be rid of her because she boosted prisoner morale, and was much gratified when many of the men wept at the news of her departure. McVay was the most despondent of all, rolling several messages across the passage that night, sensing that he'd never again watch her lips form his words. He was right—Belle lost all track of him after her release—but at least she got her trousseau; Superintendent Wood, keeping his infuriating promise to grant her every wish, shopped for her and sent each purchase on to Richmond.

The following morning, Belle and the Confederate prisoners lined up in the Old Capitol courtyard and were hustled out into the street, where throngs of secessionists gathered to cheer them, including a young mother who threw kisses to the departing soldiers using her baby's hand. Belle treated them to her broad smile and queenly wave and let herself be lifted into her carriage. She had two gold saber knots, one for General Joe Johnston and one for Stonewall Jackson, hidden beneath her skirts.

The steamship *Juniata* cast off toward the mouth of the Potomac, where it dropped anchor and spent the night, the prisoners taunting the Union officers with rebel songs and cheers for Jeff Davis ("Three cheers for the Devil!" came the ready retort). In the morning it steamed up the murky waters of the James, and the passengers spotted the rebel "Stars and Bars" waving from windows as they rounded the river bend. Belle went straight to the Ballard House, where she was serenaded by the city band and held court in the parlor alongside Rose Greenhow. For Belle the encounter was secondary only to meeting Stonewall Jackson, and she went to sleep marveling that she'd become who she wanted to be before even growing up.

Rose Greenhow and Belle Boyd stayed at the Ballard House, Richmond's finest hotel.

Elizabeth stepped out into the stifling heat, wagging a palm-leaf fan by her face, taking care to breathe through her mouth. The air so reeked of suffering and death that the city burned barrels of tar, considered an efficient fumigator, in an attempt to clean it. She spotted her next-door neighbor, R. J. White, who was hoping to patent a machine similar to the Union's "coffee mill" gun, which poured bullets, as from a hopper, at the rate of 120 per minute. He tipped his hat. She nodded in reply, feeling his gaze on her back as she crossed the street to her friend Eliza's home.

There Mary Jane was waiting with the seamstress, both of them expecting Elizabeth's visit and aware of the risks involved in her plans. They were among a growing number of servants and slaves working on behalf of the North, passing information to Union engineers, cartographers, generals, and scouts. In rural areas they acted as guides through unfamiliar country, sharing knowledge about water, game, and the habits of the enemy. Escaped slaves, armed with axes, went on expeditions through the woods, hacking at the heavy timber and preparing the way for the advance, leaving the roads strewn with the bodies of bloodhounds sent to track them down. In Port Royal, South Carolina, the great slave rescuer Harriet Tubman organized a band of Negro spies, gaining their confidence through kind words and sacred hymns.

Elizabeth explained that she and Mary Jane needed an intermediary, someone willing to pass information to and from the Confederate White House, someone whose discretion matched her valor.
They had come to the right woman; the seamstress even volunteered how it could be done. She held up a gown of purple brocade,
the sort Varina Davis wore every day while the poorer classes, suffering from the effects of the blockade, transformed their draperies
into dresses.

Cotton dresses were constructed with the skirt attached directly
to the bottom of the bodice, but the ones made of finer material,
like Mrs. Davis's, were constructed with the skirt attached to a
waistband made from lining or scrap fabric and *then* attached to the
bodice. If Mary Jane had the time and skill, she could partially dis-
assemble the garment to access the inside of the waistband, hide her
dispatches inside, stitch it back together, and bring it in for "repair."
But if she was in a rush, she should simply sew the information into
a fold in the skirt and package it before dropping it off. Elizabeth
would pick it up the following morning and prepare the information
for delivery through the lines.

It was settled, but they needed a contingency plan in case the
seamstress was unavailable or compromised, or if Mary Jane, for
whatever reason, wasn't able to leave the Confederate White House
to drop the dresses off. Elizabeth thought of Thomas McNiven,
a Scottish immigrant whom she'd met one recent Sunday at St.
John's Episcopal Church. After the service, a fellow parishioner approached, clutching a young, red-headed man by the arm. The parishioner introduced him as "a friend to be trusted."

McNiven owned a bakery at 811 North Fifth Street, less than
one mile from the Confederate White House, a frequent stop on his
route, and Mary Jane could easily arrange to slip him dispatches
and drawings during his early morning deliveries. He also had his
own burgeoning network of Union sympathizers, people far outside
Elizabeth's usual circles, mostly other Scottish immigrants and pros-
titutes in the red-light district along Locust Alley. One of his most

reliable and prolific sources was "Clara A.," who always operated alone and catered only to the "carriage trade," including politicians and both Union and Confederate commanders; "Bull Head" was reportedly a nickname for Confederate colonel Lucius B. Northrop, and "Big Belly" for secretary of state Judah P. Benjamin. She recorded the highlights from some of her sessions in a diary:

General Limpy, the food fop—he must do the undressing. Shoes too.

Big brass, big belly. Since my rule, he brings only Yankee money. Wonder where he gets it. Wonder what old sourface would say if he knew one of his plate-lickers had so much Yankee money.

The Maryland Governor? Do it bending over, bark sometimes.

Four big generals last night came together. Red beard really has red hair all over. They brought two more barrels of wine and twenty blankets. Must sell some of the blankets. Have too many.

Christ! The praying general was brought in today by Preacher H. He is rough and brutal. After I serviced him, he dropped to his knees and asked God to forgive me for my sins!

Elizabeth was pleased that her Richmond Underground was expanding its ranks, although larger numbers carried greater risks—for mistakes, for turncoats, for the lethal consequences of discovery. Her friend Charles Palmer—who, like John Minor Botts, was a wealthy Unionist ex-Whig—had connected with a man named F. W. E. Lohmann, a grocer who himself was the leader of a group of German Unionists in Richmond. Botts recruited William S. Rowley, a New York native who rented a farm on the outskirts of Richmond. He was an odd-looking man, a brunet with a bright red beard that looked like a pinecone made of hair, and he would become Elizabeth's most valuable spy; she called him "the bravest of the brave, and the truest of the true," a man of "rare perception and wonderful intuition." Three new members—ice merchant Burnham Wardwell, carpenter William Fay, and engineer Arnold Holmes—even sent their children on missions to the prisons. Some of the adults created a recognition system, carving favors made from peach pits, little swinging three-leaf clovers in their centers, that dangled from men's watch chains and ladies' pendants. When the clover was upside down, it was safe to talk; when it was right side up, it wasn't.

Elizabeth's money and family connections gave her entrée into the Confederate Army and Navy departments, where she lured numerous Northern-sympathizing clerks into her confidence. One, a "trusty Union man" in the Adjutant General's Department, had access to information about the strength of rebel regiments, brigades, divisions, and corps, along with their movements and where they were stationed. There were nameless, faceless members, too, people who were wary of associating with outspoken Unionists but wanted to do their part. They jotted down gossip and observations and snatches of overheard conversation and crept, in the thick of night, to St. John's, leaving their notes, wrapped around flowers, by certain tombstones in the graveyard.

Mary Jane, meanwhile, played her role inside the Confederate White House, entering each morning through the servants' door, polishing every inch of mahogany, addressing Mr. and Mrs. Davis as Marse and Mistress, being careful to avert her eyes. She knew which Davis servants were loyal, always looking for another traitor among the staff, and forced herself to join in their condemnation of the Yankees.

She kept the president's office spotless, taking her time shining his desk, picking up each paper and memorizing it before lowering it back into place. Davis and his top generals employed an encryption system superior to that used by Rose Greenhow. While Rose switched and substituted letters and symbols for only one alphabet, the top-level cipher utilized as many as twenty-six alphabets based

on the letters of a key phrase: MANCHESTER BLUFF, COMPLETE VIC-
TORY, or, later in the war, COME RETRIBUTION. Davis wrote drafts
of his messages in plain English, and marked sensitive portions "to
be placed in cipher," a task handled by his confidential secretary or
a clerk with the Confederate Signal Corps. Leafing through Davis's
papers, Mary Jane saw either the English version of the president's
messages or English written in over the cipher.

There was much for her to report in October. At the end of
the month, General McClellan began to cross the Potomac River
east of the Blue Ridge, leading the Army of the Potomac southward
toward Warrenton, and General Lee felt compelled to act lest he be
cut off from Richmond. He ordered Stonewall Jackson to remain
in the Valley, in position to confront the right flank of the Union
army, while he and General James Longstreet moved eastward into
a position to block the Northern advance.

President Davis summoned Lee to Richmond for a conference,
during which Mary Jane pressed her ear to the door and heard the
general plead for "every support" in the coming conflict. His ranks
were depleted by nearly sixty thousand men, one-third of whom
had gone absent without leave, "scattered broadcast over the land,"
Lee said, engaging in the destruction of private property, feigning
illness, and "deceiving the guards and evading the scouts." Many
of them were determined to make it back to distant homes, while
others simply remained "aloof" in the vicinity of the army. "Unless
something is done," Lee warned, "the army will melt away."

At the end of her workday, back in servants' quarters, Mary Jane
transcribed pages of information and maps from memory and sewed
them into the fold of a dress. She brought that garment to the seamstress the following morning. If she had crucial information to deliver right away, she hung a red shirt on the laundry line, a signal
to Elizabeth to visit the seamstress before the usually scheduled day.
The seamstress then unstitched and unfolded those dispatches and
prepared them for Elizabeth to pick up, before the final, and most
dangerous, phase of the operation.

Her brother John would spread the papers over their dining room table, identifying key points and phrases and numbers. Using a stack of books piled nearby, he carefully pricked the first letters of random words to create precise sentences, just as Elizabeth had shown him, then gathered several blank invoices and purchase orders he used for his small chain of hardware stores.

He would complete these business documents as if they were genuine, encoding information from Mary Jane and from personal observations on his own regular visits to Petersburg and Fredericksburg. Certain quantities of items corresponded with certain military terminology: 370 iron hinges meant 3,700 cavalry; 30 anvils meant 30 batteries of artillery; 40 vises meant 4,000 battle-hardened shock troops. He told friends and associates that he was going either to visit his sister in Philadelphia or on a business trip—not to make new sales, as the Confederate government would target him for trading with the enemy, but to collect on certain accounts from before the war.

He took the next train on the Richmond, Fredericksburg & Potomac Railroad. Disembarking in Fredericksburg, he spent two days at his office there, doing legitimate business and arranging for his pass to cross the lines, over the Rappahannock River. From there he took a horse to Aquia Creek and then a steamboat north to Washington. Once in the capital, he gathered his false business paperwork and pinpricked books, transcribed the information hidden therein, and headed for the War Department at Seventeenth Street and Pennsylvania Avenue.

His departure shifted something inside Elizabeth; she was all in, on every front, unable to stop what she had set in motion even if she wanted to. She had never meant for her little brother to become so enmeshed in her operation as to risk his own life, but after his wife left he'd insisted on helping the Richmond Underground. Doubtless Mary was still furious that he had taken their children. There was no telling how far her anger could reach, what she might do, what she might say and to whom. Elizabeth's sporadic kindnesses to rebel soldiers did not negate years of speaking openly about abolition. Neighbors had already threatened to shoot her dead. One of General Winder's detectives had begged to sleep on her parlor floor. Another tried to entrap her on the street.

Her mind had to adapt, acquire new patterns. She kept her notes and plans within arm's reach, prepared to destroy them quickly if Confederate detectives raided her home. She wore a simple cotton bonnet and calico dresses so that she would not be immediately recognized on the streets; sometimes she even stuffed cotton into her cheeks to distort the shape of her face. If she turned to talk to a friend, she found a detective at her elbow. She saw strange faces peeping around the columns and pillars of the back portico of her home. Every night, before turning down the lights, she lowered herself to hands and knees and checked beneath the bed, certain that someone was waiting to get her.

Belle spent as much time as she could with Rose Greenhow, accompanying her and Little Rose to Richmond's hospitals, where she cared for the wounded rebel soldiers and entertained them with stories. The work made her miss her own father, temporarily home on sick leave in Martinsburg, which was once again under Confederate control. She decided to return home for a brief visit, and knew exactly what she would do as soon as she arrived.

After she put her younger siblings to bed and kissed her mother good night, she lay down herself and let her mind craft the next day's drama, imagining the scene, the setting, the lines she would say and hear. The sun seemed to rise just for her, slanting in her direction, flattering her angles and curves. She put on her prettiest riding costume, rebel gray and cut close, with a soldier's sash encircling her waist and her palmetto pin blooming on her chest. She found the gold saber knot intended for Stonewall Jackson and slid it down her bodice. It was time.

She saddled up Fleeter and rode out to the Confederate encampment at Bunker Hill, about eight miles from Martinsburg. Stonewall had his headquarters in a stately Greek Revival–style brick building with a gabled roof and a small log slave cabin in the back. It was called the Boyd House, after a family unrelated to Belle, but she took the coincidence as a good omen.

Instead of staying inside the house, Stonewall had chosen to camp on the lawn. She approached General Jackson's tent, hoping that she wasn't interrupting his daily prayer, and that he would re-member her name and her heroic dash across the battlefield at Front Royal. An aide appeared and asked for her name and her business. She answered and waited, pressing her hand against the gold saber knot to make sure it was still there.

The aide emerged from the tent and said that the general re-fused to see her. Belle demanded to know why. She felt her eyes blur-ring, her throat caving in. She fought to maintain her expression, a cool leveling of her features, an unspoken insistence that there was surely a misunderstanding. She repeated her name, italicizing each word: *Belle. Boyd.* Daughter of Benjamin, cousin and friend to count-less rebel soldiers. The aide shrugged and said that the general was "not altogether assured" of her loyalty. Her loyalty? She had risked her life for the cause and just spent a month in a Yankee prison. The aide threw up his hands. Her mind groped for reasons: Did he fear that her quick release from prison, by people who had actual proof of her activities, was some sort of trap? Did he believe that she had secured her freedom by vowing to betray him? Had he heard the gossip that some of her associations with Union officers weren't solely for the benefit of the Confederacy? Was he equally unsettled by the newspaper mentions of her bare arms and low-cut dress, the way she told reporters of her desire to "share his dangers"?

The aide either couldn't or wouldn't say. Quietly she mounted Fleeter and peered over her shoulder, her face loaded with a mur-derous glare. "If I ever catch you in Martinsburg," she threatened the aide, "I will cut your ears off."

She rode the ten miles home, brushed past her siblings and the servants, and tumbled backward onto her bed. The sun had shifted position in her window, sharpening the shadows along her wall. She retrieved the saber knot from her corset and made a fist around it. She would try again, and the scene would unfold differently, the dialogue edited, the ending correct.

When she dismounted at Stonewall's tent he would greet her right away, gently laying his hands upon her head, his long fingers forming a crown. How pleased he was to see her once more well and free. He would advise her, sweetly, that if his troops were forced to retreat she must leave her home again, for the enemies would quickly move in, and it would be foolish for her to expose herself to the caprice or resentment of the Yankees and face more time in prison. He would promise to give her timely notice of his movements so she could plan ahead. He would bestow upon her the title of honorary aide-de-camp, and invite her to attend a review of his troops. She'd thank him, lift the saber knot from her dress, and lower it into his palm. As she turned to leave, he would whisper, "God bless you, my child." They would never meet again.

PLAYING DEAD

ON THE MARCH, VIRGINIA

Despite General Lee's fear that his army was "melting away," Allan Pinkerton's final report estimated the Confederate strength to be 130,000, nearly twice the number Lee had present for duty. But McClellan, for once, gave Lincoln what he wanted and began chasing Lee, moving eastward from the Shenandoah Valley. The Army of the Potomac aimed to reach Warrenton, just south of Manassas, and from there swing east to Richmond.

It was warm for October, the leaves just beginning to curl around the edges. The men marched from sunrise to sundown, the worn soles of their boots gaping and shutting like mouths. Emma's comrades were in fine condition, but her injuries from her mule accident hadn't healed. There were days when she could not set her left foot on the ground and she could feel herself bleeding inside, a condition she called "hemorrhaging from the lungs." But she continued her duties as mail carrier and courier, and on the third day of the march she was given a message for headquarters. She turned her horse, Frank, around and started at a brisk canter, riding mile after mile, passing train after train, and finally made her delivery.

The weather had turned since the army first set out, a stark, thick cold weighting the air, and after sundown it began to sleet. For hours she wandered along icy paths, searching for her regiment, but had no way of knowing it had shifted its route after a skirmish with the enemy. Instead she came upon a small village. She intended to knock on any strange door in the hope of finding food and shelter for the night, but spotted a band of Confederate guerrillas roving between the homes.

She had heard guerrillas were on the rise in Virginia, "irregular fighters" who indiscriminately plundered, kidnapped, and murdered Northern-sympathizing citizens, breaking into homes looking to shoot any Yankees they found. In response groups of Federal cavalry hunted them down, assisted in some towns by companies of farmers, equipped and paid by the government. One elderly local patriot who stalked rebel guerrillas wore a shirt made from a Union flag, but on account of his advanced age the Confederates left him alone.

Quietly Emma rode on, but by two in the morning her horse showed signs of giving out; neither she nor Frank had eaten in nearly twenty-four hours. She approached a farmhouse and with numb hands rapped on the door. No answer. After hitching her horse under the cover of a woodshed she lowered herself to the ground, now covered in three inches of snow. Her soaking clothes chilled her skin. She feared she might die from exposure but Frank lay down beside her, keeping her warm "by the heat of his beautiful head."

At daybreak she resumed her search for her regiment, stopping in a cornfield to feed Frank. She heard a rustling behind her and turned to find a party of Union cavalry. They were looking for a group of Confederate guerrillas, and asked if she'd seen any. Indeed she had, she said, and offered to lead them back to the village. They set off.

She was thinking she was grateful for the temporary company, for the extra sets of hooves cutting through the dusty snow, when from the corner of her eye she saw a twitch of movement, a darting streak of blue. She heard the claps of repeating rifles and the thud of two bodies hitting the ground, and realized they were under attack by rebel guerrillas in Union attire. Frank reared, her head whipped back, and she was launched into the cracked china sky.

When she crashed down her mind seemed to close. She pushed against the feeling, struggling to keep awake, to feel the sting of snow against her cheek. She waited for a current of pain, a taste of blood, something to indicate whether she'd been shot. Through slitted eyes she saw two of her Union comrades, limbs limp and splayed. Her horse lay prone, inches away, close enough for her to touch the bristled fringe of his lashes. Frank was fighting, too, gusts of dry breath smoking the air, his front hoof slapping at the earth, trying to gain purchase and lift himself up. She watched him fail. He groaned once, fell back, and threw his neck across her body, pinning it beneath him. She felt the warm slow spread of his blood along her skin, and she lay still, letting him soak her. She was soaking inside, too, her old wounds made new again, relieving themselves. A minute or an hour passed; she couldn't tell. Then came a distant sound of galloping and a glimpse of battered boots. She realized that one of the Confederate guerrillas had returned, and at once she felt wholly, terrifyingly alive.

He sank the tip of his saber into the corpse of a Union man; even playing dead might not save her. A grunt as he rolled the body on its side. She heard the rustling of material, the clink of weapons being collected. She was next.

Two hands encircled her ankles and pulled. Her bad hip jostled in its socket. The hands yanked again, sliding her body out from under her horse, stripping off a layer of its blood. He found her hips, his fingers squirming in her pockets, turning them out. She struggled to make her mind work, to remember what she was carrying: no official papers, at least, and only five dollars. At last his hands lifted and she sensed him receding. She lay still, with closed eyes, until the surviving Union men returned. One of them lent her his horse, walking alongside her until she found her regiment. She thought about her own horse, Frank, and how she would have borne yet another broken limb if she could only have him back. Inside her tent she said a prayer, thanking God for sparing her "unprofitable life."

The Army of the Potomac continued its march, reaching War-renton by the first week in November, but it was not soon enough for Lincoln. By then the Confederates, under General James Longstreet, had slipped down the Shenandoah Valley and wedged themselves between the advancing Union troops and Richmond, while Stonewall Jackson remained in the Valley on McClellan's flank. Lincoln was furious and, like many of his Republican advisers, not entirely convinced of McClellan's loyalty. To his private secretary the president explained that when McClellan kept "delaying on little pretexts of wanting this and that I began to fear that he was playing false—that he did not want to hurt the enemy." He replaced McClellan with Ambrose Burnside, a well-respected thirty-eight-year-old general whose luxuriant muttonchop whiskers gave rise to the word *sideburns*. Allan Pinkerton quit the Secret Service in a considerable huff, declaring that he "had no confidence in the ability of General Burnside" and blaming the removal of his esteemed chief on the "political cabal at Washington."

McClellan gave an informal reception in his tent for his staff officers, pouring champagne into tin cups and raising his own in a toast: "To the Army of the Potomac, and bless the day when I shall return to it." He rode from camp to camp, reveling in the salutes, the shouts, the caps tossed into the air, and then delivered an emotional farewell to his troops:

"In parting from you I cannot express the love and gratitude I bear you. As an army you have grown up under my care. In you I have never found doubt or coldness. The battles you have fought under my command will proudly live in our nation's history. The glory you have achieved, our mutual perils and fatigues, the graves of our comrades fallen in battle and by disease, the broken forms of those whom wounds and sickness have disabled—the strongest associations which can exist among men—unite us still by an indissoluble tie. We shall ever be comrades in supporting the constitution of our country and the nationality of its people."

By the end of McClellan's address, his men, Emma included, were weeping openly, and she thought about all that had changed—and even more that had not. It was a year before, exactly, when she and Jerome Robbins had stood side by side watching a fireworks display in the general's honor, and when she allowed Frank Thompson to reveal who and what he was.

Jerome, still languishing at Camp Parole, learned of McClellan's departure three days later from the newspapers, and felt "depressed indeed." He hoped the rumors of his imminent release would prove true and wanted, above all, to see Emma; lately he'd corresponded more often with "Frank" than with his girlfriend back in Michigan. "Have been somewhat busy with preparations for going away which report seems well-authenticated," he wrote in his journal. "I feel glad in the thought that I may again meet dear comrades. . . . My dear friend Frank is last on the list but not least. How very often do I think of you my dear friend, and more than ever since I learned of your illness. Oh, how I have longed to be there to care for you but God willed it otherwise and let us be content." A part of him relished being the sole keeper of Emma's secret, and wanted no more—but no less—than this strange, furtive thing they had.

While General Burnside sent troops to Falmouth, just across the Rappahannock from the town of Fredericksburg, Emma received orders to ride to Washington to collect and deliver mail, a two-day trip. Her leg still pained her but she was able to splint it. Along the way she passed through the battlefield at Manassas, where her army had twice lost to the rebels. It had been months since the last contest but the ground was still thick with the evidence of carnage, skeletons scattered like a box of spilled matchsticks, bones bleaching in the winter sun.

Emma stopped at the body of one, a cavalryman who lay together with his horse; nothing but the bones and clothing remained. One arm—or, rather, the bones and the coat sleeve—was raised straight up, its hand severed cleanly at the wrist and lying nearby on the ground. She marveled at the hand's perfect condition, the fingers and joints still fused together and arched delicately, as if poised over a piano. She dismounted and considered taking the hand with her, but found no clues as to the soldier's regiment or name, no way to track down his family. There were depraved hucksters in town selling "Yankee skulls" at ten dollars apiece, and rebel women wearing brooches made of soldiers' bones, and she did not want to make a souvenir out of a sacrifice.

At headquarters in Washington she met a twenty-six-year-old Union lieutenant named James Reid, whose regiment recently had been brigaded with hers as part of the Ninth Corps. He was the opposite of Jerome Robbins: a six-foot-three Scotsman with blond hair and the neck of a bear, someone disinclined to confide to a diary about his faith or anything else. When the war broke out, he'd enlisted as a private with the 79th New York Highlanders, a Scottish infantry regiment that arrived for training in Washington wearing kilts, an ensemble derided by the *New York Military Gazette* as "short petticoats and bare knees, in poor taste and barbarous." Back home, in Manhattan's Yorkville neighborhood, his wife, Mary, waited with their sons, three-year-old George and one-year-old William. He told Emma stories of being captured at First Bull Run and his six-month imprisonment near Richmond, impressing her with his valor and devotion to the cause, and she wondered when she'd see him again.

A "Yankee skull" goblet, a necklace of Yankee teeth, and other items—some real, some invented—of "Secesh" industry.

She returned to Falmouth, bringing along a pocketful of apples, doughnuts, and an orange for the 2nd Michigan's commander, Colonel Poe. She found the army encamped for miles along the Rappahannock. A heavy snowfall had turned the ground into slushy mud. "It looks rather dismal," wrote one soldier. "The men are ragged & short of shoes & only have shelter tents which are in fact more like a dog kennel than a habitation for men." Conflicting rumors swirled around camp: they were either going into winter quarters or advancing into Fredericksburg any day. Burnside's initial plan was to take the city while it was too weakly held to resist, but the pontoon bridges he needed to cross the river arrived late, allowing Confederate forces to occupy the city and stretch out a six-mile defensive array in the hills and ridges above it. Although Burnside would have preferred to wait until spring, he knew that Lincoln, and the Northern public in general, wanted a confrontation and wanted it now.

The general convened a meeting at his headquarters to announce his decision: the engineers would lay the pontoons across the Rappahannock under the cover of darkness, supported by heavy artillery on the near side of the river. The army would then cross in force and attack the Confederates head-on. Burnside's staff unanimously objected; the rebels occupied the high ground, and it was unconscionable to send troops charging across an open field and up a heavily fortified hill where the enemy would be waiting. Even if the men, by sheer luck, made it across the field, they would then have to climb a stone wall while under constant Confederate fire. "If you make the attack as contemplated it will be the greatest slaughter of the war," one colonel told Burnside. Another agreed, adding, "The carrying out of your plan will be murder, not warfare." But they succeeded only in annoying Burnside, who argued that his plan would work simply because it would take Lee by surprise. With that, at least, his staff agreed, if only because Lee wouldn't expect so foolish a move.

Belle made an appearance at Fredericksburg, encouraging the rebel soldiers and snooping around the enemy, making note of their positions. She noticed a barefoot young Confederate private and offered her own pair of fine cloth gaiters laced at the side and trimmed with patent leather. "If it rests his poor feet only a little while, I am repaid," she explained to another soldier. "He is not old enough to be away from his mother." Bewildered, the private held up Belle's size-two-and-a-half boots, wondering how he might make them fit.

She decided to take a chance and write to Stonewall Jackson at his headquarters, asking the general if it was safe enough for her to return to Martinsburg, which was still in and out of Union hands. To Belle's surprise and relief he replied promptly, proof he was now convinced of her loyalty. He advised her to avoid her hometown and the risk of another imprisonment and suggested she instead visit relatives in Tennessee. "Truly your friend," he signed off, "T. J. Jackson." She would do as the general said.

Lee, meanwhile, felt confident about the positions and morale of his troops. He had urged the people of Fredericksburg to flee their town before Union troops attacked it. Six thousand citizens were suddenly homeless, trudging through the grimy snow, little girls dragging their dolls behind them, slaves tugging picked chickens and bags of flour. Every move Burnside made was monitored by Confederate spies. One rebel agent, posing as a Union telegraph operator, set up his paraphernalia in the basement of a sutler's store, located conveniently near the Rappahannock. The sutler, a fellow rebel spy, sold goods to the unsuspecting men of Burnside's army and reported their conversations to the telegraph operator.

Only one thing seemed amiss. Although Lee had a direct route from Richmond for his supplies, via the Richmond, Fredericksburg & Potomac Railroad, shipments were slow and late in coming, if they ever came at all, prompting the general to send written protests against the lack of "zeal and energy" in the management of the

RF&P. What Lee did not know was that Samuel Ruth, the railroad's superintendent, was a member of Elizabeth's Richmond Underground.

Emma spent her days delivering messages to and from Union commanders. Riding along the bank of the river, she saw the rebel batteries frowning on the heights beyond the city of Fredericksburg, and the Confederate and Federal pickets within patrolling the grounds, close enough to smell each other. She watched as Union engineers began assembling six pontoon bridges, rushing to finish before dawn.

On December 11, as a church tower in the distance struck five, the dense morning fog began to break up, exposing the engineers. Emma glimpsed hundreds of Confederate guns poised on the opposite banks. There was a flash of musketry, the hissing of shot, a terrible cacophony of screams. Bodies dropped into the frigid waters of the Rappahannock. Batteries of Union cannon overlooking the river opened on the town and riverbanks, a roar that could be heard for miles. The surviving engineers hurried to finish the bridges. "The work went steadily on," Emma observed, "not withstanding that two out of every three who were engaged in laying the bridges were either killed or wounded. But as fast as one fell another took his place." Members of the 7th Michigan climbed into boats and crossed the river under a relentless barrage of fire. Those who made it to the other side stormed into Fredericksburg and began driving the rebels out of town, back toward the heights where Lee's main forces were entrenched. By nightfall the bridges were finished but the city was destroyed—shells crashing and bursting, houses crushed, smoke swirling, and flames leaping. Union soldiers looted everything in sight: books, petticoats, hats, bonnets, musical instruments, pillows, bedclothes, furniture. Dead rebels lay strewn about the streets.

Two days later, after both sides had regrouped and tended to their wounded, the armies prepared for battle. Emma learned that her old friend, Colonel Poe, had lost his orderly, and she volunteered to take the orderly's place in the coming battle. She rose before dawn and dressed in her uniform, now embellished with buff epaulets trimmed in gold braid, signifying her position as aide-de-camp. She celebrated with an entry in her journal. "I wish my friends could see me in my present uniform!" she wrote. "This division will probably charge on the enemy's works this afternoon. God grant them success! While I write the roar of cannon and musketry is almost deafening. This may be my last entry in this journal. God's will be done. I commit myself to Him, soul and body. I must close. [Poe] has mounted his horse and says Come!"

Low-lying clouds rolled in, enshrouding the field in fog. Fifes shrieked and bugles called. Burnside kept to his plan, even though, as one Confederate colonel joked, "A chicken could not live in that field when we open upon it." The general ordered Emma's division to attack. With a fierce yell three brigades, one after another, charged across the field that lay between the town and the heights. They ran with heads down, as if they could not bear to glimpse what awaited them. As soon as they came within range of the Confederate guns they began to drop, one witness said, like "grass before the scythe." All day long, brigade after brigade charged out, each one having witnessed the fate of the one before.

No one got within fifty yards of the stone wall. Many soldiers stayed on the field, crouched behind the dead or dying, trying to ignore the sound of bullets slapping the corpses of their friends. Within the first hour the Union had lost nearly three thousand men, both dead and wounded, and still they kept coming. By the end of the battle they would lose ten thousand more. General Lee watched it all from the hills above. "It is well that war is so terrible," he said, "we should grow too fond of it." Emma watched, too, mourning the "thousands of noble lives which fell upon that disastrous field," and wishing that General McClellan were still in charge.

All day long she rode up and down the lines, carrying messages and relaying orders, close enough to feel the heat of fire, to hear the cries for help and pleas for water, the dreamy, delirious voices murmuring loved ones' names. One comrade marveled at "Frank's" bravery, his riding "with a fearlessness that attracted the attention and secured the commendation of field and general officers." She witnessed a man fatally shoot himself with his own pistol and another taking aim through his side, rendering himself unfit for duty. Only once in twelve hours did she get out of the saddle, stopping to assist an officer who lay writhing in agony on the field.

Emma realized she knew him: James Reid of the 79th New York, the lieutenant she'd met while collecting the mail at headquarters, the one who was nothing like Jerome. He clutched her arm and begged for help, pinning her with his pale blue eyes. She determined he was suffering from cramps and spasms and pulled him off the field. She pried open his mouth and pushed opium pills onto his tongue, assuring him they'd help with the pain. She poured a stream of whiskey down his throat, wiped the residue from his chin. Within the hour he impressed her once again by getting back on his horse and galloping onto the battlefield.

Emma didn't want to face the truth: for the second time during the war, she was falling for a man.

[PART THREE]

1863

WHEN YOU THINK HE MAY
BE KILLED TOMORROW

FREDERICKSBURG, VIRGINIA, AND KENTUCKY

After the Battle of Fredericksburg James Reid moved into Emma's tent, where she revealed more of herself than she ever had to Jerome.

Jerome, finally released from Camp Parole, noted in his journal that his "friend Frank" had grown "extremely fond" of the married soldier, that they seemed to be "particular friends." He never inquired about the nature or extent of their relationship—he was, after all, still courting Anna Corcy—but questions and suspicions crept through his mind. It was all he thought about when he walked with Emma to get the mail at headquarters, or when she stopped by his tent to show him a new novel, *Pauline of the Potomac*, about a female spy for General McClellan, or when she left camp for a few days, picking up messages at Alexandria. "Have not had a very long chat with Frank and I feel quite lonely without him," Jerome wrote, "but I suppose he enjoys his tentmate. . . . Reid seems a fine fellow & is very fond of Frank." He began referring to Emma as Reid's "pet."

He told himself that there was nothing untoward about Emma and Reid's relationship, that they were simply "individuals who repose in the pleasantest arbor of friendship." He did not want to suspect the worst of Reid, whom he called "one of nature's noblemen," nor of Emma, his fellow devout Christian and confidante. He thought about them bunking together, working to weatherize their tent, collecting stones and bricks to build a hearth and chimney, sharing stories by their fire. He wondered if Emma had recounted the night she sheared her hair and bound her breasts and renamed herself Frank Thompson, if he was no longer the only one who knew her truth. He made sure to share the news of each love letter from Anna Corey, searching Emma's face for the impact of his words.

While away collecting mail Emma sent Jerome a letter, signing her real name for the first time since she was a girl back on the farm in Canada, doodling in her Bible:

Jan. 16/63

Dear Jerome,

In the first place, I will say that I am happy to know that you are prospering so well in matters of the heart. In spite of the ridicule which sentiment meets with everywhere, I am free to state that upon the success of our love schemes depends very much of our happiness in this world. . . . Dear Jerome, I am in earnest in my congratulations & daily realize that had I met you some years ago I might have been much happier now. But Providence has ordered it otherwise & I must be content. I would not change now if I could—if my life's happiness depended on it. I do not love you less because you love another, but rather more, for your nobleness of character displayed in your love for her—God make her worthy of so good a husband.

Your loving friend,
Emma

Perhaps signing her name let her reclaim some small vestige of truth, lost not because of Frank Thompson—he had always been an authentic part of her—but because she coveted a man who wasn't hers to have. She had allowed herself to imagine an honest life with Jerome, but this "love scheme" with Reid was a dead-end sin, bringing her equal measures of joy and misery, leaving both her and Frank with nowhere to go. For now, her lies and her lives were intertwined, two strands of a twisting double helix, but she knew the structure could not be sustained. It could unravel at any time, from either within or without.

Her comrades spread stories about women being exposed in the ranks, and such stories seemed to grow more numerous by the day. Most inadvertently divulged their sex in the course of regular army life, as was the case with "Charles Norton," a private with the 141st Pennsylvania Infantry who stole a fellow officer's boots. The resulting investigation revealed this "general favorite" to be a she, who was "speedily mustered out of the service." Two more women, serving under Union general Philip Sheridan, had been discovered just a few weeks earlier. "While out on the foraging expedition these Amazons had secured a supply of 'apple jack' by some means, got very drunk, and on the return had fallen into Stone River and had been nearly drowned," the general wrote. "After they had been fished from the water, in the process of resuscitation their sex was disclosed, though up to this time it appeared to be known only to each other."

A careless few betrayed themselves through stereotypical feminine behavior. Two women serving with the 95th Illinois Infantry were outed when an officer threw apples to them. They were dressed in full military uniform, but instinctively made a grab for the hem of their nonexistent aprons in order to catch the fruit. They were discharged immediately. Another woman who'd recently tried to enlist was suspected when a commander witnessed her giving "a quick jerk of her head that only a woman could give." A recruit in Rochester, New York, forgot how to don pants, and tried to put hers on by pulling them over her head.

In the most dramatic discovery, a member of the Army of the Potomac—a corporal from New Jersey—had recently given birth to a baby boy in a camp not far from Emma's own. She had concealed her sex and her pregnancy beneath an oversize coat, and went into labor on picket duty. The soldier, said another picket, "complained of feeling unwell, but little notice was given his complaints at first. His pain and other symptoms of severe indisposition increased, becoming so evident that his officers had him carried to a nearby farmhouse. There the worthy corporal was safely delivered of a fine, fat little recruit."

The rank and file of the Union army seemed thrilled by news of the birth, an affirmation of life in the midst of relentless destruction and death, and to Emma's surprise the men admired rather than condemned the new mother, even taking up a collection for her. Yet despite their generous reaction, Emma was unnerved. Her comrades might take a harder look at her smooth face and diminutive feet, listen for tones lurking just beneath her practiced voice. They might turn "Our Woman" from an affectionate nickname into an accusation. Despite her bravery on the battlefield and her work behind the lines, she could still be dismissed or arrested, depending on the whims of her superiors. She had worked too hard for too long to create Frank Thompson, and she wanted his demise to be her decision alone.

After the bloody defeat at Fredericksburg, President Lincoln replaced Ambrose Burnside with General Joseph Hooker, whose headquarters—a combination, it was said, "of barroom and brothel"—were so infamously wicked that his name would become synonymous with the world's oldest profession. "Beware of rashness," the president warned his new commander, "but with energy and sleepless vigilance go forward and give us victories." In mid-March "Fighting Joe" sent Emma, James Reid, Jerome Robbins, and the rest of the Ninth Corps to Kentucky as part of the Army of the Cumberland, preparing to support General Ulysses S. Grant's renewed attempt to capture Vicksburg, a fortified city that held the Mississippi River for the Confederacy. Emma reached for her journal and penned her last entry as a member of the Army of the Potomac: "The weather department is in perfect keeping with the War

Department, its policy being to make as many changes as possible, and every one worse than the last. May God bless the old Army, and save it from total annihilation."

She took the train to Louisville and then continued on south-east to Lebanon, where she received orders to infiltrate enemy lines. Confederate forces under General John Pegram, said to be the advance of General Longstreet's division, had crossed the Cumberland River and were circling Union troops at nearby Danville. Emma was to gather any information she could about numbers of cavalry, infantry, and artillery.

Before her arrival Union troops had skirmished with a group of Confederate cavalry, taking five rebels prisoner, and one of them unwittingly donated his uniform for her mission. Noncommissioned officers of the cavalry had no standard uniform, the only requirement being that the cloth was butternut in color, and she decided to pose as a Southern civilian traveling from farmhouse to farmhouse, seeking butter and eggs for the rebel army. As always, she carried her seven-shooter revolver.

Emma wandered past the lines and into a village, knocking at the first door she came to, and was surprised to find herself in the midst of a wedding party, a group of soldiers and sundry relatives gathered around a long mahogany table piled with ham, biscuits, horehound candy sticks, and a traditional wedding fruit cake, sliced open and spilling flecks of apples and raisins. Her mouth watered; her last decent meal had consisted of hardtack and beef soup at Christmas. The bride, Emma learned, was a young widow whose husband had been killed in combat a few months ago, and she'd been eager to remarry ever since, fearful that the war was fast depleting her pick of eligible men.

It was a valid concern; the war was on its way to claiming one in five Southern white men of military age, a situation that prompted frantic letters to the editor. "Having made up my mind not to be an old maid, and having only a moderate fortune and less beauty, I fear I shall find it rather difficult to accomplish my wishes," an eighteen-year-old Virginian named Hattie wrote to the *Southern Literary Messenger.* "Do you think I will be overlooked amidst this wreck of matter and crush of men and horses?" Fear of spinsterhood led to a breakdown in mores—"the blockade don't keep out babies," one South Carolina woman quipped—and to unconventional liaisons; many girls settled for marrying amputees, or even Yankees. Another captured the attitude of Southern belles as the war progressed: "One looks at a man so differently when you think he may be killed tomorrow."

Emma noticed that this bride had forgone "widow's weeds" in favor of a more festive costume: frills of lace at her wrists, pearl buttons trailing down her spine, a single spray of orange blossoms encircling her head, purple fabric in honor of the dead. Her new husband spotted Emma loitering on the periphery of the party and broke away from the crowd.

He was dark-haired and imposing, with a long, loping gait and a mustache that spiraled like the shell of a snail. He introduced himself as Captain Logan, a recruiting officer for the rebel army, and asked what business she had in the village.

Just collecting supplies for the soldiers, she replied.

The captain nodded, raised his glass to his lips. She congratulated him, apologized for interrupting the festivities, and turned to leave.

A hand curled around her shoulder and turned her back.

"See here, my lad," he said. "I think the best thing you can do is enlist and join a company which is just forming here in the village, and will leave in the morning. We are giving a bounty to all who freely enlist, and are conscripting those who refuse. Which do you propose to do? Enlist and get the bounty, or refuse and be obliged to go without anything?"

His gaze meandered down her body. It occurred to her that she would rather be discovered as a woman than as a Union spy.

"I think I shall wait for a few days before I decide," she said.

"But we can't wait for you to decide. The Yankees may be upon us any moment, for we are not far from their lines, and we will leave here either tonight or in the morning early. I will give you two hours to decide this question, and in the meantime you must be put under guard."

He clenched Emma's arm and pulled her to a corner of the room. Two soldiers positioned themselves on either side, closing her in. Her head reached only as high as their shoulders. She talked up at them, presenting herself as a simple, honest Kentuckian, worried, like anyone else, about the Yankees invading the town. She wondered how many rebel troops were currently in the area; where were they heading next? With every sip of whiskey her guards let slip another scrap of information; the rebel forces at Danville comprised 3,500 cavalry, three regiments of infantry, and six pieces of artillery. The revelers rose from the table and formed a grand chain, clasping shoulders and sliding hands, keeping time to the plink of the banjo and the clank of spoons, spinning in tight concentric circles like the works of a watch. Minutes tumbled into hours and the circles merged into one, halfhearted and teetering, voices garbling the words to "God Save the South," whiskey sloshing over glass rims. At a pause in the music Captain Logan lumbered toward Emma, and asked if she'd made her decision.

"I've concluded to wait until I'm conscripted," Emma said.

"Well, you will not have to wait long for that, so you may consider yourself a soldier of the Confederacy from this hour, and subject to military discipline." He added that he was forming a company of cavalry, and would be ready to set off in the morning.

The captain returned to the circle, leaving her to consider his words. She knew he was serious, especially as she would be required to take the oath of allegiance to the Confederate government, but she trusted in God and her own ingenuity to escape the dilemma. She would have to flee before being forced to declare her loyalty to the rebel cause; if she refused to be sworn in at the service after conscription, Logan and his soldiers would suspect her true intentions and execute her on the spot. All night long she stood, foot throbbing, her guards watching her as she watched the captain, twirling and kissing his bride.

At daylight Captain Logan reappeared. He led her to a group of waiting soldiers and pointed out her horse. Within moments they were trotting briskly over the country, the captain complimenting her riding skills. You will be grateful, he said, when the war is over and the South has gained her independence. You will be proud to have been a Confederate soldier, to have driven the vandals from their soil and steeped your saber in Yankee blood. "Then," he added, "you will thank me for the interest I have taken in you, and for the gentle persuasion which I made use of to stir up your patriotism and remind you of your duty to your country." As he finished this speech a Union reconnoitering party rose in sight atop a hill and galloped toward them, cavalry in advance and infantry in the rear.

The Confederate captain ordered Emma and his men to charge forward, weapons drawn. A chorus of rebel yells roared fiercely in her ears. They were upon each other then, thrashing and rearing, and she found herself on the Federal side of the line. The Union captain recognized her and signaled for her to fall in next to him, a position that put her directly across from Captain Logan, the man to whom she owed "such a debt of gratitude." This was her chance, she realized, "to cancel all obligations in that direction."

Her brain directed her arm to rise and her arm obeyed, aiming her revolver at his head. She wanted not to kill him, but merely to spoil the "graceful curve of his mustache," to render him, permanently, a lesser version of himself.

She pulled the trigger.

BREAD OR BLOOD

Illustration of the Richmond bread riot, April 1863.

RICHMOND

On March 27 Jefferson Davis declared one of his numerous "fast days," in which he called upon the citizens of the Confederacy to attend church and abstain from food and drink. On such occasions Elizabeth and her family instead enjoyed a better dinner than usual and even sweet tea and dessert; brother John frequently returned from his trips north with sugar, a rare delicacy in these long, lean days of the blockade. Guilt always accompanied this act of defiance, as Elizabeth heard stories and witnessed firsthand how Southerners were starving. In Atlanta a dozen women held up a butcher shop, brandishing guns and escaping with armfuls of meat. In Richmond a war clerk complained that he had lost twenty pounds, and his wife and children were emaciated. In Vicksburg, Mississippi, people turned to eating rats, dogs, and cats, and a local newspaper wrote of "the luxury of mule meat and fricasseed kitten." One mother fed her daughter soup containing the remains of the girl's pet bird.

"Alas for the suffering of the very poor!" Elizabeth wrote in her journal. "Women are begging with tears in their eyes, and a different class from ordinary beggars . . . There is a starvation panic on the people."

Such circumstances, to her thinking, had but one positive effect: declining morale among the populace aggravated declining morale among the entire Confederate army.

The events of April 2, less than a week after the last fast day, unfolded as if Elizabeth had choreographed every moment, starting with a young girl waiting on a bench in Richmond's Capitol Square, her body a rigid structure of stark angles and jutting bones. She turned to an old lady sitting next to her, said she was too weak to stand, and raised a skeletal arm to adjust her bonnet. The old lady gasped, fluttering a hand over her mouth, and the girl gave a hard laugh. "That's all there's left of me!" she said.

People started filing into the square, hundreds and hundreds of them, taking their places amid the blooming forget-me-nots and sweetly trilling birds. They wielded axes, hatchets, clubs, knives, hammers, guns, stones. A black maid hurried to the square to collect her wandering charge, worried that the child might "catch something from them poor white folks." A chorus of voices shouted "Bread or blood! Bread or blood!" a rolling drumbeat that matched the pace of their steps.

More rioters joined in, thousands of them now, punching their fists toward the sky, funneling through the alleys, charging

into stores and grabbing everything they found, bread and turnips and hams and shoes, hoisting them into carts and wagons
lined along the curb, where female getaway drivers waited in
stern complicity. They smashed locked doors with their hatchets
and hammers and stones, working with a terrible earnestness,
looting both necessities and luxuries: bonnets, tools, cornmeal,
piles of bacon, barrels of flour, strings of glinting jewels. One
merchant rolled up his sleeves and stood guard at the door of his
shop, waving a six-shooter, vowing to defend his property at the
risk of his life. Another tossed out boxes of needles, hoping that
would be enough to placate the crowd.

Within the hour the Public Guard filed up Seventeenth Street,
Jefferson Davis close behind. The president clambered into a cart and
looked out over the crowd, which fell instantly and utterly silent. In
a mournful tone he urged the people to "abstain from their lawless
acts" and wondered why they had stolen jewelry and finery when they
claimed to be starving. No response. The president turned out his
pockets and gathered a thin roll of Confederate graybacks in his fists.

"You say you are hungry and have no money," he said. "Here is
all I have; it is not much, but take it."

"The Union!" the mob screamed. "No more starvation!"

He tossed the money and watched the people dive to catch it.
"We do not desire to injure anyone, but this lawlessness must stop.
I will give you five minutes to disperse. Otherwise you will be fired
upon."

The crowd murmured and rustled but stayed rooted in place.

"Captain Gay," Davis intoned, "order your men to load with
ball cartridges."

"Load!" the captain shouted.

There was a swift clattering of ramrods as the soldiers obeyed,
stuffing powder and balls down the muzzles of their muskets.

"Captain Gay," he said again, "if this street is not cleared
within five minutes, order your men to fire down the street until it
is cleared."

The president lifted his pocket watch into the air to count the tick of the hands.

All at once the rioters gave up. They broke off in pairs and threes, hatchets and clubs dangling by their sides, stones dropping to the ground, loot lying forgotten in the carts and strewn across sidewalks, broken glass crunching under their plain brown work shoes. Police tracked down and arrested forty-seven of them.

By nightfall all of Richmond seethed with suspicion, convinced that the "bread riot" was the work of Yankee spies. "This demonstration was made use of by the disaffected in our midst," one resident declared, "for the misrepresentation and exaggeration of our real condition." The city council ruled that whoever planned the riot had "devilish and secret motives," and the War Department ordered newspapers to stifle coverage so that the Confederacy would not be further humiliated. The *Richmond Examiner* nevertheless condemned the rioters as "emissaries of the Federal Government" and found it "glaringly incongruous" that so many of the supposedly impoverished suspects were able to post bail and retain expensive counsel, paying as much as $500 cash—money supplied by Elizabeth and a few wealthy Unionist friends.

The Richmond Underground made sure news of the bread riot reached the Union army and the Northern press, along with personal updates about the Confederate president. The current year, in the words of Varina Davis, had opened "drearily" for her husband. He was still morose over the fall of numerous Southern cities in the western theater: Donelson, Tennessee (the battle that earned Ulysses S. Grant the nickname "Unconditional Surrender"); Nashville (whose citizens, the Northern press reported, were all "hard up"); and Memphis, where, said one correspondent, "secession had made havoc of all female charms and graces," infecting women like a "contagious disease." Davis's health, always poor, was worsening by the day. He suffered from gastric distress, bouts of malaria, and constant inflammation in his left eye, which Varina treated with mercury drops. Chronic neuralgia left even his face in constant agony.

Often he was too sick to walk the half mile to his official office near Capitol Square and instead worked at home, Mary Jane lingering with her duster, cleaning the adjacent children's nursery until she could get to his desk. She was there after he penned a response to the bread riot, reprinted in the Yankee press, in which he appealed to Southerners to "respond to the call of patriotism" with continued sacrifice and prayer.

Most of Richmond's prominent families responded, stripping their beds of blankets, sheets, and pillows; packing up the contents of their pantries; gathering coats and shoes; and sending everything off to Confederate soldiers. Some even offered their horses to replenish the army's dwindling supply; thousands of mounts, on both sides, were lost to convalescent camps, where they recovered from battle fatigue and the equine equivalent of "soldier's heart," an early term for post-traumatic stress disorder. General Lee, in an April 16 letter to Davis, confided his anxiety over "the present immobility of the army, owing to the condition of our horses and the scarcity of forage and provisions." Civilians who didn't willingly contribute risked a visit from rebel "impressment agents," charged with confiscating food, fuel, slaves, horses, and anything else the army might need. Elizabeth witnessed the agents seize horses from bread carts and unsuspecting peddlers arriving at the market. Most of the livery stables on Franklin and other streets had been completely swept of their steeds. Some residents, in anticipation of the coming agents, mounted diminutive slaves on their horses and sent them, under whip and spur, in the opposite direction.

The agents already had come to Elizabeth's door three times, shaking papers from the War Department in her face and claiming they had the right to her horses. They did not pay her on the spot as they had been instructed to do. Three of her brilliant white horses were gone, never to be seen again, and she stashed her last remaining horse in the smokehouse. There she considered him safe until agents again made the rounds in her neighborhood, creeping behind the mansions and peering inside stables. One night, when all was clear, she rolled up her Oriental rugs and led her horse inside her home. She wrapped his hooves with rags and spread straw across the floor of her study.

She implored the horse to stay quiet, scratching his withers and sending her nieces with baskets of fresh apples to keep him company. To her amazement he stayed in the study for days at a time, until the officers gave up, never neighing or stomping loud enough to be heard. It was, she marveled, as though he "thoroughly understood matters"—that in the South's "dear young Government" there was no safety for anyone, not even an innocent beast.

A WEAN THAT'S BORN TO BE HUNG

LEBANON, KENTUCKY

When the last bullet had been fired and the smoke cleared Emma could see what she had done to Captain Logan, now writhing on the ground. There was a hole where the right side of his nose used to be, his once fine face caving in on itself. A pink ribbon of lip clung to his mouth by the thinnest thread of skin. She thought of his new bride, and how she would no longer "rejoice in the beauty of that manly face." He was alive, at least for now, but "sadly spoiled." She had one second to brace herself for a counterattack.

The rebels rushed at her at once, each of them wanting the pleasure of killing her. They swiped their sabers at her neck, nicked a shallow gash across her horse's throat. The Union men hurled themselves in the middle, protecting her, driving the enemy back, pouring volley after volley until half of the rebels lay dead. They gathered their prisoners and their wounded and returned to camp, where her commanding officers commended her "coolness." They also told her she could not go undercover again as long as her regiment remained in Lebanon. "I would not be permitted to go out again in that vicinity, in the capacity of spy," she wrote, "as I would most assuredly meet with some of those who had seen me desert their ranks, and I would consequently be hung up to the nearest tree."

Her disappointment about the end of her espionage career was soon eclipsed by news about James Reid. He had resigned his commission, planning to take his children and wife, who had suddenly become gravely ill, back to Scotland. At night Emma grew aware of an intensifying internal heat, her brain burning beneath her skull. Her body couldn't settle on a temperature and a sheen of sweat formed along her goosebumped arms. She allowed herself to admit what was happening: a recurrence of the malaria she'd contracted nearly a year before, on the banks of the Chickahominy River near Richmond.

She dreamed of her mother, who'd once confessed to a Scottish clergyman her fear that her youngest daughter would meet with a violent death; the clergyman had advised her mother not to worry. "It is an auld saying," he replied, "an' I believe a true one. A wean that's born to be hung will never be drooned." And now here she was, having survived her childhood, the bloodiest battles of the war, and several narrow escapes from the enemy, prepared to die in a most anonymous and unspectacular manner, shivering on a cot in a tent. She yearned for the touch of her mother's cool hand. She applied for a medical furlough and was denied, with no reason given. Discovery, she thought, was "far worse than death." She refused to go to the hospital for reasons only James Reid and Jerome Robbins knew. The latter sat next to her now, feeding her sips of whiskey spiked with quinine sulfate, somewhat ambivalent about the job; he alone understood that Emma's malaise was partly due to matters of the heart.

Outside her tent, there was a short ripping sound followed by a boom. Then screaming, moaning, heavy huffs of breath. Jerome took her outstretched hand and lifted her. She willed her legs to support her weight and leaned against him, shuffle-dragging her way to the opening of the tent. A shell had exploded randomly, accidentally, near a group of soldiers sitting together. Through swirls of smoke and dust she saw fingers dangling from hands, hands torn from wrists, faces half missing, cubes of unidentifiable flesh, blood leaking from ears. A deck of cards was strewn about the ground, hands half played.

As soon as she felt herself falling she felt herself caught. Jerome led her back to her tent, to her cot, and lowered her down. He stayed a moment, sitting next to her. The carnage was nothing she hadn't seen before, but yet she felt a sinking inside, a collapse of every last reserve.

She recognized the immediate and acute loss of her "soldierly qualities," and loathed what she had become: "a poor, cowardly, nervous, whining woman, and as if to make up for lost time, and to give vent to my long-pent-up feelings, I could do nothing but weep hour after hour, until it would seem that my head was literally a fountain of tears and my heart was one great burden of sorrow." Jerome hurried to tend to the wounded, leaving her alone, both of them aware of words left unsaid.

He said them to himself later that night, alone in his own tent. "I passed most of the day with Frank, who has been and is unwell," he wrote. He pressed the pencil harder, as if the force of his hand could redirect her attention from James Reid back to him, and what he thought they'd had: "How strange are some of the incidents of life. . . . It is unpleasant to awaken to the conviction that one dear as a friend can forget, in their selfish interest, that others may not be void of the finer sensibilities of the human heart. It is a sad reality to which we awaken when we learn that others are receiving the devotion of one from whom we only claim friendship's attention."

She had kept other thoughts from Jerome, as well. She was despondent about the imminent departure of General Poe, her friend and mentor, who was also leaving the regiment. Without Poe she might lose her position as mail carrier, and with it, the freedom—those long, solitary rides—that gave Frank Thompson respite and helped him keep his secret. Many of the men who had enlisted with her in Michigan back in the spring of '61 were gone, too, transferred to other regiments or wounded or killed. And she privately mourned the loss of someone she'd never met, a woman whose fate warranted one line in the *Louisville Daily Democrat*: "A young woman wearing soldier's apparel, and belonging to the Fourteenth Iowa regiment, shot herself in Cairo on Sunday night because her sex was discovered."

The story made Emma feel a sudden absence, a distancing from herself, as if she were living in the third person instead of the first, caught between pronouns. She had to decide which one of her selves to kill.

Once she made her choice she did so in a manner befitting Frank Thompson, who would not want her to weep or waver or pine. Three weeks after her last mission, on or around the night of April 17, she defied her fever and dressed in full uniform, slipping out of her tent and away from camp, slashing her way through the woods, saying good-bye to no one, each step as liberating as it was lonely. In the morning, at roll call, they would be looking for Frank Thompson and determine that he had deserted, a crime punishable by death. They would be ready to hang someone who had only briefly existed, a man she could make vanish just as easily as she conjured him.

A DREADFUL BLOW

VIRGINIA AND WASHINGTON, DC

On May 2, after the Battle of Chancellorsville in northeastern Virginia—a Confederate victory against all odds, as Lee's army was half the size of Hooker's—Stonewall Jackson was struck by three balls: one passed through his right hand; another through his left arm between elbow and wrist; and the last two inches below the left shoulder, shattering the bone and serving the brachial artery. He fell from Little Sorrel and was caught by a Confederate captain, to whom he remarked, "All my wounds are by my own men." Field surgeons amputated the arm and were about to toss it onto a pile of limbs, but Jackson's chaplain saved it, giving it a proper Christian burial in a nearby cemetery. "If it was in my power to replace my arm," Jackson told the chaplain, "I would not dare do it unless I could know it was the will of my Heavenly Father." One of the general's staff officers would mark the area with a tombstone inscribed:

ARM OF STONEWALL JACKSON
MAY 3, 1863

A week later pneumonia had set in, and Jackson was in critical condition. At the Confederate White House, Mary Jane watched Jefferson Davis wait for the inevitable, rubbing his temples with bony fingers, pacing with the tick of the clock. He did no work, confessing that he was "unable to think of anything but the impending calamity." Two of Mary Jane's fellow servants were dispatched to wait for news, one to the telegraph office and the other to the Richmond depot, where hundreds of people gathered for each arriving train, shouting up to the engineer, "How is he? Is he better?"

Stonewall Jackson died on a Sunday, just as he'd always wanted. His body arrived in Richmond the following morning, where it would lie in state before being buried in Lexington, Virginia, more than a hundred miles away from his arm. Mary Jane was there that evening when the president told a visitor, "You must excuse me. I am still staggering from a dreadful blow. I cannot think." A week later, General Lee came to Richmond to confer with Davis about his next move: he would embark on his Gettysburg campaign, striking across the Potomac to invade the North for a second time, putting himself in a position to threaten Philadelphia, Baltimore, and Washington.

Belle was visiting family friends in Alabama when she learned of the tragic news. As a public sign of mourning she wrapped a black crepe band around the sleeve of her riding habit, directly over the scar on her arm. She would wear the band for thirty days, the time allotted for the outward sign of a soldier's sorrow, but privately she mourned the general longer, invoking him in her dreams and drafting a brief elegy: "It is not for me to trace the career and paint the virtues of 'Stonewall' Jackson," she wrote. "That task is reserved for an abler pen; but I may be permitted to record my poignant grief. . . . The sorrow of the South is unmitigated and inextinguishable. . . . [His glory] had neither dawn nor twilight. It rose and set in meridian splendor."

She had enjoyed her five-month tour though the South, although the memories were now bittersweet, since Stonewall had been the one to advise her to go. In Culpeper, Virginia, she spent a night at a hotel popular with blockade runners and impressed her fellow boarders by using perfume instead of water for her bath. "She was a brilliant talker," said one, "and soon everybody in the room was attracted to her, especially the men. She talked chiefly to the men—

indeed, I am afraid she did not care particularly for the women."
In Knoxville, Tennessee, while she was visiting with Confederate
general Joseph Johnston, the Florida Brass Band serenaded her and
a crowd exhorted her to make a speech. Belle urged the general
to appear in her stead, but when he refused she poked her head
through the window: "Gentlemen!" she called. "Like General Johnston, I can fight but I cannot make speeches. You have my heartfelt
thanks for the compliment."

In Montgomery, Alabama, she heard a rumor that Stonewall had
been shot, but that the wound was minor. She was not prepared for
the reality, delivered shortly thereafter via telegram from the office
of Governor John Gill Shorter. She quickly sent a wire to Virginia
governor John Letcher for confirmation: "Please telegh if Gen. Jackson is dead. If so save me a lock of his hair. Yours truly, Belle Boyd."

Wearing her mourning armband, she made one last stop
in Charleston, South Carolina, where she dined with General
Beauregard and his staff officers. Beauregard, the once-rising military star of First Manassas, had since clashed with Jefferson Davis
and been banished to the Palmetto City. One of the general's officers presented Belle with fruit and a parrot, both fresh from Nassau
via a blockade runner. Belle conspired to take the parrot with her as
she headed home, and planned to teach it to say Stonewall's name.
The press traced her journey back to the Shenandoah Valley, reporting, strangely, a pit stop in Philadelphia, where she allegedly
stayed in a brothel on Twelfth Street, dressing in male attire. The
story prompted Union authorities to pursue her for entering Northern territory, a violation of the conditions of her parole.

By the time Belle arrived in Martinsburg in mid-June, she found
that home had changed irrecoverably. At a convention in
Wheeling, Virginia, the northwest portion of the state had seceded
to become its own state, West Virginia, the thirty-fifth in the Union.
Berkeley County, which included Martinsburg, voted in favor of an-

nexation into the Union 645 to 7. Confederate soldiers who had missed the vote refused to recognize West Virginia's legitimacy, as did Jefferson Davis and the editorial boards of Southern newspapers. "The last scene in a stupid farce was played at Wheeling," scoffed the *Richmond Whig*, "in the pretended inauguration of 'West Virginia' as a State. We put the account on record for the present amusement of our readers, but more especially for the future instruction of prosecuting attorneys, when they come to try some of these smart fellows for treason." The new West Virginia constitution freed all slaves born after July 4, 1863, and freed all others on their twenty-fifth birthdays. Belle would no longer own Eliza, but the former slave would always remain loyal to her; decades later, the headline for her obituary read: "Was Property of Belle Boyd."

As vexing as Belle found the concept of "West Virginia," she had more immediate concerns. Her mother, now thirty-seven, was about to give birth to another child, and her father was once again on sick leave from the army, having suffered from exhaustion since Stonewall's final campaign. General Lee's Gettysburg Campaign had ended with staggering Confederate losses, and cars traveling southward along the newly repaired Baltimore & Ohio Railroad carried thousands of wounded men. All private homes, including Belle's, served as makeshift hospitals, with the men on cots scattered throughout parlors and studies and hallways. Whenever Belle wasn't tending to her parents she acted as nurse, boiling linens, changing bandages, and offering sips from her private stash of whiskey.

Fear of another arrest clawed at the back of Belle's mind. The Confederates had once again retreated from Martinsburg, leaving it vulnerable to Union occupation. Even if she hadn't gone to Philadelphia, West Virginia was now a Union state, and a visit home was a clear violation of her parole. She hoped, amid the aftermath of the battle, that she could remain inconspicuous, but certain adversaries in the Shenandoah Valley had snidely noted her return. "'Tis said Belle Boyd is in town tonight," wrote Lucy Buck. "What next?"

Belle's baby sister was just three days old when, as she sat in her mother's room, she heard Eliza exclaim, "Oh, here comes de Yankees!" She crept to the window, peering out with one cautious eye. An entire brigade looped around the front of her home. There were the familiar hard raps on the door; she dreaded descending the stairs. Major Nathan Goff, future West Virginia congressman, secretary of the US navy, and Federal judge, stood in her drawing room, on the very spot where, two years earlier, she'd shot the Yankee soldier dead. Belle took note of his blandly earnest expression and arranged her features to match it.

"Miss Boyd," he said, "General Kelly commanded me to call and see if you really had remained home, such a report having reached headquarters. But he did not credit it, so I have come to ascertain the truth."

She tilted her head and replied, "Major Goff, what is there so peculiarly strange in my remaining in my own home with my parents?"

Goff looked stumped. "But do you not think it rather dangerous? Are you then not really afraid of being arrested?"

"Oh no," Belle gasped, "for I don't know why they should do so. I am no criminal!"

"Yes, true," Goff conceded, "but you are a rebel and will do more harm to our cause than half the men could do."

"But there are other rebels besides myself."

"Yes, but then not so dangerous as yourself."

Belle took the words as a compliment and couldn't bring herself to challenge them. They stared wordlessly at each other, Belle's baby sister wailing in the background. "Good morning," the major said finally, tipping his hat. He showed himself out and led his brigade in a slow march down South Queen Street.

For the next few days the view outside the Boyd home was free of Yankee officers, and Belle hoped she would not be bothered again. Her mother had fallen ill, unable to sit up or eat solid food, her only nourishment broth sipped through a china feeding cup. Her midsection was tightly swathed in a damp warm towel, a practice known as "abdominal binding" that would mold her figure back into shape. Belle spent hours massaging her mother's stomach, trying to rid her body of any remaining placenta or blood. The July heat smothered the room. Belle was terrified that Mary might have puerperal fever, a bacterial infection for which there was no cure, often caused by attending doctors who didn't bother to wash their hands. As one prominent obstetrician put it, "Doctors are gentlemen, and gentlemen's hands are clean."

But one of Captain Goff's men soon returned, bearing an order for Belle's arrest. Benjamin Boyd met him in the drawing room. His wife was poorly, he explained, and if they took her daughter now, it might literally kill her. Would they consider letting Belle stay at home, at least until his wife's condition wasn't so dire? The officer acquiesced, stationing a chain of guards around her house, and Belle made the most of her confinement. At night she kept her window open to listen to their conversations, and during the day she chatted with them on the porch, asking about the bloody draft riots in New York, her eyes peering above the wooden slats of her fan. One soldier proved particularly attentive until his comrade came along and said, with an air of prophetic warning, "Know who that is? Why, that's Belle Boyd. First thing you know she will chloroform you; better keep a watch out for her." Another feared she might "send a dagger through their hearts."

Her reprieve was short-lived. A few weeks later Captain Horace Kellogg of the 123rd Ohio rang the bell of the Boyd home, announcing that Secretary of War Edwin Stanton wanted Belle in Washington by eleven o'clock the following morning. At the news Mary Boyd suffered a nervous relapse, surely thinking of her nephew, William Boyd Compton, who four months earlier had been captured by Union officials and sentenced to be hanged.

Belle wiped her mother's brow and told her not to worry; she could take care of herself. A year before, her prison sentence had seemed romantic and brave, but the prospect now struck her as "dreary." Her mother was ill, her father was injured, her little siblings were frightened and helpless. She had no desire to burnish or expand her reputation as the Secesh Cleopatra, at least not at the moment.

Her father accompanied her on the train, and for a hundred miles she wept into the sleeve of his muslin shirt, fearing she would never see her family again. At the prison gates he leaned to kiss the crown of her head and told her good-bye.

Inside Carroll Prison, a row of town houses alongside the Old Capitol Prison, guards led her past cells crammed with rebels, prisoners of state, hostages, blockade runners, smugglers, spies, and numerous Federal officers convicted of defrauding the government. She was given what was colloquially known as the "room for distinguished guests," a small honor that failed to elevate her mood. Despite the lofty moniker, her cell resembled all the others: ten by twelve feet, its single window streaked with grime. A broken chair and splintered table perched at the foot of a rusty iron bed. The table held a chipped washbowl and pitcher set, and she checked her reflection in jagged shards of mirror set inside a wooden frame.

She took an instant dislike to one of her guards, perceiving an insult behind his every word and expression. In retaliation she loosened a brick from her windowsill and hid it beneath her dress, waiting. The next time he paced beneath her cell, she pushed her hands through the bars and aimed it squarely at his head.

A few days into her imprisonment Belle got new neighbors, including a "Miss Ida P.," charged with being a rebel mail carrier. Superintendent Wood, hoping Belle and Ida might discuss methods, routes, and members of the Confederate mail service, allowed them to visit each other's cells, always under the close watch of Private Lyons Wakeman of the 153rd New York. When Belle blew kisses to the blue-eyed, five-foot-tall soldier she was unwittingly flirting with a woman: twenty-year-old Sarah Rosetta Wakeman, who had left her home in upstate New York a year earlier and reinvented herself as a man. She signed many of her letters home "Rosetta," confident that her true identity would remain secret as long as she needed it to be.

A fter fleeing her camp in Kentucky, Emma took the Chesapeake and Ohio Railroad to a random destination: Oberlin, Ohio, where she checked into a boardinghouse, ridden with chills and fever, still dressed and living as Frank Thompson. For the first time Frank felt like a disguise, an identity she hid inside rather than owned, and she needed to summon the energy to shed him for good. Over the next few weeks, as her fever abated, she did, boarding a train for Washington, DC, where she traded her frock coat and trousers for a corset and petticoats and let her curls grow out.

"I never for a moment considered myself a deserter," she wrote. "I left because I could hold out no longer"—words that applied both to leaving the army and to leaving her old identity behind. She took solace in the knowledge that she retained the best of Frank Thompson: his daring, his cunning, his seamless adaptability—the parts that had been hers all along. How strange, she thought, that she'd allowed two men, a species she once considered the enemy, to know both sides of her equally, and even stranger that they'd kept her revelations to themselves.

She had no idea what happened at camp after she'd deserted, that James Reid, before leaving himself, had paid a visit to Jerome Robbins's tent. The men exchanged words about Emma, words that, either directly or obliquely, confirmed what Jerome didn't want to believe—Reid and Emma, *his* "Frank," were lovers. Enraged, Jerome reached for his diary: "Do you know I have learned another lesson in the great book of human nature?" He paused for a moment, and decided to protect his friend even as he disparaged her. "Frank," he continued, "has deserted for which I do not blame him. . . . He prepared me for his departure in part. . . . Yet he did not prepare me for his ingratitude and utter disregard for the finer sensibilities of others. Of all others whom I trusted as friends he was the last I deemed capable of the petty baseness which was betrayed by his friend R at the last moment."

He felt powerless to record the details, the tangents of his spinning thoughts. He had concealed Emma's identity for nearly three years, proved his willingness to both see her for what she was and look discreetly away. Had she ever acknowledged his sacrifice? Did she realize the perfidy of her actions, the hypocrisy of her prayers? Jerome had been complicit, a blind and eager fool, and now the burden of her secret wasn't his alone. Everyone had heard about the confrontation between him and James and now his comrades were gossiping in their tents, snickering around their fires, mocking a confidence for which he had risked his life. "We are having quite a time at the expense of our brigade postmaster," wrote William Boston, a soldier with the 20th Michigan. "He turns out to be a girl, and has deserted when his lover, Inspector Read [*sic*], and General Poe, resigned. She went by the name Frank and was a pretty girl."

Emma Edmondson, as herself.

In Washington, Emma took a room at a boardinghouse, paying rent with what was left of her soldier's salary. She delighted in having female friends once again and planned to become a missionary. She dropped the last syllable of her family surname and sat for photographs, her first as a woman in five years. She thought often of James Reid, recalling in particular his words that Jerome had been her "only friend" in camp. On one Sunday afternoon, after church, she sat down in the parlor and composed a letter to her friend:

> *Dear Jerome,*
>
> *This is Sabbath afternoon . . . anxious to drop you a line before going to visit some of the hospitals where so many of Hooker's poor wounded soldiers are. . . . I intend to start on Wednesday for New York City. . . . I dare not write you the particulars of anything more until I hear from you and know where you are for fear it might fall into other hands. . . .*
>
> *Oh Jerome, <u>I do miss you so much.</u> There is no person living whose presence would be so agreeable to me this afternoon as yours.*
>
> *How is "Anna"? I hope you have received a favorable answer to your letter sent just before I came away. I always remember you, and sometimes her, at the Throne of Grace. May God bless you both and make you faithful to him and to each other. . . . I will write you from New York as soon as I make arrangements there. . . .*
>
> > *Goodbye my dear boy.*
> >
> > *E. Edmonds*

Jerome took a brief furlough home to Michigan to see Anna and then rejoined the army in eastern Tennessee, where he received Emma's letter. He was still recovering from his trip, during which Anna had told him she intended to marry someone else. To his surprise Emma's words somehow cheered him, distracting him from his heartache, and once again he opened his diary. "I received a letter from my friend 'Emma,'" he wrote, "whose history . . . is stranger than fiction." He forgave her, finally, for sharing it with him.

No One Ignorant of the Danger

RICHMOND AND EN ROUTE TO EUROPE

Rose met with Jefferson Davis often during her year in Richmond, taking a carriage from her hotel to the Confederate White House, bringing Little Rose to play with the president's daughter, Maggie. Normally Davis preferred to discuss the war only in written correspondence and steered visitors toward more frivolous banter. Over wine and juleps, he would reminisce about his days on the frontier, about meeting Andrew Jackson as a boy, about planting cotton, about anything he did when he was younger and free from this awful responsibility. During dinner with a Confederate colonel he mused for hours about "the instincts of cows, and calves, and horses and dogs and sheep and pigs." The colonel politely excused himself around ten o'clock, fearing that by midnight the conversation would have evolved into a monologue on cats.

But that summer's events forced Davis to broach weightier topics. During the Battle of Gettysburg the Confederate army lost twenty-three thousand men, nearly a third of its numbers killed, wounded, or missing. Lee fell back toward Virginia with his caravan of casualties stretching fourteen miles. The simultaneous fall of Vicksburg was equally disastrous, yielding control of the Mississippi River to Union forces. Southern newspapers and politicians criticized Lee's performance in Pennsylvania, calling the campaign "foolish and disastrous" and opining about his "utter want of generalship." The losses also affected the Confederacy's position abroad, as Davis and Lee had long dreamed of earning international recognition for their country through prowess on the battlefield. With each defeat that hope diminished, but the president had a last-ditch idea: he would send Rose Greenhow as his emissary to court the French and British elite, in the hope she might rally support for the Confederacy.

It would be an unconventional, even unprecedented, move. The wives of US officials had occasionally joined their husbands on foreign assignments, but never before had an American president sent a woman abroad to represent her government. Davis considered Rose ideally suited for the job: she was, as his secretary of the navy said, "a clever woman . . . equally at home with Ministers of state or their doorkeepers, with leaders and the led," and, just as important, "very slender—beautiful." She was articulate, fairly fluent in French, and as imperious as any royal. And her imprisonment in a Union "Bastille" had stirred sympathy among Europe's leading citizens, who considered it an example of Yankee vulgarity and oppression.

Davis, Rose wrote, "affords me every facility in carrying out my mischief."

In her room at the Ballard House, Rose set her travel trunk atop her bed and filled it with necessities: pantalettes, corsets, hoop slips, crinolines, button-up boots, white kid gloves that barely hid the bones of her wrists. She had lost all her finery during her imprisonment and planned to shop in London and Paris, trading her mourning attire for silk ball gowns with Grecian corsages and double puff sleeves and cascading, knife-pleated skirts. She packed Little Rose's favorite papier-mâché doll and her best gingham and wool dresses, intending to enroll her daughter in a Paris convent. She tucked a fresh leather journal in alongside her completed memoir, *My Imprisonment and the First Year of Abolition Rule,* and filled her gold-plated case with calling cards. She mustn't forget two items for James Mason, the Confederate envoy in London: a box of fine Southern tobacco and a letter from Jefferson Davis, explaining why his services were no longer necessary in England.

Rose knew the mission would be difficult, despite her considerable and deftly honed charms. Mason and his counterpart in Paris, John Slidell, had been lobbying for the Confederacy overseas since the beginning of 1862, hoping that the dearth of Southern cotton would compel Europe to confront the United States over its blockade and sympathize with the Confederacy. Although England and France fretted over the blockade they found ways to circumvent it, buying "white gold" from India, Egypt, and Brazil. Most of all, neither country wanted to support a loser, and they had both decided to wait and see whether the South proved its might on the battlefield (the bloody defeats at Antietam and, more recently, Gettysburg seemed to answer that question, at least for the moment). But Lincoln's Emancipation Proclamation constituted Rose's greatest challenge. The war was no longer about opposing ideologies or the oppressive Yankee government, but about ending slavery. As soon as word of the proclamation reached England, the masses held hundreds of meetings across the country to hail Lincoln's action, a reaction that dissuaded the British cabinet from making any radical move benefiting the Confederacy.

To pay for her trip, the Confederate government gave her 540 bales of cotton to be converted for a premium price in England. Before leaving Richmond she met with General Lee, who despite the recent losses impressed her with his calm, confident tone. She originally planned to go straight to Wilmington, North Carolina, where she and Little Rose would catch a blockade runner leaving for Bermuda, but instead took an impromptu detour to Charleston, currently under siege. She did not even stop to wash her face. Traveling south along the coast, she passed train cars loaded with cotton and even carriages and horses, all heading inland away from the surging Union troops, many of them regiments composed entirely of Negroes—a sight that made her recoil. Porters passed handbills warning about the imminent peril of the city and calling for three thousand slaves to work on the defenses. Nearing the city she heard the leaden boom of the guns.

She stayed in Charleston long enough to meet with her old contact General Beauregard, who asked her to pass on a message to the president. Rose complied, sketching a diagram of rebel forces and composing a seventeen-page letter to Davis that underscored the gravity of the situation: Beauregard needed heavy guns and mortars to combat Union troops around Morris Island in the harbor; without them, Charleston would fall. "Be assured that every body is wide awake just now," she wrote, "and no one ignorant of the danger to the Palmetto City." She promised to send another update soon.

Davis read the letter at his desk in his home office. He had been working there exclusively lately, his health so poor that rumors in the War Department countenanced even the possibility of his death. Much of the cause may have been anxiety, fed both by his army's precarious situation and by the suspicion that there was a mole in his midst, someone relating Confederate concerns and plans to the enemy as soon as they materialized. One month earlier he had expressed this fear to Confederate brigadier general G. J. Rains, insisting that "no printed paper could be kept secret."

Outside, just behind the door, Mary Jane gathered the children's toys to return them to the nursery, making sure to leave nothing behind.

On August 5, Rose and Little Rose set sail from Wilmington on the *Phantom*, a sharp, slim needle of a steamship, 193 feet long and only 22 feet wide, painted light gray to render her invisible on moonless nights. Every element of her design maximized her ability to slip past the Union blockade. She had two engines for speed and two hinged masts that could be dropped on deck to lower her profile. Her seven-foot draft allowed her to coast over the shallow bar at New Inlet, which presented a problem for bulkier Union warships.

She burned hard, smokeless anthracite coal. Sailors wrapped her paddles with tarps to muffle the sound of the churning water. No one was permitted to smoke or speak on her deck. She normally didn't take passengers, but Rose, on official government business, was an exception. Captain Stephen Porter, a furloughed naval officer, escorted her and Little Rose onboard and oversaw the loading of Rose's bales of cotton, weighing from 500 to 700 pounds apiece. He had nine hours until the sky was wholly black and the tide high; then he could make a run for it. It was still light enough to glimpse the Yankee ships offshore, nosed in their direction, and many more were out there, drifting unseen.

With pocket watch in hand, the captain counted the minutes. Two other blockade runners, the *Pet* and the *Hebe*, pushed off first. "There they go ahead of me," he said, and Rose sighed at his "rather piteous" tone. "Oh never mind," she snapped, "wait 'til the full tide is in." She hid her own anxiety beneath a steadfastly serene expression, and busied herself putting Little Rose to bed in their stateroom. The girl was exhausted from the adventure and nearly asleep, the rise and fall of her chest matching the soft sways of the ship.

And then the *Phantom* was off, whooshing past the *Pet* and the *Hebe*, slinking through a shallow channel, picking her way toward New Inlet. At half past midnight a Union blockader spotted her and launched a signal rocket, the descending flare's imprint lingering on spots of sky. Four other Yankee ships took note and swarmed in, guns manned, firing rocket after rocket to mark the *Phantom*'s course.

Rose watched from the deck, gripping the railing, narrowing her body as if it were one with the ship. The crew fueled the *Phantom*'s furnace with sides of bacon and turpentine-soaked cotton. They picked up speed, reaching sixteen knots, tearing through the V of blockaders, pushing and pushing and extending the distance until the enemy was no longer in sight. As soon as Rose was alone again, far out in the rocking sea, she ran to her room and vomited. When she recovered she penned an entry in her journal: "How my blood boils to think that even in the wide ocean we are not free from the despicable foe who seeks by numerical strength to crush us. . . . Yet God is just, so can this thing continue?"

Little Rose awakened and insisted on accompanying her back to the deck. The crew spread a mattress between the cotton bales. Rose lowered herself onto it, fixing her eyes on the waning moon. Little Rose wobbled over, hands clasped against her temples. "Mama," she cried, "I can't stand to have my head bumped from side to side." Rose locked the girl in the crook of her arm and twined her fingers in the bagging of the cotton bales, wishing for everything to be quiet and still. She kept a bucket by her side for the entire trip.

By the end of their fourth day at sea they reached Bermuda, a British colony and neutral port. Confederate agents were allowed to transfer goods from blockade runners to larger ships bound for England, and Union vessels were forbidden to fire at rebel ships in Bermuda waters. Captain Porter escorted Rose and Little Rose to Mrs. Haywood's, the lone lodging house on the island, where most guests were fellow Confederate expats waiting out the war. They all assembled in the parlor to welcome Rose and thank her for her service and sacrifice. She heard church bells ringing and considered attending Mass, but, feeling another wave of nausea, instead went straight to her room.

During the next few weeks Rose took her daughter for long walks, admiring the whitewashed homes and riotous gardens, excursions marred only by encounters with the island's black population. "The negroes are lazy, vicious, and insubordinate," she complained to her journal. "The negro in this as other English Islands were emancipated in 1834, since which time the material wealth and prosperity of the Island has been constantly diminishing. . . . Few negro women reach the age of puberty without becoming mothers and it is an established etiquette that the first child shall be white in order to make indisputable the claim of the mother to good society. After this it does not damage her position should her succeeding progeny be black." She dined often with Major Norman S. Walker, the Confederate agent in charge of military shipments, and discussed how he might send her coded dispatches from Richmond.

They left for England at the end of August, arriving in mid-September, Rose having clung to her bucket all the way. Her first stop was Liverpool, where she planned to sell her cotton, and then it was on to London to meet various politicians and dignitaries. Despite the recent disappointments back home, she was confident about her chances. In all of her fifty years so few had dared to disagree with her, and the ones who had seldom failed to regret it.

On a September afternoon back in Richmond, one of Elizabeth's servants opened the door of the mansion to admit a pair of Confederate officers into the parlor. They identified themselves as detectives for General John Winder, provost marshal of Richmond, and requested Elizabeth's permission to search the premises. Her brain whirred and clicked, deciding on a strategy. She smiled, inviting them inside as if they were old friends. Of course, she said, she knew General Winder well—what a decent and kind man, a true Christian, who had facilitated her efforts to aid the prisoners.

If she mentioned the prisoners directly, perhaps the officers would be less likely to suspect that she was, at that very moment, hiding several Union escapees inside her secret room.

She turned, motioning for the men to follow her, and saw her nieces at the foot of the stairs. A year had passed since her brother John had taken the girls away from their mother, and they hadn't seen or heard from her since, a silence both comforting and disquieting. John was in the North now, delivering dispatches about Confederate strategy after the profound disappointment at Gettysburg. When there were disruptions on the RF&P railroad, Union raiding parties tearing up the lines, he made the hundred-mile journey on horseback.

"These good men are tired of the war and would like to see our house," she said, directing her words at Annie. "Please take your sister to your room and play until I bring you cookies."

Annie took Eliza by the hand and led her out of sight.

Another Van Lew servant named Hannah appeared, and Elizabeth asked for some applesauce cookies and sweetened tea for the soldiers. Hannah returned to the parlor with tea and said she would start the cookies right away.

Elizabeth suggested they begin their tour; there were fourteen rooms and the detectives were welcome to all of them. They strolled through the dining room and the parlor and descended into the basement kitchen, where Hannah was mixing sugar and eggs, keeping her eyes on the bowl. The detectives peered into cupboards, the fireplace, the windows granting a vista of the cascading back gardens, sloping down to the bank of the James. They lingered in the study, perusing newspapers stacked on the desk, taking books from shelves, flipping through pages. Elizabeth watched silently from the doorway, praying that the detectives would miss the pinpricked letters, that she and John hadn't left any evidence behind.

She was as annoyed by the intrusion as she was afraid of its possible consequences. Her brother often returned from his trips north bearing gifts for Winder: angle braces, metal strapping, carriage bolts, screws, nails, hinges—items useful in repairing battered freight wagons for the rebel army. This search seemed a violation of her unspoken deal with the Confederate general: she would continue to supply him with such goods (along with the occasional basket of fresh vegetables from her farm), and he would refrain from prying too deeply into her life.

The detectives sipped their tea, taking their time; they hadn't enjoyed many sweets since the war began. She glanced at the mantel. The second hand on the clock moved with agonizing lethargy. The silk of her chemise clung to her damp back.

They came toward her and she stepped backward, letting them out of the study. She exhaled silently behind their backs.

Two more stories to go.

Elizabeth walked ahead of the detectives up the stairs, her breath thick in her throat. The men's boots landed heavily on the wood. They reached the top of the stairs and started down the hall, working as a unit, pushing aside dresses in armoires, checking under walnut beds. They peeked into Annie's room and returned her cheerful wave. At the end of the hall another set of stairs led to the top floor, and at the end of that hall, behind the coat of whitewash and the splintering old dresser, was the door to the secret room.

Four Union soldiers huddled inside there now, having escaped from the prison on Belle Isle, where inmates were sheltered only by tents, exposed to the elements. One prisoner reported that fifteen to twenty-five men died there every day and were buried on the grounds, tucked inside thin sheets of canvas instead of coffins. After Elizabeth secured Confederate uniforms for the men, and when she deemed it safe, she would lead them to a wagon and tell them to lie on the bed, facedown. A servant would cover them with a tarp and pile horse manure on top, just enough to make a credible load. The men would breathe through strategically placed holes in the bottom, settling in for the ride to the Van Lew farm.

From there they would make their way to Fort Monroe and relate everything they'd seen and heard to Union authorities: enemy troops were heading toward Charleston; the fortifications at Petersburg were formidable; there was a "most exciting" rumor that Lee was deserting the entire state of Virginia to reinforce Confederate troops in Tennessee. She knew that Northern papers would report the escapees' arrival, including the detail that they came disguised in rebel uniforms, and that General Winder would want to know how they had gotten them.

She and the detectives were nearly at the end of the eighty-foot hall, at the entrance to the secret room. She hoped that the prisoners would hear snippets of the detectives' banter, that they would recognize there were three sets of approaching footsteps instead of just one. She willed them to stay quiet behind the sliding slab of a door.

The detectives paused, standing in the window's dusty light.

La Belle Rebelle

WASHINGTON, DC

One night in late September, around nine o'clock, Belle sat at the window of her cell inside Carroll Prison, singing "Take Me Back to My Own Sunny South," watching the Yankee passersby stop and gape upward. Two months into her second stint in jail, and she was undeniably, inescapably bored. Her rebel companion, Ida P., had been released after vowing to do no more harm against the Northern government. Prolonged flirtations with various prisoners and guards had yielded neither true love nor actionable information. And an illness was creeping upon her, slogging its way through her system. She trilled the final refrain of her song, rested her head against the bars, and closed her eyes, contemplating whether to sleep or cry. Before she could decide an object whizzed past her head, splitting the air beside her ear, thwacking the opposite wall.

The Yankees were trying to kill her.

She felt her way toward the gas and turned it on, following the pale strip of light. On the floor lay not a bullet but an arrow, an envelope affixed to its end. She tore it open and read the letter inside:

> *Poor girl! You have the deepest sympathy of all the best community in Washington City, and there are many who would lay down their lives for you, but they are powerless to act or aid you at present. <u>You have many very warm friends,</u> and we daily watch the journals to see if*

there is any news of you. If you will listen attentively to the instruc-
tions that I give you, you will be able to correspond with and hear
from your friends outside.

On Thursdays and Saturdays, in the evening, just after twi-
light, I will come into the square opposite the prison. When you
hear someone whistling "'Twas within a mile of Edinbro' town,"
if alone and all is safe, lower the gas as a signal and leave the
window. I will then shoot an arrow into your room, as I have done
this evening, with a letter attached. Do not be alarmed, as I am a
good shot. . . .

Do not be afraid. I am really your friend.
C.H.

At first Belle dismissed the note, certain it was a Yankee hoping to collect evidence against her, but caution yielded to curiosity. She bribed a sentinel, giving him oranges and apples in exchange for an india-rubber ball, which she split into halves. On the appointed nights, after hearing the lyrics "and each shepherd woo'd his dear," Belle waited for the arrow and hurled the ball, her missive hidden inside.

Her mysterious pen pal's letters focused on politics rather than passion, offering updates about the Union army. Her old consort, Union general Nathaniel Banks, was attempting to gain a foothold in Texas, and General Grant left Louisville for Chattanooga, Tennessee, where his officers were starving under a Confederate siege. Occasionally a miniature Confederate flag came attached to the arrow, little tokens from the secessionist ladies of Washington. She slipped one into her bodice and pressed herself against the bars of her cell, waiting for the sentries to notice. One did, right away, and reached out to snatch it. Belle stopped him by raising a revolver to his temple.

"The discomfited Yankee," one witness reported, "suddenly let go of his rifle and stood, with eyes staring and mouth open, the embodiment of a person who had seen a supernatural apparition."

Belle cleared her throat until she had the attention of every prisoner on her floor, and declared, "These are the kind of men Lee whips."

She couldn't shake her illness. Her stomach curdled, her nose bled, a crop of red spots bloomed on her skin. Her friend C.H.'s missives went unanswered. Another prisoner, a kindly old Confederate doctor, made the diagnosis: typhoid fever. Her mother tried to have her transferred to a hospital but was rebuffed by Secretary of War Stanton. "She is a damn rebel," he reportedly responded. "Let her die there!" A Negro nurse sat on the edge of her bed, patting her forehead. She longed for her own Negro maid, Eliza, back home in the Valley.

After three weeks the fever passed, and she felt well enough to write to General John Martindale, commander of the forces in Washington, requesting permission to walk daily in Capitol Square. To Belle's surprise he agreed, on one condition: she must vow that, on her word of honor as a lady, she would communicate with no one, either by letter or by word of mouth. Her every step would be tracked by a corporal and a guard with loaded muskets. This precaution seemed warranted in light of recent newspaper reports stating she had "several official lovers in Washington."

Belle agreed, so desperate for a breath of pure air that she intended to keep her promise. But she couldn't prevent throngs of Southern sympathizers, both ladies and gentlemen, from lining up in the square to watch her stroll past. During one walk a group of teenage girls, scarcely younger than Belle herself, laid a gift at her feet: a square piece of bristol board bearing a Confederate flag and her name, both brightly stitched in worsted. The Union corporal ordered the girls to leave the square immediately and confiscated the flag. Belle's walking privileges were revoked, but the guard let her keep the present in exchange for five Federal dollars.

She cradled it in her lap and traced the letters of her name, marveling at how many people knew it; the French had even bestowed their own title on her, "La Belle Rebelle." She no longer belonged wholly to herself, a realization she promptly recorded for posterity. "Had I been a queen, or a reigning princess," she wrote, "my every movement could not have been more faithfully chronicled at this period of my imprisonment. My health was bulletined for the gratification of the public; and if I walked or was indisposed, it was announced after the most approved fashion by the newspapers. Thus, from the force of my circumstances, and not through any desire of my own, I became a celebrity." She encouraged rumors that she had been sentenced to death, a fate reversed only by President Lincoln's clemency.

Without her daily walks Belle's fever returned, or at least she reported as much to Superintendent Wood. She fainted at the slightest provocation, landing prettily, one open hand to her brow. Her father, still ill and on leave from the Confederate army, traveled to Washington to negotiate her release. Belle wasn't allowed to see him even after she learned he'd saved her: the Yankee government agreed to commute her sentence and banish her to the South. She would leave for Fort Monroe on the first day of December, and from there board an exchange boat for Richmond.

She began preparing for her release, sending word through the prison that she needed certain items from the outside. A friendly sentry, made friendlier by a generous offer of Federal greenbacks, delivered the goods: a uniform meant for a certain Confederate major general (whom Belle declined to name); a few pairs of army gauntlets and felt hats; $20,000 in Confederate notes, $5,000 in Federal dollars, and nearly $1,000 in gold; several letters of introduction to Confederate officials in Richmond; and her most prized possession, a pair of field glasses she claimed had belonged to Stonewall Jackson. She packed away the clothing and glasses but hid the other items on her body, tying the bills to her hoops and sliding the letters into her drawers.

Before Belle left Washington she begged to see her father, still sick and staying nearby with a niece, and too weak to come to her. Her request was refused, and by the time she departed she was in a foul and petulant mood, exacerbated by the "annoying and ungentlemanly" conduct of Captain James Mix, her escort and a former bodyguard of Lincoln's, who had achieved some renown in the North by saving the president from an unruly horse. Upon her arrival in Fort Monroe twenty-four hours later her spirits had not improved, and in this mind-set she prepared to see Union general Benjamin Butler, the commander of the Department of Virginia and North Carolina. She was to be kept under Butler's watch until her departure for Richmond, and although she had never met the general, she already thoroughly detested him.

Belle was not alone in her enmity for the "Beast," as he was called throughout the Confederacy, and every Southerner could readily catalog his offenses. The first occurred just after the start of the war, when he acted without orders and seized Baltimore, imposing martial law on the city and quelling secessionist ardor in Maryland. For this the Union army promoted him to major general and sent him to Fort Monroe, where he defied the 1850 Fugitive Slave Act, refusing to return runaway slaves to their owners and considering them contraband of war, a policy subsequently adopted by other Union officials. But it was his brief tenure in New Orleans that most enraged Belle, especially his General Order No. 28—the "Woman Order," as Butler inelegantly called it—which avowed that any female who by word, gesture, or movement insulted or showed contempt for any officer of the United States should be arrested and treated as a prostitute.

Southerners of every rank and tier protested Butler's reign of tyranny. Jefferson Davis labeled the general a felon, sentenced him to death in absentia, and commented that Northerners were "the only people on earth who do not blush to think he wears the human form." A South Carolina man offered a $10,000 reward for Butler's capture or delivery, dead or alive. Merchants sold chamber pots emblazoned with his face (and what a face: mothers taught their toddler children to "Make like Butler" by distorting their features and crossing their eyes).

General Benjamin "Beast" Butler.

Belle waited to enter Butler's headquarters, hands on hips, one boot tapping. She considered not only the words she wished to say to the general but how she would later recall them; her memory of the encounter might be the only surviving one, and she needed perfect, precise lines. For inspiration she considered the infamous exchange between Butler and Eugenia Phillips, one of the spies in Rose Greenhow's ring, after Phillips was arrested and sent to New Orleans. "I expect to be killed before I leave the South, by either you or Mrs. Greenhow," Butler said. Eugenia let one eyebrow creep upward and replied, "We usually order our Negroes to kill our swine."

Belle stood at the doorway, waiting for one of Butler's small and muddy eyes to rise; the other was weighed down by a drooping lid.

"Ah, so this is Miss Boyd, the famous rebel spy," he said. His smile employed only one side of his mouth and revealed a crooked picket fence of teeth. "Please be seated."

"Thank you, General Butler, but I prefer to stand." She felt herself shaking—in anger, not in fear, she assured herself. The very sight of him—the sloping forehead, the brick of a neck, the intricate web of purplish veins etched beneath his skin—plucked a quiver in her chest.

"Pray be seated," he said again. "But why do you tremble so? Are you frightened?"

"No," Belle said, and then corrected herself, reaching for the words she'd so deftly arranged. "Ah! That is, yes, General Butler; I must acknowledge that I do feel frightened in the presence of a man of such world-wide reputation as yourself."

This pleased Butler, she could tell. He touched the tips of squat fingers together. "What do you mean," he asked, "when you say that I am widely known?"

Belle took a step closer.

"I mean, General Butler, that you are a man whose atrocious conduct and brutality, especially to Southern ladies, is so infamous that even the English Parliament commented upon it. I naturally feel alarmed at being in your presence."

His smile collapsed as Belle's widened. With glee she noted the "rage depicted on every lineament of his features," and was already halfway out the door when he ordered her to leave.

On Wednesday evening, December 2, officers arrived to escort Belle to the provost marshal's office in Norfolk, where she found her luggage, two Saratoga trunks and a bonnet box, waiting for inspection. A man and two female guards set to work ransacking their contents, sullying her private underpinnings with their grubby Yankee hands.

Belle assured them that such thorough searching was unnecessary, since she had just come from prison, but they continued digging, launching chemises and drawers into the air until they hit the bottom. They held each smuggled item aloft as if displaying a prize catch. Belle shrugged and said she had no idea how those things came into her possession. When the guards told her that General Butler would be the new owner of Stonewall Jackson's field glasses, she dropped her face in her hands and wept.

Three days later Belle boarded the Southern flag-of-truce vessel at City Point and began the twelve-hour journey up the James River to Richmond. In the Confederate capital she checked into the Spotswood Hotel, one of the city's finest, where a drunken rebel soldier once rode his horse into the lobby and through the bar—an escapade that reminded Belle of her younger self, charging into her parents' dinner party saddled on Fleeter, demanding to be included.

Belle had been in Richmond for just a week when she learned from a Confederate captain that her father had died. She secluded herself in her room, waiting for word from her mother. When the letter came, it said that Benjamin Boyd spoke often of his oldest daughter in his last, lucid moments and exclaimed that she was "being torn" from him, a lamentation as universal as it was private, sung and recognized by people on both sides of the lines.

WOMEN MAKE WAR UPON EACH OTHER

LONDON AND PARIS

After arriving in London, Rose accepted an invitation for tea with Thomas Carlyle, the eminent Scottish writer and historian whose essay "Occasional Discourse on the Nigger Question" assailed the notion of racial equality and argued that emancipation in the British West Indies had been an economic disaster. He viewed the American Civil War as an unnecessary conflict, dismissing it as "a smoky chimney which had taken fire," and she was counting on him to be an early ally in her mission.

From her boardinghouse in Mayfair she took a hansom cab to Chelsea, taking in the tidy, flag-studded streets and sun-stippled views of the Thames, the heavy smells of shipping and tar. Carlyle lived in a redbrick house at 5 Cheyne Row, thickly corniced and wainscoted to the ceiling, each crevice and shelf toppling with books. He was taller than Rose had expected, with a slight, sinewy figure and grizzled hair. He greeted her wearing a dressing gown and slippers, and his brogue seemed to soften with each sentence he spoke.

He raised his cup to his lips and asked, "What sort of looking animal is Lincoln?"

As she launched into a detailed description of "the beanpole," Carlyle rose abruptly, shook himself like a great dog, and declared "the flat nosed Negro of Haiti and Abraham Lincoln, the rail splitter of the United States, as a worthy pair to stand side by side in history."

Rose murmured her agreement. Carlyle, appeased, sat down again and asked for a description of Jefferson Davis. He clamped his fingers over his eyes. "I see him," he murmured. "God has made the situation for the man."

For hours they spoke about the "crimes and imbecility of the North" (in Rose's words), and at midnight he walked her to the door. They bowed toward each other.

"I will do anything for your country," he said.

"An article or few words from you, sir, will carry weight and would be deeply gratifying to our President and our people."

Carlyle promised to consider it.

One of Rose's most pressing concerns was finding a publisher for her memoir, which she was certain would rouse interest in the Southern cause. She met with Richard Bentley, "Publisher in Ordinary to Her Majesty," as he preferred to be called, who offered her a contract stipulating that she split the profits for all sales in Great Britain, on the Continent, and in the United States. Rose agreed, reserving for herself the right of publication within the Confederate states, and dedicated the book to "the brave soldiers who have fought and bled in this, our glorious struggle for freedom."

The book attracted widespread attention, in both Europe and America. The Southern-sympathizing *Standard* gleefully recounted Rose's unflattering portraits of Mary Todd Lincoln and observed, "Men may not war on women, but short of death—actual or social— women may, and habitually do, make war upon each other." The *Morning Post*, which supported the Confederacy even as it denounced slavery, spent most of its review critiquing Rose rather than her work: "She boasts that she made herself as obnoxious as any man to the masters of the situation by her acts; and her words, as she reports them, leave nothing to be desired as to the rancour of her tongue. She unsexes herself, and then abuses her captors for want of observance and consideration of her sex." Back in the United States, the *New York Times* said exactly what Rose had expected, calling the book "as bitter as a woman's hate can make it" and concluding that "many may wonder, not that she was treated with such severity, but that she got off so easily."

She paid equal attention to war reports from abroad, despairing over the news of the Confederates' defeat at Lookout Mountain in Chattanooga. Hoping that the Northern press had exaggerated the details—the capture of eight thousand rebel soldiers, in particular— she begged her friend, the Confederate politician Alexander Boteler, for a more impartial version. "All the accounts come through the Yankee press," she wrote. "The effect is most depressing. . . . My friend you know not the importance of sending correct information, which can be used so as to counteract the Yankee accounts. I believe that all classes here except the Abolitionists sympathize with us and are only held back from recognizing us for fear of war with the United States."

She closed by saying she wished she could write more freely but feared her letters would be intercepted. Perhaps he might use her name and get a cipher from Judah P. Benjamin, the Confederate secretary of state, so they could correspond without the burden of self-censorship. She promised to write from Paris, where she presumed the "educated and thinking" classes would be on their side.

R ose and Little Rose left London on December 12, taking the Dover, Chatham, & Kent Railway, passing the magnificent cathedral of Canterbury, watching the smoke and murk of the city recede in the distance. The train arrived in Dover in time for them to catch the 9:40 a.m. ferry to Calais, a ninety-minute crossing and the quickest Rose could find. She hoped to minimize her chances of seasickness—"the greatest evil," she wrote, "to which poor human nature can be exposed"—but found herself at once crouching on the deck, face in hands, the crew gathered around laughing at her, all of them fortunate enough to avoid paying "the same tribute to Neptune." It felt as if days had passed before Little Rose touched her shoulder. "O, Mama," she said in her sweet twang, "here is our flag." Rose raised her head and there it was: the Stars and Bars, flying from a sleek little ship bobbing along the shoreline. Her daughter danced from foot to foot, pulling at her hand to hurry her along.

Rose checked in at the Grand Hotel on the rue Scribe in central Paris and spent the next ten days sightseeing with Little Rose. They gaped at the Campana Collection in the Louvre and took carriage rides around the Bois de Boulogne, which had been the royal hunting forest before Napoléon III transformed it into a public park. On one excursion they spotted the emperor himself, skating on the Grand Lac with his wife, the Spanish-born empress Eugénie, clad entirely in black, from her petticoats to her cuffs to the heron's plume on her hat. "She fell 4 times," Rose noted. "Ill-natured people sayed [sic] it was to show her feet, which are pretty. . . . When she fell no white was visible."

On Christmas Eve she took Little Rose to visit the Pensionnat du Sacré-Coeur, a convent on rue de Varenne that operated a Catholic boarding school. Her daughter fell silent, hiding her face in the folds of Rose's skirts; there was no sign at all of the bold little rebel who had helped Rose smuggle information and talked back to prison guards. Rose tried to console her daughter: she'd meet students from all over the world, learn to speak French, receive her First Holy Communion. But Little Rose wept at the thought of being separated from her mother. "My little one is very shy," Rose wrote, "and she does not like the idea of being placed here. Nevertheless it is to be her destiny." She wished for her daughter the rigorous formal schooling she'd never had herself.

With that settled, Rose resumed her business, meeting with Confederate diplomats James Mason and John Slidell. An old Washington neighbor left his calling card at her hotel. She scanned the papers (the damned Yankee press, again) for developments back home, reading of Lincoln's Proclamation of Amnesty and Reconstruction, which pardoned those who "directly or by implication, participated in the existing rebellion" if they took an oath to the Union. He also issued his annual message to Congress, crowing that public opinion at home and abroad was favorable and that "the crisis which threatened to divide the friends of the Union is past." The *New York Times* reprinted gloomy excerpts from the *Richmond Whig*: "Why is the attitude of European Powers—England especially— now inimical to us? Because the superior diplomacy of the United States has made it appear that it is in their interest to be so."

She compiled a twelve-page letter to Jefferson Davis, acknowledging her "dismal" mood but assuring him she still believed she could "do some good." She would not leave Paris until she'd pleaded the Confederate cause to the emperor of France himself.

PLEASE GIVE US SOME OF YOUR BLOOD

RICHMOND

The rebel detectives stood outside the secret room, one with an elbow propped on the dresser that hid its entrance. Elizabeth listened to their meandering chatter, certain they were just testing her, waiting for the moment when she would push the dresser aside, throw open the door, and expose her deed, like the madman in "The Tell-Tale Heart." She did not breathe again until they began the long walk down the hall.

Downstairs, at her dining table, the detectives ate cookies and drank tea, as if they'd forgotten their main purpose for being there. A few days later the escapees, disguised as Confederates, would make it north via Fort Monroe, just as planned. Her niece Annie thought the whole episode great fun and told the detectives to come back soon, a request Elizabeth suspected they would heed.

One December afternoon a fifteen-year-old girl named Jose-phine Holmes smiled at the Confederate guards and entered the Libby prison hospital. She passed rows of men suffering from scurvy and diarrhea and fever, some of them dead in their beds. Any patient able to lift his head did so and watched her pass, enjoying the brief and rare sight of a pretty girl, her skirts swishing and braids looped up like lassos. She aimed herself toward the bed of one

John R. McCullough, an assistant surgeon for the 1st Wisconsin Infantry who had been captured in Tennessee and sent to the hospital for a minor illness. When she reached his side, she pressed a small bag between his hands. Inside he found a mound of fine Virginia tobacco and a cryptic note: "Would you be free? Then be prepared to act." Josephine—the daughter of a member of the Richmond Underground—returned the following day and whispered the plan: McCullough would escape by faking his own death.

He wouldn't be the first prisoner to attempt this ruse. Some "dead" escapees had allowed themselves to be stacked in piles of bodies near the gates and sprinted to freedom after dark. Others were loaded into wagons and hauled off to the burial ground, stealing away once the wagons rambled out of sight. In one infamous case, a prisoner who worked on burial detail was sent out on a job under the watch of several guards. When he dropped the first shovelful of dirt on the body, it yelped in protest—or so the guards, who all fled in horror, thought. The prisoner, who sprinted in the opposite direction, just happened to be a gifted ventriloquist.

A few hours after Josephine's visit McCullough suffered his sudden "death." Four of his friends smeared powder on his face to give it an ashen pallor and bound his hands and feet; that way, the guards could hold the ropes when they transported the body, and would be unlikely to notice that it still was warm and had a pulse. McCullough stayed in the dead house, from midday until dusk, lying as still as he could, ignoring the itch of powder in his eyes, the smell of flesh just beginning to turn. At the designated moment his comrades staged a sham fight to distract the guards, and he rose from among the truly dead and ran as fast as he could. One fellow prisoner, Captain Harry Howard, a scout for General Benjamin Butler, followed him to the outside, where they found Josephine waving a white handkerchief.

"Follow me at a distance," she instructed, and took them to a Union safe house.

They didn't have much time. Elizabeth coordinated with Samuel

Ruth, the Unionist railroad superintendent, who paid $3,000 in Confederate money for a pair of passes out of the city. Female members of the Richmond Underground scurried to the mansion to sew gray uniforms made of rebel blankets. While the women worked, heads bent over their needles and thread, Elizabeth retreated to her study to write a note to General Butler.

Mary Jane had sent reports about the Confederate army's desperation in the face of dwindling manpower. Rebel troops in both the eastern and the western theaters were abandoning their ranks in growing numbers. The assistant secretary of war estimated that from 50,000 to 100,000 Southern soldiers were absent without leave after the fall of Vicksburg and the defeat at Gettysburg. General Lee wrote long missives to Jefferson Davis, suggesting ways to increase the army's strength, including revoking draft exemptions. Elizabeth worried that her brother John would be conscripted into the Confederate army, a situation that would put both his life and the Richmond Underground at risk.

At the moment she forced herself to be stoic and practical, focusing not on John's personal danger but on making alternative arrangements for the Richmond Underground; she had to establish another route to deliver dispatches North. She crafted a letter to Butler, explaining, as obliquely as her direct nature would allow, her desire to help the Union. The general should know that there were many of them, dozens, hundreds—even she couldn't recite all of their names or place all of their faces—living in the Confederate capital, citizens who had become traitors with the firing of shots and the stroke of a pen. He should not take her word for it, she stressed, but listen to the testimony of John McCullough, the escaped prisoner, the man bearing this letter. He should consult someone else, too, a mutual friend: Commander Charles O. Boutelle, a Massachusetts native and topographical engineer with the US Coast Survey Office in Washington. As an accompaniment to the letter she prepared a bouquet of geraniums and winter honeysuckle; surely even "The Beast" would appreciate something of beauty.

Ten days later, after rebel guards gave up their search, Mc-
Cullough and Howard, disguised as Confederates, emerged from
hiding and began hiking through the lines to the Potomac River,
heading for the provost marshal's office in Washington. The Southern press finally caught on. "The 'dead' Yankee has arrived in the
North, in company with a live Yankee," reported the *Richmond Examiner.* "The 'Union people' of Richmond, who made the suits of
clothing and obtained passports for McCullough and Howard, will
rejoice to learn of their successful journey." The article concluded
with a question that seemed aimed directly at Elizabeth: "By the
way, are not some of them known to the detectives?"

The Union's espionage operations had undergone significant change since the days of General McClellan and Allan
Pinkerton. The Secret Service under McClellan's successor, Ambrose Burnside, also failed to connect with local Unionists in rele-
vant towns and cities, a misstep that contributed to the disastrous
Federal defeat during the Fredericksburg campaign of the previous
winter. Burnside's replacement, Joseph Hooker, was the first Union
commander to realize the liabilities of the army's informal, decen-
tralized approach, whereby each general was left to his own de-
vices, expected to run his personal espionage operations as he chose
without any oversight or overlapping of information or sources (little
wonder, then, that Butler hadn't yet heard of Elizabeth or the Richmond Underground).

Grasping the advantage of having a centralized, cohesive authority, Hooker created a Bureau of Military Information in early
1863. Its chief, a young lawyer and colonel named George H.
Sharpe, merged Pinkerton's spying and interrogation functions
with reports gleaned from other types of intelligence gathering activities, such as cavalry reconnaissance, scouting, and interception
of Confederate flag messages. It was a groundbreaking achievement: for the first time, the army commander had as comprehen-

sive a picture of the enemy's situation as sources allowed. As General George Meade, Hooker's successor, prepared to meet rebel forces at Gettysburg, fourteen locally organized groups of citizen spies aided him.

When McCullough delivered Elizabeth's letter (the bouquet, sadly, had wilted en route), Butler envisioned tapping into a similar spy network in Richmond. He did as Elizabeth suggested and contacted their mutual friend.

"My Dear Boutelle," Butler wrote, "You will find enclosed a letter from a dear friend of yours in Richmond. I am informed by the bearer that Miss Van Lieu [*sic*] is a true Union woman as true as steel. She sent me a bouquet, so says the letter carrier." Butler explained that he needed a knowledgeable correspondent in Richmond, and wanted to confirm that Elizabeth would be up to the job. "I can place my first and only letter in her hands for her directions, but I also place the man's life in her hands who delivers the letter. Is it safe to do so?" He could entice Elizabeth with monetary rewards, Butler concluded, but "from what I hear of her I should prefer not to do it, as I think she would be actuated to do what she does by patriotic motives alone."

Boutelle told Butler to trust Elizabeth and proceed.

Back in Richmond, Elizabeth waited—for what, exactly, she wasn't sure: the delivery of a book with doctored sentences, the clandestine drop of a letter by her feet, or, if McCullough had been caught along the way, the knock of Confederate detectives at her door. The response came, finally, in the form of William S. Rowley, her most trusted spy, pulling his carriage up to the mansion, cool and composed even in his excitement. Alongside him sat Captain Howard, the scout and escaped prisoner, who had risked his life to cross the lines back into Richmond. Victory: Boutelle had vouched for her and Butler had listened.

The three of them huddled close in her study, the fire spitting at their backs. She read Butler's dispatch, lips moving noiselessly with each word:

My Dear Aunt:

I suppose you have been wondering why your nephew has not written before, but we have been uncertain whether we should be able to send a letter. The Yankees steal all the letters that have any money in them through Flag of Truce, so that we thought we would wait until we got a safe chance.

I am glad to write that Mary is a great deal better. Her cough has improved, and the doctor has some hope. Your niece Jennie sends love, and says she wishes you could come north, but I suppose that is impossible. Mother tells me to say that she has given up all hopes of meeting you, until we all meet in heaven.

Yours Affectionately, James ap Jones

From his pocket Captain Howard retrieved a vial of tannic acid, which he tipped and spread across the page. He motioned for Elizabeth to move it closer to the fire. The heat lapped at her fingertips. Silently she coaxed the alternative letters to appear, watching the strokes of ink stretch and curve and morph into an entirely different message:

My Dear Miss:

The doctor who came through and spoke to me said that you would be willing to aid the Union cause by furnishing me with information if I would devise a means. You can write through Flag of Truce, directed to James ap Jones, Norfolk, the letter being written as this is, and with the means furnished by the messenger who brings this. I cannot refrain from saying to you, although personally unknown, how much I am rejoiced to hear of the strong feelings for the Union which exists in your own breast and among some of the ladies of Richmond. I have the honor to be,

Very respectfully, Your obedient servant

Captain Howard produced another vial filled with what looked like water, an odorless and colorless liquid—"S.S. Fluid," he called it. She was to use it for all future correspondence. He dipped his finger into a bottle of milk and smeared it across the invisible fluid, demonstrating how it turned black. Next he pressed into her hand a small square piece of paper featuring six rows and six columns of letters, rimmed by a right angle of numbers—a cipher based on the Polybius checkerboard. She kept the key to the cipher in the case of her watch, which she wore at all times, even to sleep. The enemy would find the key only if she weren't alive to prevent it.

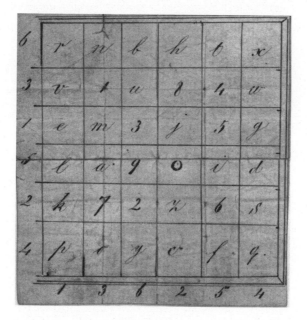

Elizabeth Van Lew's cipher.

Sometime after Butler made contact Elizabeth received another letter, this one delivered through the mail. The words lurched across the page in a deeply pressed and erratic script, as if scrawled by someone very young or very old. She filed it away, evidence for a crime yet to be committed: "They are coming at night. Look out! Look out! Look out! Your house is going. FIRE. Old Maid, is your house insured? Please give us some of your blood to write with."

It was signed, "White Caps," and concluded with a crude rendering of a skull and crossbones.

Fragments of a threatening note from the "White Caps" to Elizabeth Van Lew.

[PART FOUR]

1864

THAT UNHAPPY COUNTRY

PARIS

After a round of exchanges with Napoléon III's cabinet, Rose was invited to visit the Tuileries Palace, on the right bank of the Seine, at half past one on January 22, a Friday afternoon. She was led through a web of hallways opening into improbably grand spaces: reception parlors, a chapel, a theater, dining rooms covered in wall panels illustrated with hunting and mythological scenes, acres and acres of gilding. The last stop, after several flights of a spiral staircase, was the audience chamber, where the duc de Bassano, an imperial aide, greeted her and invited her to sit. "His majesty will receive you in just a few moments," he promised. She glanced up at the soaring dome ceiling and was briefly transported to the Senate chamber back home in Washington, where, exactly three years before, she'd listened to then senator Jefferson Davis reluctantly make the case for secession and bid his Northern colleagues a "final adieu."

The imperial aide returned and bowed low. "Entrez, Madame, dans le cabinet de l'Empereur," he said, and closed the door behind him.

Napoléon III stood like an ornament in the middle of the room, wearing a large diamond eagle in the clasp of his kerchief. He was fifty-five years old and known for equally vigorous political and sexual appetites. Nearly an inch shorter than his uncle, Napoléon

Bonaparte, he seemed too small for his striking face, dominated by thick lips and a mustache with bayonet-sharp tips. His gray eyes appeared expressionless and somehow veiled, as if his gaze were directed inward rather than to the outside world. It was rumored that he had his first love affair at the age of thirteen. As emperor he pursued an eclectic array of women, delegating the responsibility of arranging trysts to his beleaguered social secretary. Once his wife, Empress Eugénie, produced an heir, she reportedly refused to have sex with him again, finding the whole business "disgusting." He reminded Rose, in a way, of Confederate general Earl Van Dorn, a notorious lothario known as "the terror of ugly husbands," murdered the previous year by a man who claimed that the general had bedded his wife.

Rose curtsied deeply, holding the pose as Napoléon approached and extended his hand.

"Vous parlez français, madame?" he asked.

In nearly flawless French, which she'd learned from her late husband, Rose responded, "Non sire, je ne parle pas assez pour me faire comprendre, mais je sais que Votre Majesté parle parfaitement anglais [No sir, I do not speak well enough to make myself understood, but I know Your Majesty speaks English perfectly]."

He smiled, his mustache tips slicing upward, and held Rose's hand as he led her to a chair. He sat opposite her and leaned forward, intertwining long fingers.

"You are from the South," he said.

"Yes sir, from that unhappy country."

She wasted no time making her case, urging the question of recognition. The Confederacy was entitled to it, she argued, and the moral strength it would provide.

Napoléon was inclined to agree, for various reasons. He was a nationalist who sympathized with the aspirations of people for national self-determination. The Union blockade had disrupted the French economy, crashing the cotton textile industry and hindering the export of such luxury goods as wine, silks, clothes, and perfumes, and influential Parisian merchants were demanding that the government relieve their distress. Above all he wished to expand his empire overseas in Mexico, where he had troops stationed for two years, a plan the Lincoln administration vehemently opposed. And on a personal note, Napoléon shared Rose's view of the American first family as boorish and unsophisticated. In August 1861 his cousin and close adviser, Prince Napoléon Joseph Charles Paul Bonaparte, "Plon-Plon" for short, attended a state dinner at the White House, disdainfully noting the president's "large, hairy hands" and Mary Todd Lincoln's attempt to speak French in an atrocious Kentucky accent. "Mrs. Lincoln," the guest of honor reported, "was dressed in the French mode without any taste; she has the manner of a petit bourgeois and wears tin jewelry."

Now Napoléon hedged his bets, telling Rose he had hoped to support the South but could not do so alone. He had, in fact, made frequent overtures to England on the subject, which the country consistently "evaded or rejected." Please, he urged, assure President Davis of his sympathy and his untiring efforts.

Rose opened her mouth to respond, but the emperor wasn't finished.

"Tell the President that I have thoughts on his military plans," he added. "He has not concentrated enough. The Yankees have also made true blunders. If instead of throwing all your strength upon Vicksburg, you could have left that to its fate, and strengthened Lee so as to have taken Washington, the war would have ended. England would have been obliged to recognize you, as I should, of course."

Rose felt her throat pulse against her neck. It didn't matter that he was the emperor of France, or that she was a guest in the royal palace; *no one* would get away with criticizing the Confederacy in her presence.

"The President is fully convinced of the wisdom of such a movement," she said, the words hot in her mouth. "But there were grave political reasons for pursuing the course we have pursued in order to prevent the alienation of our own territory." She raised an eyebrow and lowered her voice, adding, "Besides, you can have no just conception of the war and of our military operations. The State of Virginia is as great as this mighty Empire, and to show disregard to any portion of the country would excite feelings injurious in a crisis like this."

Napoléon conceded the point and changed the subject, asking about General Lee, and she took the opportunity to boast: "Sir, he is worthy to be one of your marshals."

With that she was finished and stood to go. The emperor again reached for her hand.

"I wish you would remain in France," he said.

She allowed his hand to linger over hers and replied, "Even the attractions of your mighty capital cannot keep me."

"I know your history. The women of the South have excited the admiration of the world." They walked toward the door, still touching. "I wish you a prosperous voyage, and tell President Davis that my admiration and my sympathy are with him and his people."

She turned to face him and tried one last time. "Ah, Sire," she said, "I wish you would bid me tell him that you would recognize us as one."

"I wish to God I could. But I cannot do it without England. . . . But you may assure the president that I will make renewed efforts to serve him."

They said good-bye and Rose returned to her hotel, where she recorded the meeting in her diary—"So much for my interview with this ruler of the destiny of Europe"—and drafted letters to both Jefferson Davis and her friend Alexander Boteler. "I had the honor of an audience with the Emperor," she wrote. "The French people are brutal ignorant and depraved to a degree beyond description and have no appreciation of our struggle."

Despite Rose's disdain for the French she accepted every social invitation she received, taking particular delight in returning to the Tuileries Palace for a masked ball, where Napoléon III greeted her wearing a plain dark suit festooned with a broad red sash (a subdued choice of attire for the emperor, who usually dressed for such events as a seventeenth-century Venetian noble). She pinned red roses in her hair and jewels at her breast and felt, for the first time in nearly a decade, that her mourning had passed, that she had permission to separate her business from her pleasure, that not every glance or word or touch was a piece of currency waiting to be exchanged.

The evening was like a scene from *Arabian Nights*, which she had just seen with Little Rose at the Théâtre Impérial, with armed guards lining the grand stairs "like so many petrified steel clad warriors" and the salon a wild blur of pomp and color: sword-wielding gentlemen in embroidered court dress; diplomats swathed in sashes and medals; ladies in gowns with exaggerated volume only at the back thanks to a modified crinoline called the crinolette, the precursor to the bustle. At midnight sharp the doors to the dining room swung open, but only the foreign ambassadors were invited to dine with the royal family. One ambassador asked Rose to accompany him, and she quickly accepted, thrusting her arm through his. The duc de Morny, Napoléon's half-brother, engaged her right away. Rose deemed him "very like the emperor, only younger, handsomer, and not so intellectual looking."

She was less impressed with the women. Princess Mathilde, Napoléon's first cousin and former fiancée, wore a magnificent tiara of diamonds but was herself "fat and vulgar looking." Princess Clotilde, the wife of Prince Napoléon, Matilde's brother, was "thin with sharp features, turned up nose and very unnatural looking." The empress wore a dress of white tulle garnished with a smattering of brown velvet butterflies, and her thick neck was entirely obscured by looping rows of pearls. The entire room knew that her husband was currently having an affair with a dancer. "She is not at all pretty," Rose concluded, "nor distinguished in appearance." She stood next to the empress for as long as possible, a tacit invitation for everyone to compare.

After the ball Rose felt "no desire to enter further with the gaieties of Paris" and longed for news from home. A letter from a Confederate friend in London buoyed up her spirits slightly—"I think things are looking better for 'our' side," he wrote—but reports in the Yankee press suggested otherwise. The papers spoke of daily bread riots; food was so scarce in Richmond that every rat, mouse, and pigeon had deserted the city. A single hotel meal cost nine Confederate soldiers $600, and a common kitchen utensil cost $1,000. One woman sold her hair for $100 in order to buy four pounds of flour. The masses were turning against Jefferson Davis, suggesting that if he thought $25,000 and the presidential mansion weren't sufficient compensation, then he had best resign.

The military situation was equally troubling. Soldiers had to kill their best mules for sustenance. Davis was so desperate for new recruits that he urged passage of a law to employ free blacks and slaves in noncombatant duties normally performed by soldiers, so as to utilize every white man possible on the battlefield. Richmond was bound to the South only by a single narrow line of communication, and the city was in imminent danger. "As soon as General Butler has a sufficient force (and we know he is increasing it day by day)," read one report, "we trust he will do more than harass and threaten the rebels."

Rose still believed in her mission and in a Confederate triumph, but it was now a question, she wrote, "of hope deferred." She did not know how much time she had left.

DESPICABLE REMEDIES

RICHMOND

In early January both President Davis's manservant and Varina's personal maid, Betsy, fled to Washington and divulged sensitive information. "The condition of our servants began to be unsettled," Varina noted, suspecting that they were being paid to go North. Ten days later, during a reception at the Confederate White House, someone kindled a fire in a woodpile in the basement in an attempt to burn it down; smoke led to its discovery just in time. In the excitement two more slaves escaped and the mansion was robbed, all in the same night. Speculation abounded: it was the work of "Yankee plotters," of escaped Union prisoners, of the president's other servants, of all of the above.

Despite the increased risk Mary Jane continued to spy, being extra obsequious as she waited on the president and his visitors. She incorporated her reports into Varina's dresses and hung a red shirt on the line, signaling to Elizabeth that a delivery was waiting at the seamstress's home. Elizabeth gathered information from other sources, too, mainly her friend Charles Palmer, characterized by rebel officials as a "loud-talking, violent" opponent of secession, and William S. Rowley, who used the code name "Quaker." She interpreted, edited, and enciphered their notes, checking her Polybius square carefully, each letter represented by a two-digit number. Her latest dispatch contained 146 words in all, or approximately 652 letters, requiring 1,304 laborious strokes of her pen to convey everything on her mind.

At sundown on January 30 she handed the report to Merritt Rowley, "Quaker's" seventeen-year-old son, watching him tuck it inside the hollowed-out heel of his boot. The boy had stayed with her a week in preparation for his journey, and she treated him as if he were her own, arranging for a guide to deliver him safely to Northern lines and giving him a copper Civil War "token" that would identify him as a loyal Unionist. Such tokens had been in circulation since 1862, when people—panicked about the possibility of being driven from their homes with all of their worldly possessions—started hoarding "real" money, rendering commerce all but impossible. Private merchants, in both North and South, began minting these tokens in order to fill the void. Tokens fell into two major categories: "store cards," which advertised a business; and "patriotics," which featured nationalistic slogans or images—"Union Forever" and "Death to Traitors" were popular choices. The Bureau of Military Information procured a number of tokens featuring the same pro-Union design and smuggled them out to spies in Southern territory. In case of capture, they could maintain plausible deniability since nearly everyone had a few tokens in a pocket or a purse.

Elizabeth had one final piece of information to pass on, a warning as imperative as it was ironic. "Tell Butler," she said, "that all women ought to be kept from passing from Baltimore to Richmond. They do a great deal of harm." She knew of one who carried treasonous mail for the rebels, making the journey in a wagon while pretending to sell corn, and Elizabeth hoped the general would catch her. "General Butler will take care of you," she promised, and watched from her window until Merritt Rowley slipped from sight.

Despite being paid $1,000 in Confederate money, the guide shirked his duty, bringing the boy only as far as the Chickahominy River before abandoning him. Merritt emptied his pockets, coming up with a few graybacks to buy a ride across the river, crouching down in the boat to avoid the rebel pickets, their torches blinking

Tokens used by Union spies to identify one another.

like fireflies through the trees. It took him five days to journey the seventy-five miles southeast to Fort Monroe, where guards brought him to General Butler. The boy opened his palm to reveal the Union token and retrieved the letter from his muddy boot.

An aide to Butler swiped acid across the page, waiting for the ink to reveal itself, for the innocuous message from "Eliza A. Jones" to become something else. Once the symbols appeared, the aide checked them against the cipher code and began transcribing the contents. Butler decided to test the boy, who stood before him, hat in hand.

"Well, my boy," he said, "where did you get that letter from?"

The response came in a nervous tumble: "Miss Van Lew gave it to me. I stayed for a week with Miss Van Lew before I came away. Miss Lizzie said she wanted to send you a letter, and I said I would bring it. Miss Lizzie said you would take care of me. I left there last Saturday night. Miss Lizzie told me what to tell you."

"Well, what did she tell you to say? You need have no fear here."

"She told me to tell you of the situation of the army," Merritt said, and summarized, from memory, the main points of Elizabeth's letter: There was a plan to alleviate overcrowding in the city's prisons by shipping the inmates off to Georgia, far removed from the battlefields of Virginia and from any potential rescue operations by the Union army. Lee had about twenty-five thousand men, and there was a rumor he was in secret session at Richmond. The city could be taken more easily now than at any other time since the war began.

"Miss Van Lew," he concluded, "said not to undervalue Lee's force." He relayed Elizabeth's warning about seemingly innocent female travelers hoping to pass through the lines, and Butler sent him on his way.

Alone, the general reread Elizabeth's words: "Beware of new and rash council! Beware! . . . Do not underrate their strength and desperation." He sent her report to Secretary of War Stanton with the appeal, "Now, or never, is the time to strike." With a force of six thousand men, Butler proposed to "make a dash" for Richmond and free thousands of Union prisoners, men who would then aid in the second objective of his plan: the capture of the Confederate Cabinet and President Davis.

At midnight on February 7, Elizabeth awakened to the terrifying clamor of the alarm bells. Lifting the hem of her nightgown, she rushed out to the parapet, craning her neck and scouring the skyline. She heard the familiar cry—"To arms! To arms! The Yankees are coming!"—the hauling of cannon through the streets, children running to and fro, the tramp of armed men in all directions. The time had come: Butler was acting on her information, striking just as she had advised.

But the Confederates were ready, charging out to Bottom's Bridge, ten miles from the city, digging earthworks and rifle pits and blocking the path with felled trees. Union forces retreated back to Fort Monroe with a loss of nine killed and wounded, wondering how their surprise raid had gone so wrong. Butler blamed the defeat on the "corruption and faithlessness" of a Yankee deserter who had tipped the enemy off.

Elizabeth had little time to dwell on the disappointment. For weeks she had been receiving cryptic reports from informants inside Libby Prison, warning that there was to be an exit and she should prepare, and so she did, nailing dark blankets over the windows in her parlors and burning the gas day and night to heat the rooms. "We were so ready for them," she wrote, but she had no advance notice of the specific date or exactly how it would be done.

Working between the hours of 10:00 p.m. and 4:00 a.m., using nothing but chisels and spoons, two inmates hacked through the back of a kitchen fireplace, creating a portal to the basement. From there a group of them, numbering about seventy, worked in shifts, excavating a fifty-foot tunnel to an enclosed yard across Twenty-First Street. The yard led to a gate fastened by a swinging bar, the only exit. Twice a day Libby's clerk, Erasmus Ross, conducted a prisoner count. While five of them were digging below, their accomplices sneaked to the end of the line, hoping to be counted twice to conceal the absences. Once, eager to play a joke on Ross, some of the prisoners decided to imitate the repeaters, tiptoeing to the end of the line. Ross first counted too many, then too few, and then, flummoxed, unleashed a furious tirade against the "damned Yankees," who laughed so hard for so long that the clerk gave in and joined them. Never once did the prisoners suspect that Ross was secretly on their side.

On the night of February 9 a group of 109 men, two or three at a time, made their way through the tunnel and across the yard, free at last. Several of them headed straight for Elizabeth's mansion, pounding at the servants' entry door and begging to be let in. Others huddled across the street, watching from the lawn of St. John's Episcopal Church. The Van Lews' driver answered the door. He studied the men, faces smeared with dirt, the chilled air turning their breath to smoke. He caught snatches of their explanation of who they were and why they had come. He knew that Elizabeth hid prisoners in her secret room, and that she had prepared the windows and parlors for the possibility of sudden and desperate guests. But he also knew that rebel officers had disguised themselves as friends in the past, setting traps for Elizabeth that she had so far managed to elude. He refused to be the one who facilitated her capture, and on that night—of all nights—she was not there to speak for herself.

The servant told the men they were not welcome and closed the door.

Earlier that evening, just hours before the great escape, Elizabeth donned her usual disguise: a "coarse" homespun dress and sunbonnet, a basket of cakes on her arm, and cotton stuffed in her cheeks. She rode, unrecognized, to a farmhouse on the outskirts of Richmond, where a Unionist family was hiding her brother John. As she feared might happen, he'd been drafted by Confederate authorities. John had officially deserted, a status punishable by death.

In the South, where the Confederate army was loath to execute potential soldiers, public humiliation and torture were the preferred penalties for deserters. Some were forced to wear boards imprinted with the word *coward*, or were branded on the cheek with the letter *D*. Others suffered the brutal practice of "bucking," being bound in a contorted position and left that way for days. Desperate, Jefferson Davis begged the women of the South to shame deserters into returning to the army. Despite the potential consequences and the president's pleas, desertion continued unabated. Some men, propelled by a desire to leave "this one horse barefooted naked famine stricken Southern Confederacy," went over to the enemy and took an oath of allegiance to the United States. John hoped to evade Confederate detectives and join them.

When Elizabeth arrived at the farmhouse, she embraced her brother and sat up with him for hours, talking to the point of exhaustion. She went to bed, lying just a few feet away from the farmer's wife, who rocked back and forth in her chair, puffing on her pipe, blowing curlicues of sweet smoke. For hours Elizabeth tossed and turned, afflicted by some strange nervousness she couldn't quite name. In the morning, when a servant arrived with supplies for John's trip north, he mentioned the escapees that had been refused at the door. Elizabeth wept, "greatly distressed" that she hadn't been able to offer the prisoners refuge.

She was even more concerned about John's welfare. Now her brother had to give up all hope of fleeing to Union lines, since the streets would be teeming with Confederate detectives searching for the escapees. She told John to stay put, resumed her disguise, and quickly rushed back home. Turning onto Grace Street, she saw a chain of guards posted around the perimeter of the mansion. She spat the cotton from her mouth and approached the door. They parted, giving way, and she smiled at them as she let herself in.

Desperate situations sometimes require despicable remedies, she thought, and changed clothing, trading her homespun frock for a more dignified day dress in a lightweight worsted wool. She set out for provost marshal John Winder's new office on Broad Street, weaving through cavalrymen patrolling the streets in somber pairs.

Winder took the matter of desertion seriously, combing the hospitals for stragglers and issuing public appeals for help (he was especially pleased when a fourteen-year-old boy apprehended a deserter and brought him in, giving the kid $30 and a job as a messenger). Depending on his mood, he had the power to help Elizabeth's brother or deliver him straight to the front lines. To her relief he received her kindly, inviting her to sit down and asking how he might help. Neither of them mentioned the day his detectives almost found the secret room in her home.

She offered a selective version of the truth, admitting that her brother had deserted but insisting he had done so only because he suffered from a painful and debilitating injury. She hoped, as she spoke, that Winder wasn't aware of the report by the Confederate army doctor who had examined John, conceding that he was "of a delicate appearance" but concluding nonetheless that he was fit for active duty.

Winder ran a hand through his wild white thatch of hair, the hair Elizabeth once suggested belonged in the Temple of Janus, and said, "Bring him to me tomorrow morning and I will do what I can for him."

Rising, she placed her hand over her heart and thanked the general. She prayed that his words weren't setting another trap.

He stopped her and suggested he might be able to do her another favor. He understood that she was being harassed, that certain characters had made threats against her home and family and person, threats that had only escalated since the beginning of the war. If she were amenable he could board temporarily at her home, keeping a very close watch on her. She would be doing him a favor, too; his wife had fled the volatility of Richmond to live with relatives in Hillsborough, North Carolina, and now he was all alone.

Elizabeth knew his primary intent was to spy on her, but she had to tell him yes.

THE SANCTUARY OF A MODEST GIRL

RICHMOND AND EN ROUTE TO EUROPE

After the death of her father Belle longed to return to Martinsburg, even just for a visit, and launched a campaign to convince the Federal government to lift her banishment from Northern territory. Several Southern senators petitioned Washington on her behalf, and she herself wrote letters to Lincoln and Secretary of War Stanton, appealing to them as the daughter of a brother Mason. Not one of them responded. Next she tried Ward Hill Lamon, the bodyguard of Lincoln's and old family friend who had helped her back in 1861, after she was caught spying for the first time. "You knew my Father, my Mother, & you know me," she wrote. "For god's sake use your influence (for I know you have it) with Mr. Lincoln, for me to return to ma. I know she is nearly broken hearted. Relieve my grief. My father dead. My Mother, nearly wild with grief & I an *exile*. Oh, God! 'tis too hard. I pray you will listen to my entreaties. Anser [*sic*] soon."

Ultimately Belle was not permitted to go home. Instead she decided to travel as far away as possible, following Rose Greenhow's path and running the blockade to Europe. She told the press of her plans, and was gratified by the prediction that she would receive "great attention as one of the heroines of the war." She did not mention that she'd sent a note to Jefferson Davis offering to carry Confederate dispatches to England, and now she sat in her room at the

Spotswood Hotel, awaiting his reply. The president's approval would confirm her value to the cause, elevating her even in the minds of those who still dismissed her as an "accomplished prostitute" or, even worse, "absurd."

A knock at the door announced the arrival of a clerk from the State Department, bearing $500 in gold to cover her expenses, letters of introduction signed by Secretary of State Judah P. Benjamin, and several dispatches to British officials, the contents of which Belle never divulged. When she arrived in London, she was to report to Henry Hotze, a Swiss-born Southern sympathizer serving as a Confederate propagandist and commercial agent. She invented an alias, "Mrs. Lewis," and created a pretext for her mission: she was traveling as a private citizen who was suffering from ill health and needed to recuperate. True, her face and figure were as recognizable as her name, but the precaution still gave her a small measure of security. A civilian passenger caught on a contraband-carrying ship would most likely be detained as a witness for court proceedings and not as a prisoner of war, but a Confederate courier could expect to be treated only as an active foe.

On a Sunday evening in March, at dusk, Belle departed Richmond and took the train to Wilmington. There she boarded a steamer named the *Greyhound*—a striking vessel, she thought: a brand-new, three-masted propeller, its dark lead hull brightened by a sleek red streak. At the mouth of the Cape Fear River they dropped anchor, waiting to push off until the moon disappeared. Outside the bar, about six miles in the distance, she saw vessels from the Federal fleet cruising in quiet, inquisitive circles. She batted away her fear, focusing instead on the importance of her mission. She was comforted, too, by the ship's commander, Captain George H. Bier, a former US naval officer who had switched sides to serve with Stonewall Jackson. He also used a pseudonym, "Captain Henry." He had known her father and would take good care of her. She told him she had gold coins concealed all about her person and offered to similarly conceal his. They showed each other their scars.

The *Greyhound* idled off the coast at Fort Fisher, waiting while the Confederate Signal Corps blinked lights in sliding boxes, revealing the location of channels. Belle acquainted herself with the other civilian passengers, among them Mr. Pollard, the esteemed editor of the *Richmond Examiner.* He had also taken an alias, "E. A. Parkinson," and was gallant enough to call her Mrs. Lewis, although she couldn't imagine he didn't know her true identity. The ship's crew began a routine search and discovered four stowaways, all of them fugitive deserters who were told to choose between returning to the Confederate army or being tossed overboard.

Around 10:00 p.m. Captain Henry changed into a black suit. Every light was extinguished, every word spoken in a whisper. The anchor was raised and the ship glided off. Belle caught the captain's muffled commands of "Steady" and, more frantically, "Cut off your smoke!" She made her way to the deck, which was piled high with cotton, young officers perched on the tallest bales to keep watch for the Yankees. Her body, pulled by the weight of the gold, rocked with the rise and dip of the waves. Silently she prayed. No one considered sleep. The dread of capture stretched through the night, ending, at last, when the sun rose to reveal no enemy sail in sight.

Her relief gave way to seasickness, and Belle retched over the side of the deck. As soon as she lifted her head she saw a Union ship named the *Connecticut* bearing rapidly down, and heard the startled cry of the lookout: "Sail ho!" The *Greyhound* picked up speed, slicing through the water, but the enemy ship closed the distance foot by foot, her masts rising higher and higher, her hull looming larger and larger. Belle sifted through a tumble of thoughts: "Unless Providence interposes directly in our behalf, we shall be overhauled and captured; and then what follows? I shall suffer a third rigorous imprisonment . . . and every kind of indignity." The chase continued, the Yankee ship gaining. She knew what was coming, and she held her breath, waiting for it: the thin white curl of smoke spiraling through the air, the horrifying hiss of the shell, the deep, slushy report of its explosion underwater.

The crewmen began rolling bale after bale of cotton and heaving them overboard. "By damn!" one called. "There's another they shall not get." Captain Henry, revolver in hand, reminded them that he was the master of the vessel and ordered them back to the wheel. Belle watched the shots come faster now, too fast to count, pocking the surface of the sea or bursting high like falling stars; all of them seemed aimed directly at her. Captain Henry paced and checked his compass and screamed, "More steam! More steam! Give her more steam!" He spun around and clutched Belle's arm.

"Miss Belle," he said, "I declare to you that, but for your presence on board, I would burn her to the water's edge rather than those infernal scoundrels should reap the benefit of a single bale of our cargo."

Belle straightened and stripped her face of expression; she was as much of a soldier as everyone else on this ship. "Captain Henry," she replied, screaming over the clamor in the sky, "act without reference to me—do what you think your duty. For my part, sir, I concur with you. Burn her by all means—I am not afraid. I have made up my mind, and am indifferent to my fate, if only the Federals do not get the vessel."

The crewmen were still lobbing bales of cotton and other valuable cargo out to sea: barrels of tobacco and medicinal turpentine, used both for whooping cough and as a diuretic; a keg containing around $30,000; fat packages of Confederate bonds. Mr. Pollard hurled items from a letter satchel, until the white waves were dancing with ambrotypes and souvenirs. Belle found the engine room below, and emptied her bodice, skirts, and purse of all government dispatches, threw them into the firebox, and watched them curl into themselves and vanish. She tossed in everything but the gold coins on her body and the pistol in her belt.

She hurried back to the deck and heard the captain's resigned shout: "It is too late to burn her now. The Yankee is almost on board of us. We must surrender!" The engine gasped to a stop. Belle heard a low, dense humming and looked up to see a missile sear the air above her head. "By Jove!" the captain shouted to no one in particular. "Don't they intend to give us quarter, or show us some mercy at any rate? I have surrendered."

A voice boomed over from the enemy ship: "Steamer ahoy! Haul down that flag, or we will pour a broadside into you!"

Captain Henry nodded and the flag made a jerky descent. Small boats coasted in all directions, carrying crewmen dressed in pristine uniforms of gold and blue. The Yankee officers climbed aboard the *Greyhound* in an orderly flurry, and their captain, Lieutenant Louis Kempff, sauntered over to Captain Henry. Belle stood in the doorway of her cabin, close enough to listen to their conversation.

"Good day to you, Captain," Kempff said. "I am glad to see you. This is a very fine vessel, and a valuable one. Will you be good enough to let me see your papers?"

"Good day to yourself, sir, but as to my being happy to see you, I cannot really say that I am. I have no papers."

Kempff led Henry onboard the Federal ship, leaving an ensign, William M. Swasey, in charge of the *Greyhound*. He wore lavender kid gloves and used his fingers to smooth his greasy, center-parted hair. Belle loathed him on sight, drawing a silent conclusion: "An officer as unfit for authority as any who has ever trodden the decks of a man-of-war." He caught her eye and seemed to intuit a threat.

"Sergeant of the guard!" he called. "Put a man in front of this door, and give him orders to stab this woman if she dares come out." He took a step closer to Belle and said, "Now, ain't ye skeared?"

She reared up on her toes. "No, I am not. I was never frightened at a Yankee in my life!"

Swasey seemed surprised by this response. The Yankee sailors invited themselves into Belle's cabin, a brazen transgression that surprised her; she had hoped her quarters "would have been respected as the sanctuary of a modest girl." One approached her and said with studied insolence: "Do you know that it was I who fired the shot that passed close over your head?"

"Was it?" Belle asked, head tilted, hand on hip. "Should you like to know what I said of the gunner?"

The officer nodded. "I should like to know."

"That man, whoever he may be, is an arrant coward to fire on a defenseless ship after her surrender."

He stared at her, but her eye was drawn to another Union officer who had just dropped over the side of the ship, crossing the deck by the wheel. Everything about him was immediately and inexplicably significant—she hadn't been this riveted by a man since her late, beloved Stonewall Jackson—and she composed a mental file, marking the way his dark hair swept his shoulders, the eerie brilliance of his large and bright eyes. He was not "strictly handsome," she acknowledged to herself, and "neither Phidias nor Praxiteles would have chosen the subject for a model of Grecian grace," but he moved like a refined gentleman, a refined *Southern* gentleman, and she felt her rebel proclivities relax and fall away, like a corset undone: "Oh," she thought, "what a good fellow that must be." She forgot every previous flirtation with a Union soldier, making room in her mind for what was to come.

To Belle's dismay he passed by her cabin without inquiring about her, and so she took the initiative, turning to an officer standing nearby.

"What is the name of the new arrival in this party of pleasure?" she asked.

"Lieutenant Hardinge," he replied.

She repeated the name to herself, taking turns stressing different syllables, turning it into a private refrain. He came back into view, approaching the disagreeable Ensign Swasey. She angled closer, eager to hear their conversation, or even just the sound of Hardinge's voice.

"Hallo, Hardinge, anything up?" Swasey asked. "What is it?"

"Yes sir," Hardinge said, and told him that by order of Captain Almy, the commanding officer of the *Connecticut*, he had come to relieve him of the command of this vessel.

As soon as Swasey left, Hardinge, now in charge, summoned all of the men from Belle's cabin. He stepped closer, filling her doorway, propping his hands on the frame. She realized she didn't even yet know his full name. He bowed deeply, letting fall the long brown fringe of hair. She found sly, secret meanings beneath his words—"I am now in command of this vessel, and I beg you will consider yourself a passenger, not a prisoner"—and she couldn't consent fast enough when he asked permission to enter her room.

You Are Very Poor Company

PARIS AND LONDON

For once Rose exercised caution in her affairs. There would be no stack of illicit love letters left for detectives to find, no whispers about late-night callers or torrid flirtations with married men.

Her primary and most public suitor was Granville George Leveson-Gower, leader of the House of Lords and a widower. He was eager to wed again and had fixed on Rose, escorting her to balls and parties and debating with her about the war, a pastime she enjoyed despite his troubling positions. Two years earlier, he had written a widely circulated memorandum—one that was said to have influenced the prime minister, Lord Palmerston—against intervention, arguing that recognition "would not by itself remove the blockade" or supply England with cotton and would mostly serve to stimulate the North. Rose believed that only she could change Lord Granville's mind.

"Our people earnestly desire recognition," she reminded him one night. "The opinion of a nation who had showed such wonderful capacity for resistance and self-government was entitled to grave consideration."

They continued the conversation behind the closed doors of her carriage.

But she was more focused on work than romance, shuttling between Paris and London, suffering from seasickness and *la grippe*, as the French called it, her heart breaking every time she had to leave Little Rose. "Took my child back to the convent and left her sobbing bitterly," she wrote. "It was a heavy trial to say goodbye. God bless her. My heart is very sad." News from home depressed her further. While Lee's army bided its time in winter quarters in Virginia, waiting for spring to break, Union general William Tecumseh Sherman tore through the South, capturing Meridian, Mississippi, and destroying depots, railroads, bridges, locomotives, storefronts, arsenals, hospitals, offices, hotels, and cantonments until the city ceased to exist. It didn't help, either, that her efforts were often wasted on daft and senseless people, such as Edward George Stanley, fourteenth Lord Derby, who "showed that he was utterly ignorant of everything, save that we had once formed a part of the old United States." (Her opinion was somewhat rectified when he deemed Rose "the best diplomatist I have ever seen.")

She was growing tired and short on patience; everything, simply put, began to seem "stupid." Dinner at Westbourne Terrace was "stupid." Dinner at the table d'hôte was not only "very stupid" but also "disagreeable." She endured the tedious company of "le creme de la creme de societie" during a "very stupid" trip to the London zoo. "The most interesting object was a very intelligent but malignant monkey," she recounted, "which approached more nearly in intelligence to man than was at all comfortable. In fact, I think the Yankee President would have had a goodly feeling of fellowship."

Another pleasant meeting with Thomas Carlyle during which he praised her "gallant countrymen" was followed by "3 stupid visits." A visit with Mr. and Mrs. George Watson-Taylor, owners of the island of Montecristo, and their friends left her bored and unimpressed. "They are all stupid," Rose concluded, "and [I was] right glad when it was time to come home." An afternoon at the House of Commons was especially disappointing—nothing like her time spent in the Senate gallery back in Washington: "The place assigned for ladies is small and dark and screened from observation by a bronze grating. It is almost impossible to hear. Written up on the wall in the only place where a ray of light penetrates: 'Silence is especially enjoined,' which seemed the most absurd idea for a ladies gallery." Even a charity event at the Grand Hotel was "dull and lifeless," as the men mingled on one side of the ballroom and the women on the other, with "no conversation going on between the sexes."

Alone in her hotel room Rose wrote melancholy diary entries, fat tears blurring the ink: "What trifles color life and make it dark as night? Blessed are they who let no human feeling stir their lives. I know I ought not to be miserable and yet I am and tears which I try to keep back flow down my cheek and blind me." Abruptly she changed tone, scolding herself for revealing too much, even on the page. "Well," she signed off, "I will put up my paper and hope that tomorrow's sun will disperse the cloud, which is now heavy upon my soul. "*Je suis très misérable ce soir. Au revoir, Mrs. Greenhow, vous êtes très mauvaise compagnie* [I am very miserable tonight. Good-bye, Mrs. Greenhow, you are very poor company]."

In public she relied on her default emotions of anger and disgust, as exhibited one night during a party attended by a number of London abolitionists. One guest innocently posed a question: Who was the superior man, President Lincoln or President Davis?

Rose turned to face him and replied, "Sir, if you accept the scientific weight rather than the religious one—and believe man in the beginning was a baboon or an orangutan, and that successive ages of improvement has brought him to this present high state of perfection, almost equaling the God head—I will assume Mr. Lincoln is the beginning of the specimen, Mr. Davis the end."

The man backed away and did not approach her again.

Another guest, the Reverend Newman Hall, a prominent supporter of the Union, seized the chance to lecture Rose about slavery. She listened for a moment, sipping her sherry, until she could no longer tolerate his unintelligible nonsense.

"Your remarks are so absurd," she declared, "that I could almost suppose that you could have derived your argument from the romance of Mrs. Beecher Stowe."

A crowd had gathered around them, turning its collective head to follow the volley of words.

Hall confessed that Mrs. Stowe was, in fact, an inspiration, at which point Rose refused further conversation. "I consider you a subject for compassionate tolerance rather than argument," she murmured in mock sympathy, "and a candidate for the strait jacket."

The crowd tittered and hissed.

"What about the massacre at Fort Pillow?" Hall persisted, referring to a recent battle along the Mississippi River in Tennessee, where Confederate troops slaughtered more than three hundred soldiers of the US Colored Troops after they had surrendered. From the North came reports of unimaginable butchery: one man nailed to the boards of a tent so he couldn't escape when it was set afire; Negro children as young as seven forced to stand and face a rebel firing squad. "If I were a Negro," he argued, "I would have taken arms!"

The crowd fell silent, making room for Rose's reply. She hurled her words: "Then I would have shot you with as little compassion as I would a dog. You must excuse me but I do not consider the opinions of a man who confesses Mrs. Stowe as his authority worthy of reputation."

"The sympathy of England is worth something," he protested.

"The sympathy of the enlightened classes are all with us. Besides, we attach far less importance to that now than you seem to think. When it is to your interest to recognize us, you will. Our destiny is entirely in our own hands and the events of this war have removed from us all anxiety upon the slavery question. The fate of the slave rests with his Southern Masters, as the Masters with God."

She allowed Hall one last line: "But will you free him?"

"Never! Either extermination or eternal slavery is his lot, according to the lights before me."

The crowd parted, letting her pass, flinging whispers at her back. She took a carriage back to her room and turned in for the night, too tired even to cry.

BE PRUDENT AND NEVER COME AGAIN

RICHMOND

On April 7, in an outbuilding behind a farm on the edge of the city, Elizabeth stood over an open metallic coffin, staring at a month-old corpse. The body was that of Ulric Dahlgren, a Union colonel just twenty-one years old, who had been killed during General Butler's latest raid on the city, a second bid to liberate Union prisoners that had gone spectacularly wrong. Again the alarm bells had rung late in the night, coaxing Richmond's motley home defense brigade—clerks, factory workers, politicians, merchants, schoolboys, and elderly men—out of bed and into five battalions, trudging their way to the battle front. Elizabeth had seen them drilling earlier that afternoon, spotting many Northern men among them—"Some I know to be Unionists," she wrote. "So potent is fear to blind conscience!"

Union troops, led by Colonel Dahlgren, advanced on the city, but were betrayed again by a Northern deserter. Dahlgren's men failed to connect with another body of Federal troops under H. Judson Kilpatrick, and the Confederates took advantage of the confusion. Kilpatrick was forced to withdraw toward the Peninsula and the safety of Union lines, while Dahlgren was left to contend with Richmond's local militia and reserves. He, too, tried to retreat, but fell into an ambush and was fired upon and killed. The Confederates claimed to have found on Dahlgren's body papers detailing a plan to execute Jefferson Davis and his entire cabinet, but Elizabeth believed they were forged—nothing more than an attempt to "irritate and inflame" the Southern people and defame a noble man. After defacing Dahlgren's corpse and putting it on display at the train depot, they buried it in a shallow, muddy hole below Oakwood Cemetery.

Elizabeth bribed a Negro gravedigger to divulge the exact location of the grave, and three members of the Richmond Underground—including a current candidate for mayor—dug up Dahlgren's body and conveyed it to the farmhouse in a mule-drawn wagon. That morning, as soon as her new boarder, General Winder, finished eating breakfast with her mother Eliza (he seemed to have a soft spot for the sixty-six-year-old widow, just two years older than he) the women called for a carriage, planning to meet their Unionist friends at the farmhouse. Winder professed to be on his way to the office, but Elizabeth suspected he first would spend some time rifling through her papers and drawers, both exasperated and pleased by the lack of damning evidence.

Now, at the farmhouse, Elizabeth lowered herself closer to the coffin. The young colonel wore an unbleached cotton shirt and the pantaloons of a Confederate, with a tattered camp blanket for a shroud. The pinky finger of his left hand had been cut off for its golden ring, a memento from a deceased sister, and his prosthetic wooden leg had been sent to service a wounded rebel soldier. He had not been embalmed, nor treated with carbolic acid to ward off the odor of decomposition, but she was surprised by his state of preservation. Everything except his head, where he had been shot, was "fair, fine, and firm," with just an occasional purple blotch, like mildew, marring his skin. "The comeliness of the young face was gone," she saw, "yet the features seemed regular and there was a wonderful look, firmness or energy stamped upon them." Gently she let her hands drift across his chest, feeling for wounds and finding none. She cut a lock of his red hair, intending to smuggle it to his father, a rear admiral in the US Navy, through the Underground.

They were taking an incredible risk, but she believed Dahlgren deserved a proper viewing and burial in keeping with the peacetime ideal of the "good death"—the desire of every person, male or female, Northern or Southern, to die well and, in the tradition of *ars moriendi*, even artfully: surrendering the soul gracefully, meeting the devil's temptations, being laid to rest in a way befitting the life already led. War made a good death difficult. Men died far from home, leaving the fate of their remains in the hands of strangers or even enemies, as in the case of Ulric Dahlgren. In reburying the colonel, Elizabeth and the Richmond Underground were taking control of his death and making it, after the fact, a good one.

The men hoisted the casket and slid it back into the wagon, covering it with a load of young peach trees to fool the rebel pickets. They headed to another friend's farm just outside the city limits, planted a peach tree over Dahlgren's grave, and reported to Elizabeth upon their return to Richmond. "Every true Union heart who knew of this day's work," she wrote in her diary, "felt happier for having charge of this precious dust."

She returned the diary to its hiding place and headed out again, crossing the street to Eliza Carrington's home to read Mary Jane's latest report. There was much work to do before General Winder returned for the night.

Even Winder hadn't been able to save Elizabeth's brother John from conscription, but he did place John in his own regiment, the 18th Virginia, and promised to protect him and keep him from active duty. John didn't even have to wear a uniform, and he won the affection of his company by sharing his generous supply of whiskey. When Elizabeth offered Winder $6,000 in Confederate bills for the favor, the general seemed insulted, explaining that he had helped because John's conscription was a "clear case of personal animosity to the family." Elizabeth had known the general since the beginning of the war, had deftly played the mouse to his cat. She studied his face, gauged the inflection in his voice, and concluded he was being genuine.

After nearly three years of dealing with Winder, throughout all of their delicate negotiations and volatile truces, she finally understood: he recognized her as a true patriot, someone who didn't want to denigrate the South so much as nudge it back to where it belonged, a citizen stuck in a prodigal country. Because she was a woman—a wealthy, socially prominent woman at that—he tempered his suspicions with decency and Southern manners. He respected her dedication to her cause even as it diverged from his own, and the constant monitoring and attempts at entrapment were merely requirements of his job.

Elizabeth suspected, though, that his understanding extended only so far, that his willingness to look the other way was based entirely on what he might find.

With that in mind, as long as Winder was her boarder, she conducted her correspondence with Union officials from Eliza's home. At General Butler's request, she sent her spies through the lines with reports about Confederate plans to capture Norfolk and the skimpy defenses around the city: "nothing in Richmond except Home Guards & the men in Batteries & a cavalry force of 1500 men." Butler responded, addressing his queries to his "dear niece"—a unintended departure from his first letter to his "dear aunt" and a mistake that could have been deadly if detected by Confederate officials. "My Dear Niece," read one,

> Your Aunt Mary has decidedly improved in health, and will be so far helped by the spring air and warm weather as to make her quite well.
>
> Your old acquaintance, the Quaker, called on me two or three days ago, and is quite well and very happy to have escaped so luckily a visitation of the prevailing fever in his town, of which you have heard. He is going back to the North soon. He says your sisters are all quarreling over the question whose baby is the prettiest, but I decided in favor of Emily's perhaps because it is the fattest. All are well and send love.

I should like to tell you about the Negro soldiers here, but I
suppose if I did they would not let the letters go through. Keep up good
heart, Eliza. I hope we shall soon get through our trials, and meet in
a better country where all will be peace and happiness.
                    Your affectionate Uncle, Thomas ap Jones

Alone in Eliza's study, curtains drawn, Elizabeth applied acid
to the letter and held it close to the gaslight, waiting for the real
message to emerge:

Give what account you can of the rebel rams. Letter about messenger
received. Does messenger need money? If so, give him all he wants, and
it shall be repaid. Arrests will be made. Will there be an attack in North
Carolina? How many troops are there? Will Richmond be evacuated? If
any thought of it, send word at once at my expense. Give all possible facts.

As payment for her spying, Elizabeth requested shoes, gunpowder tea, and a "muff of the latest style"—items that had grown
increasingly expensive and hard to find, especially since John could
no longer travel to the North. Butler also sent $50,000 in Confeder-
ate money for her to use in recruiting others ("employ . . . only those
you know to be faithful, brave, and true," he cautioned), and on
one occasion sent her to an informant in the last place she'd expect:
General Winder's office. For two years, Captain Philip Cashmeyer
had been Winder's chief aide and "pet," and Elizabeth found it hard
to believe that he was actually a Union spy.

His loyalty to the Confederacy had seemed unimpeachable save
for one recent incident: he had been caught giving an exchanged
Federal prisoner a package containing several letters, written in his
native German, and copies of orders issued from Winder's office.
He defended himself, claiming that he wanted to impress his wife
in Baltimore with evidence of his importance as a Confederate official, but was arrested and imprisoned in Castle Thunder until the

letters could be translated and examined. The investigating officer informed Winder that Cashmeyer had made "a contemptible ass" of himself but that he was no traitor. He was, however, still under suspicion, which made Elizabeth's mission all the more dangerous.

She carried in her bosom a confidential letter from Butler to Cashmeyer, imploring him to meet another Unionist at New Kent Court House. The challenge would be calling on Cashmeyer without attracting Winder's suspicion; if she had business in that office, the general would surely wonder why she hadn't come straight to him. Thinking the Winder Building's side entrance on Tenth Street might be safer, she ducked in, creeping past the offices of various aides and adjutants and quartermasters. She found Cashmeyer alone, sitting at his desk. Her heart tripped.

If General Butler was wrong about him—if Cashmeyer really was a devoted Confederate—this incident would be evidence that Winder couldn't ignore. He would arrest her and send her to prison, if her neighbors didn't kill her first.

Without a word she dipped her hand down her bodice and produced the letter, and without a word he took it. She watched as he read, his skin turning "deadly pale." Abruptly he stood and then listed to one side, grasping his desk for purchase. She turned to leave, and he followed her, speaking in broken English.

"Be prudent and never come again," he warned. "I will come to see you."

On the last Saturday in April, in the dead of night, Elizabeth was pulled from her sleep by her driver whisper-screaming her name: *Miss Elizabeth*. He stood in her doorway, lantern in hand, and told her that three Union soldiers, just escaped from Libby, had knocked on the servants' entry door. They were still there, waiting outside, hoping to come in.

Following him on tiptoe, she tried to make herself weightless as she passed Winder's room, descended the stairs, and turned toward the back of the house. She opened the door, pressed a finger against closed lips, and motioned for the men to enter. She could not risk bringing them upstairs, past Winder's floor, to the secret room, and instead led them down to the kitchen in the basement. There they hid in dank, dark corners behind towering bags of turnips and potatoes. She told them she would be back with food and to direct them to the outhouse, and emphasized that under no circumstances were they to come upstairs.

For the next few days she strove to maintain composure when both she and Winder were home. Fortunately the general was preoccupied with work, dealing with the case of Dr. Mary E. Walker, a rare female surgeon for the Union army whose costume of men's pants, boots, cloak, and broad-brimmed beaver hat made her even more of a curiosity. She had been arrested in Georgia under suspicion of being a Union spy and sent to Castle Thunder, where she gave testy interviews to Richmond reporters. "I am a lady, gentlemen," she stressed, "and I dare any man to insult me." For emphasis she stroked a small knife resting on her lap.

On Saturday morning, just hours before the Libby prisoners escaped, Dr. Walker—or "this disgusting production of Yankeeland," as the *Whig* preferred to describe her—was marched from her cell to Winder's office, presumably for a conference about sending her north under a flag of truce. The general had been kept busy organizing the logistics of her exchange, as well as the transportation of a hundred Libby prisoners from Richmond to Danville, Virginia, to relieve overcrowding. Elizabeth was frantic herself, trying to round up three Confederate uniforms for her escapees and arrange for their journey north, and did not even consider the possibility of Winder's detectives showing up at her door until the moment they were actually there.

She couldn't recall if they were the same men who had searched her home before, but she approached their intrusion in the same manner, smiling and ushering them inside as if they were invited guests. She again instructed her nieces to go and play quietly. She suggested they might enjoy some sweet tea and homemade cookies. She began the tour of fourteen rooms—library, parlor, dining room, their boss's temporary chamber. She led them upstairs and down the hall, stopping at the entrance to the secret room, facing the wall to close her eyes and moderate her breath. Only the basement kitchen was left.

She started down the first set of stairs and then the second, the men inches from her back, a roaring quiet in her ears. She circled the wooden table and benches, long enough to seat a staff of twenty, and walked past the spit for roasting pigs. She approached the far corners and hoped the heaping bags of turnips and potatoes wouldn't suddenly and inexplicably move.

The detectives ate their cookies, drank her tea, and thanked her for her time.

They were finished, at least for now, but Elizabeth was not. She walked the thirteen blocks to Winder's office, thinking of nothing but what she would say once she arrived. She used the front entrance this time and threw open his door, as if it were as much her territory as his.

"Sir," she said, "your ordering your underlying officers to search my home for evidence to convict me in league with the enemy is beneath the conduct of an officer and a gentleman."

He looked up, startled, and she added, "It is an insult, sir, to unprotected ladies."

She let herself out before he could respond.

That night, at dinner, Winder did not mention the incident and seemed suitably chagrined. He would be leaving Richmond soon, heading to Georgia to manage the infamous military prison at Andersonville, where Union inmates were starving to death, their rations consisting of no more than a cup and a half of flour per week. She did her best to "talk Southern Confederacy," words that Winder received like salve on a wound, but she feared his successor might not be so willing to play her game.

Good-Bye, Mrs. Greenhow

LONDON

During the early days of May, Rose sequestered herself in her room, refusing all callers—even her beau, Lord Granville—and declining all invitations. She declared herself "sick with anxiety," waiting for word of the fighting back home. The first two days of battle in the Wilderness, a seventy-square-mile patch of gnarled underbrush about fifty miles north of Richmond, had cost General Grant seventeen thousand men, and at night he wept in his tent as brush fires raged, burning some two hundred of his wounded alive. The armies met again closer to the city, near Spotsylvania Court House, attacking and counterattacking, bodies piling up four layers deep, the soldiers stuffing their nostrils with leaves to insulate themselves from the stench. Lee telegraphed triumphant reports to the Confederate secretary of war, boasting that he'd "handsomely driven back" Grant's men, and Rose received the news by steamer. "Thank God it is good," she wrote. "Lee repulsed the enemy . . . if Grant is routed, I believe the time proper to press recognition."

She resumed her rigorous social schedule, accepting a congratulatory visit from Lady Abinger, lunching with the Countess of Chesterfield, dining with Earl and Countess Hilton ("he 60, she 24," she noted), and attending a party hosted by Edward Montagu Granville Stuart Wortley, better known as Lord Wharncliffe. After dinner Wharncliffe stood and raised his glass. "I am going to propose a sentiment which will be acceptable to everyone here, I am sure: the success of the Confederate cause and Mrs. Greenhow."

The woman curtsied and the men bowed, the entire room sinking in one cohesive movement, bright cravats peeping out from jackets, sprigs of flowers bobbing atop hats. She held the moment still, imprinting it in her mind, recalling her life back in Washington before it was ruined by death, and more death, and war. She did not know how many such moments she would have left, and found herself responding with an uncharacteristic economy of words: "Thank you all, for my country and for myself."

The fighting raged just twelve miles from Richmond, so close that Jefferson Davis listened from his office and Elizabeth watched from the parapet of her mansion, peering through binoculars and asking one servant, Uncle Nelson, if he could distinguish between Yankee and Confederate guns; "Yes Missis, them deep ones," he replied, leaving her to guess which one he meant. Benjamin Butler sent large numbers of troops to reinforce Grant while Lee waited for men from the Shenandoah Valley. "Confederate loss is heavy," Rose wrote. "God grant that this is the last of the bloody fighting." It only got bloodier when the two armies met at Cold Harbor east of the city. Union troops engaged in a hapless frontal assault against fortified Confederate troops and suffered nearly thirteen thousand casualties, a throttling of such magnitude that one Richmond diarist quipped, "Grant intends to *stink* Lee out of his position, if nothing else will suffice."

The good news inspired Rose to write, regarding the Yankees, "God grant that those vandals may be destroyed, exterminated so that the vile race may no longer cumber the earth. Impatiently I wait to see the bitter chalice placed at their lips, the assassin's knife at their throats and the torch of the incendiary applied to their homes." The Confederate victory failed to change Europe's position on recognition, but she had one last chance to serve her country abroad when Confederate envoy Mason summoned her for a meeting. He told her that Raphael Semmes, captain of the CSS *Alabama*, a rebel warship that had been trapped by a Union sloop-of-war off the coast of Cherbourg, France, needed her help. The captain was concerned about the mental state of third lieutenant Joseph D. Wilson, one of the *Alabama*'s officers who'd been taken prisoner.

"The very person we want," Mason remarked to the captain as Rose approached. "She can get him off if anybody can." He stood and shook her hand. "Good morning. Madam, we want somebody to do something."

"I'll do it—what is it?" Her tone was flat, all business, hiding her pleasure at the flattery. All of those meetings and parties and arguments had amounted to nothing, but there was a still a task suited only for her.

They explained: she needed to go to Charles Francis Adams, the son of John Quincy Adams and the US minister to the Court of St. James's, and "get this poor fellow off his parole, so that he can be taken care of." Since both Mason and Semmes held official positions in the Confederacy, Adams would refuse to see them; but Rose, having no title, would be at an advantage. She took a carriage straight from the meeting to the American embassy and presented her card: "MRS. ROSE GREENHOW, of Richmond."

Adams received her with "great courtesy," and refrained from mentioning a long-ago testy exchange between Rose and his wife, Abigail. During a dinner party at Rose's home in 1859, the conversation turned to abolitionist John Brown, who had just been hanged for inciting a slave revolt at a Federal armory in Virginia. Abigail leaned across the table, stared pointedly at her hostess, and called him a "holy saint and martyr." Rose didn't hesitate with her retort: "He was a traitor, and met a traitor's doom." An awkward silence settled over the gathering, and prominent Washingtonians gossiped about the incident for months.

Rose sat across from Adams, surmising his thoughts. While it was true that his wife had displayed poor manners at Rose's dinner party, there was also the obvious and pertinent fact that his government had imprisoned her as a spy and sent her into exile. He could either feel compelled to make amends or deny and dismiss her. She reminded him that Captain Semmes had taken more than 250 Union prisoners and pardoned all of them for exchange, and she hoped that Mr. Adams would follow this humane precedent.

Four days later, Joseph "Fighting Joe" Wilson was released, and he called on Rose to thank her. "Poor fellow," she said. "So happy and grateful for his release." He was just twenty-two years old and striking, with a small, beautiful mouth and a sweep of feathery dark hair. Rose forgot both their nearly thirty-year age difference and Lord Granville. She spent many of the rest of her days with Fighting Joe, even sitting with him for a formal photograph in a pose typically reserved for married couples: she in a chair and he standing behind her, one arm grazing her back. She wanted him to accompany her when she sailed back home.

The time had come, and she applied the same fervor to saying good-bye as she did to everything else. She received the sacrament of Confirmation (something she'd never gotten the chance to do as a child), considered a last-ditch effort to enlist the help of Pope Pius IX, and met with Lord Palmerston who made the argument she was so tired of hearing: recognition would only hurt the South by uniting the North, bringing together various factions—Copperheads, Irish Catholics, and about half the Democratic Party—who currently opposed the war.

Rose retorted, "Does it never occur to you that you probably bring upon yourself the very evil which you deprecate?"

She bade farewell to Mason, who consoled her by saying she had aided in "stimulating the slim English mind." She was reunited with her oldest daughter, Florence, who had arrived unexpectedly on a ship from New York. They hadn't seen each other since before the war, when Florence warned her against becoming a spy. "O how sad has been this terrible war in its effect upon families," Rose lamented. "Mine has been torn asunder." She was struck and saddened by her daughter's appearance—the sunken cheeks, the blade-edged bones, the veins shooting through the milk of her skin—but Florence was still so "very lovely." She went to Paris one last time to visit Little Rose, her heart growing "sorely tired" when the girl begged her not to leave. "Alas," Rose wrote, "inexorable destiny seems to impel me on. My heart yearns to stay and also to go. . . . The desperate struggle in which my people are engaged is ever present, and I long to be near to share in the triumph or be burned under the ruins." She embraced her youngest daughter as if for the very last time.

She packed her ball gowns, a leather money belt stuffed with $2,000 in gold, and her European diary, in which she made one final entry on August 10: "A sad sick feeling crept over me, of parting perhaps forever, from many dear to me." Her ship would leave that day from Greenock, Scotland, and she prayed that the journey back would be without incident. The business of blockade running had grown increasingly treacherous, with the Union Navy capturing or destroying nearly 92 percent of the South's blockade vessels, sending entire crews to prison, using deadly force if faced with resistance.

Rose made a pact with herself: she would sooner die than lose her freedom again.

This Verdict of Lunacy

EN ROUTE TO EUROPE

Things moved fast even by Belle's standards, starting from the moment Lieutenant Hardinge entered her cabin and made her his prisoner. Afterward he took her hand and pulled her toward the wheel of the ship. He pointed to the sky, admiring the perfect silver rind of moon, of how its light seemed to shine only on her. He murmured verses from Byron and Shakespeare—*Here we will sit, and let the sounds of music creep in our ears; soft stillness, and the night, become the touches of sweet harmony*—to the ceaseless rhythm of the waves.

Belle applauded with gloved hands, and asked for more.

He stepped closer, brought his face inches from hers, and told her he "endeavored to paint the home to which, if love could but fulfill its prayers, this heart would lead thee!"

The following night, sitting at the same spot, "Mr. Hardinge," as Belle still called him, asked the question she'd been waiting to hear, one that proved he was as much her prisoner as she was his.

"Will you be my wife?"

She admitted to herself that she loved him, and yet she couldn't stop a practical thought from drifting through her brain: "If he felt all that he professed to feel for me, he might in future be useful."

She let her expression go solemn and compressed her voice into a whisper. "Your question involves serious consequences," she told him, and insisted she could not give him an answer until they arrived in Boston—the *Greyhound*'s final destination, if not her own.

Samuel quickly proved his usefulness. While he was off giving orders, Belle stayed in her cabin with the *Greyhound*'s commander, Captain Henry, and two Union officers keeping her company. She offered the Yankees some wine, encouraging them as they swiftly made their way through the bottle. When Samuel returned he ignored the sight of his drunken officers and asked Belle if she'd seen certain papers about the ship. She pretended to think for a moment and then replied, "They must be in the lower cabin, where you've been dressing yourself."

He left to fetch them, suspecting nothing.

Belle nodded at Captain Henry, who donned his hat and walked out on deck, making his way to the harbor boat bobbing alongside the ship—the same harbor boat Samuel intended to take ashore to report his arrival. When Samuel discovered that his boat was gone, he assumed that the waterman had grown tired of waiting and pulled off. He called another and sailed to the Boston shore, returning, three hours later, with US marshal John Keyes. Together they appeared at the doorway of Belle's cabin, the marshal wild-eyed and panting his words: "Captain Henry has escaped!"

"What?" she gasped. "It is impossible! Only a few moments ago he was here!" *Again I have got the better of the Yankees*, she thought, laughing as they fumigated the ship in an attempt to smoke the captain out. They never found Captain Henry, but Samuel finally got his answer from Belle: Yes, she said, she would be his wife. She believed that God had intended them to "meet and love," and that He had purposely sent her a Yankee, a Union boy from Brooklyn. "Women," she reasoned, "can sometimes work wonders; and may not he, who is of Northern birth, come by degrees to love, for my sake, the ill-used South?"

Upon their arrival in Boston, Marshal Keyes told Belle he had procured rooms for her at the Tremont House, the city's preeminent hotel and the first in the country to offer water closets. She was to remain there, under watch of Federal guards, until he received word from Washington about her fate. He would either take her to Canada, home of numerous Confederate exiles, or deliver her to the Union commandant of Fort Warren, at the entrance to Boston Harbor.

Belle's guards reported her numerous shopping trips, during which she paid for purchases by dramatically retrieving gold coins—money the Confederacy had provided to finance her trip—from a pouch around her waist. The pouch rested atop a skirt belt holding what one observer called a "sufficiently persuasive" pistol, a piece Belle was happy to flash at anyone who cared to look. The pistol was necessary, she explained, to protect herself against the Federal spies who were stalking her. "She was proving a troublesome customer as she was overrun by curiosity seekers," Keyes concluded, "and had no discretion herself." He was eager to get rid of "her ladyship."

Samuel had gone to Washington to lobby for her release, bringing with him several letters of introduction to influential men, and Belle herself wrote to Gideon Welles, secretary of the US Navy, telling him that she wished to go to Canada—a request she believed the Federal government was unlikely to grant. Given her past activities and imprisonments, and the fact that she'd been caught on a blockade runner, the proper course would be to detain her for the duration of the war. If the North proved foolish enough to let her go, she could still sail to England, meet with Confederate officials, and orally convey the contents of the dispatches she'd destroyed.

To Belle's surprise, while Samuel was still in Washington, Marshal Keyes received a telegram ordering him to escort her and her servants to Canada. She must leave within twenty-four hours, and if she was caught again in the United States she would be shot. One Washington newspaper suggested that the government had released her because she was "insane," a characterization that Belle, for once, appreciated: "For this verdict of lunacy I thank them, if it contributed in any degree to mitigate my sentence." But Belle suspected another motivation: her claim that she was "in possession of a vast amount of information implicating certain high officials at Washington both in public and private *scandals*"—information she would withhold so long as they did what she asked. For now, she would say only that she felt "much obliged" to members of Congress and others who used their influence on her behalf.

Shortly before her scheduled departure the following day, Belle received an unsettling letter from her betrothed:

My dear Miss Belle:
It is all up with me. Mr. Hall, the engineers and myself, are prisoners charged with complicity in the escape of Captain Henry. The Admiral says it looks bad for us; so I have adopted a very good motto, viz: "Face the music!" and, come what may, the officers under me shall be spared. I have asked permission of the Admiral to come and bid you goodbye. I hope that his answer will be in the affirmative.

Belle was distraught. True, her behavior had caused Samuel's unfortunate predicament, but how was she supposed to plan and execute a wedding when her fiancé was stuck in a Federal prison? When he arrived to say good-bye he implored her not to worry. He would meet her in Canada or in London, and make her his wife.

Belle stopped at Montreal on her way through Quebec, trailed by those same Federal spies all the way, finally eluding them when she boarded her ship for England. She arrived in early August, just as Rose Greenhow was leaving, and reported directly to the Confederate propagandist Henry Hotze. She informed him that she'd destroyed communications from Richmond so that they wouldn't fall into enemy hands, and recited what she could recall of their contents. Hotze, in turn, handed her a letter, and told her it was personal.

It was from Samuel. He had come to England and, failing to find her, went on to Paris to continue his search. Guessing that she might have been delayed en route, he left word for her in London. Belle was "deeply touched at this new proof of his honest attachment" and telegraphed a message, telling him to come back right away. Their reunion was made all the more joyful by the news that Samuel was no longer officially a Yankee; the Navy Department had dismissed him for "neglect of duty" for permitting the captain of the *Greyhound* to escape. Belle would make him a proper rebel yet.

She busied herself with wedding plans, sparing no expense on flowers, cake, or dress, using the Confederacy's gold coins to pay for all of it, hoping the press was watching every move. Invitations were printed on bands of white ribbon, mounted on white parchment, folded, and sealed. Servants delivered them by hand, alerting guests to a morning ceremony on Thursday, August 25, at St. James Church in Piccadilly. Belle carried a bouquet dotted with orange blossoms to represent purity, chastity, and fertility.

Among the attendees were several distinguished representatives of the Confederate government, including the Honorable James Williams, former US minister at Constantinople under President James Buchanan; the Honorable John O'Sullivan, a former US minister at Lisbon who'd coined the term "manifest destiny" to describe American imperialism; and the aforementioned propagandist Hotze. An elegant breakfast reception followed at the Brunswick Hotel on Jermyn Street, where various guests proposed the health of Mr. and Mrs. Hardinge, eulogized Belle's work for the South, and toasted President Davis and General Lee.

News of the wedding quickly reached the United States, where Northern journals scoffed at Samuel for tolerating his wife's demand that he serve the Confederate Army. "Belle," declared the *Boston Post*, "has made a fool of him." The newlyweds spent a week together on the coast, strolling along New Brighton beach and getting lost amid the remains of Liverpool Castle. They discussed what came next. She wanted him to go to West Virginia to introduce himself properly to her family, distribute wedding cake to her cousins, and sleep in her childhood room. He would set sail on a blockade runner soon after their honeymoon.

While Samuel made preparations for his trip, Belle flailed about for a purpose. She had no official business in London, no clandestine meetings on behalf of the Confederacy, no reason for spies to trail her or for reporters to ask questions. She felt like a party everyone had left. She had to do something, say something, before no one cared what she did or said. Irrelevancy was the one enemy she truly feared.

She fetched one of the press clippings from her wedding, a long and laudatory article from the *London Morning Post*, and composed a letter to the Confederate president:

Brunswick Hotel,
Jermyn Street, London.
Piccadilly.
Sept. 22, 1864

Hon. Jefferson Davis,
Dear Sir;
　　I suppose that the news of my marriage has been rec'd in the Confederacy. I send you a paper containing the English account of my wedding. My husband will soon be in the South where I trust he will meet a warm reception, and all will forget that he was once in the ranks of the enemy. I trust from my having married a man of Northern birth my Country will not doubt my loyalty. Though I loved him I asked the advice of Mr. Hotze and other Confederates here before I took the step, fearing that my country would judge me wrong. Mr. Hardinge has given up all property and everything. His father is a Republican and has disinherited him for joining the Southern Cause and marrying a Rebel.

Do you think there will soon be peace? England wishes for it,
and all here sympathize with the South. I have been met so kindly. Of
yourself, and Stonewall Jackson and Gen'l Lee, the English have the
greatest admiration and respect. If at any time I can be of benefit to
my Country, command me.

Respectfully,
Your o'b't serv't,
Belle Boyd Hardinge,
CSA

Back in Richmond, Jefferson Davis could not have been less concerned with Belle Boyd or the question of her husband's loyalties. His five-year-old son, Joe, had died after falling fifteen feet from a porch at the Confederate White House, and there were ugly rumors that another son, Jeff Jr., had pushed him. Davis arrived just in time to hold his son's hand as he died. Jeff Jr. stood nearby, watching. "I have said all the prayers I know how," the boy insisted, "but God will not wake Joe."

The president had little time to grieve. Although Lee had held his own during the Overland Campaign—those spring battles against Grant in the Wilderness, at Spotsylvania Court House, and at Cold Harbor—the Union army had since maneuvered into a siege of Petersburg, some twenty miles south of Richmond, crossing the James River and cutting off the Confederate lines of supply. Union general William Tecumseh Sherman had just captured Atlanta, setting up headquarters and evacuating the civilian population. There was a great exile, an endless stampede of hogs and sheep, female slaves and their children, whites crammed like chattel into wagons. Those left behind became refugees in their own city, living in empty railroad cars and eating nothing but yams. Confederate ranks continued to plummet, and these losses were now exacerbated by Grant's Circular No. 31, which rewarded deserters with money and transportation home, so long as they took an oath to the Union. And Davis still had the disconcerting sense that his confidential communications were being read by the wrong eyes and repeated to the wrong ears—the most recent example being a plan to raid a Maryland prison camp to free Southern inmates. When he heard details about the raid being spoken of on the streets, he had to cancel the entire operation.

A mile away, Elizabeth penned a diary entry praising her spy inside the Confederate president's home: "I say to the servant, 'What news, Mary?'" she wrote, "and my caterer never fails!"

The Delicacy of the Situation

RICHMOND

No matter how tightly she drew the curtains or how low she turned the gaslight, Elizabeth always feared that someone was watching, aware that even her simplest movements were not what they seemed. General Winder had left her home and Richmond for Andersonville, Georgia, but his detectives remained vigilant, especially since her brother John had finally been forced to the front lines and deserted during the Battle of Cold Harbor, escaping to Philadelphia to live with their sister Anna. His friends told the Richmond press and officials that he'd been captured and carried off by a Yankee raiding party, but rumors persisted that he'd fled north, and that "Beast" Butler had hired him as a spy. "Van Lew rode out from his friend's in Hanover with a nigger in a buggy," the Confederate police chief observed. "The nigger and the buggy came back, but Van Lew didn't. It is damned strange if the Yankee raiders took Van Lew that they didn't take the nigger and buggy, too."

Elizabeth began varying her methods of delivering dispatches, both to throw off the detectives and to meet increased demand from Union officials. In one note she told them that her spying "puts me to a great disadvantage and I do not wish to do it unless it is received welcomely." It was, especially with General Grant now in charge of the eastern theater, headquartered twenty miles south of Richmond. She mined her contacts in the Confederate government, deployed members of the Richmond Underground to gather information, and picked up Mary Jane's reports from the seamstress, sitting in her study and interpreting each piece of information. Three days per week, after either encrypting her dispatches or writing them in invisible ink, she closed her fist around them and walked slowly, casually, to the ornamented iron fireplace at the far end of the room.

On either side of the fireplace grate stood two slender columns, each topped by a brass figurine of a lion preparing to pounce. She had loosened one of these so it could be raised like the lid of a mailbox. With her back to the mantel, she slipped the dispatches beneath the crouching lion and left the room; her part was finished. In a few moments one of the housemaids would enter and go about her work, seemingly intent on polishing each brass curve and mahogany spire, humming as she made her way to the lion and lifted the lid, tucking the letters down her homespun dress.

The maid took it from there, looping a basket around her arm and strolling down to Old Market Hall. She passed by the vegetable benches and butcher stalls, where ribbons of moist pink flesh dangled from strings; bacon was now $20 per pound, about $300 in today's dollars and a 6,700 percent increase from before the war. She stopped at a cart where she recognized the merchants: two men, also Van Lew servants, who'd come from the family farm in Henrico County. The maid feigned interest in the contents of the cart, sifting through the pile of corn and potatoes and peas, picking a few pieces for her basket. Instead of money she paid with dispatches, which the servants discreetly plugged inside the soles of their shoes. The shoes were great strong brogans made by a Richmond cobbler working with the Underground, and he fashioned the soles so they were double and hollow, capable of holding letters and even maps. Shoes, like everything else, had become exorbitantly expensive—high-laced shoes cost $100, and a new pair of boots $225—but Elizabeth's servants all had two pairs each and changed them every day, never wearing out of Richmond the same shoes they wore into the city each morning.

When the soles were filled to capacity, the men hollowed out an egg or a turnip, compressing the notes into slender scrolls and sliding them inside, hiding the dummy goods among the real thing. They departed after dusk, connecting with a Union scout at one of five Underground rendezvous points between Richmond and Grant's headquarters at City Point, at the junction of the James and Appomattox Rivers. The servants' preferred route was Charles City Road, which meandered southeast to the banks of the James, but if it was closed to civilian traffic they abandoned their carts and continued on foot, bushwhacking their way through thirty-five miles of woods, a journey of two days. Either way the guards passed them through the lines without question or incident; they were just slaves obeying orders, running errands for their mistress, not so much above suspicion as below it. Before daylight the Union scout, dispatches in hand, reported to headquarters, sometimes bringing along a Richmond newspaper and flowers from Elizabeth's garden, still fresh when they arrived on the general's breakfast table.

On the evening of September 27 an old family friend, Miss Lucy King, called at the mansion. She stood in the doorway, hands shielding her face, and whispered, "Do you think I am being watched?"

Elizabeth glanced left and right and hustled her inside.

Lucy told a story: one recent morning, while she lay in bed sick, her maid came to her room and said a detective was asking for her. She sent word she was ill. The detective returned the following day and left a handwritten message requesting her to report immediately to Eleventh and Broad Streets. Captain Thomas Walker Doswell, the new provost marshal, was seeking testimony against the Van Lew family, and her cooperation would be most appreciated.

Miss King felt she had no choice but to go.

Doswell pulled out a chair for her and smiled, the tip of his full brown beard grazing his cravat. He assured her she had nothing

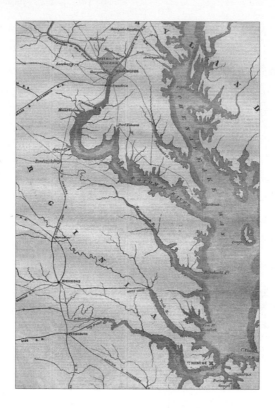

Routes of the Richmond Underground: first, seventy-five miles southeast to "Beast Butler" at Fort Monroe; and later, tweny miles southeast to Grant at City Point.

to fear. He, like his predecessor, General Winder, was well acquainted with the Van Lews. He had been a dinner guest in their home. He knew Elizabeth and her mother were upstanding Christian women, and he certainly did not want them to be lured into danger by the wrong sort of people, risking their lives for false and foolish ideals. Perhaps she was privy to information about the Van Lews? Information that, if they were on a wayward course, might help set them straight? He understood the "delicacy" of the situation, and promised that her name would not be mentioned to the Van Lews.

He leaned forward, waiting, lips pushing to maintain his smile.

"I am not with them as a spy," Miss King responded.

Doswell let her go, suggesting she keep his words in mind.

Elizabeth was relieved her friend had not succumbed to the pressure, and hoped that the new provost marshal would give up.

He didn't.

Next he sent detectives to interview various neighbors and associates. A neighbor had no direct knowledge of disloyalty but heard his sister damn the Van Lews as "good Yankees." A businessman distinctly recalled Elizabeth saying, "I wish the Yankees would whip the Confederates." The minister of the Third Presbyterian Church had heard Elizabeth praise the "violent and outrageous federal raiding parties." A Richmond grocer remembered Elizabeth's advice to "go to the Yankees and take the oath of allegiance."

The detectives were particularly interested in this last bit of testimony, given Elizabeth's brother John's suspicious absence from his Confederate regiment. They had one last person on their list: John's estranged wife, Mary, who surely harbored many incriminating secrets about the family, and who would be all too willing to share them.

General Ulysses S. Grant at City Point, fall 1864.

Not at All Changed by Death

ON THE *CONDOR*, HEADING HOME

Rose knew the South would be different from what it was when she'd left it, a country damaged and diminished and yielding itself piece by piece, just as it had been formed. The Yankees closed the vital port of Mobile Bay, Alabama, completing the blockade east of the Mississippi River. Atlanta had fallen, and now Sherman was conducting his relentless march toward Milledgeville, where women hid jewelry inside their dresses and took advice from the few haggard rebel soldiers who remained: "Lock your doors. Keep inside. If the Yankees come, unlock the door, stand in the doorway, be polite, and ask for a guard. You will not be mistreated, I hope." In the Confederate capital, Lee's army was hindered by a spy system so "complete and efficient," fretted the *Enquirer*, that "the Richmond evening papers reach Grant's headquarters before 3 o'clock the next morning." President Davis embarked on a publicity campaign, appealing to every man able to bear arms to rally to the front. In Palmetto, Georgia, he reviewed Confederate troops, drawing obligatory salutes from the soldiers but not one cheer.

But Rose was anxious to reach Southern soil, to shed the "cold isolation" that had enveloped her during her final days in Europe, and by 3:00 a.m. on Saturday, October 1, she had almost made it. The *Condor* drifted toward the entrance of the Cape Fear River, a two-hundred-mile black-water river that flowed into the Atlantic

Ocean. She was a sleek, elegant, 270-foot iron-hulled sidewheeler, painted "elusive white" to make her difficult to spot and built for tremendous speed; the British company that built the *Condor* believed she had no superior among blockade-running steamers. This was her maiden voyage.

She had a crew of forty and a few notable passengers, including James Holcombe, a judge and onetime Confederate spy in Canada, and Joseph D. Wilson, the young rebel officer whom Rose had saved from imprisonment and vowed to take home with her. A thirty-year-old British naval officer named William Nathan Wrighte Hewett served as captain, hoping to make a fortune running military supplies and coal through the Union blockade. The pilot, twenty-six-year-old Thomas Brinkman, had brought along his Newfoundland puppy. Rose carried a valuable cargo of her own: her European diary (which contained no mention of her burgeoning romance with Lieutenant Wilson); dispatches from Henry Hotze, the Confederate propagandist, for Richmond; a letter from Confederate emissary James Mason to President Davis praising her diplomatic efforts abroad; and four hundred British sovereigns, worth about $2,000—book royalties she planned to donate to a Southern relief fund. She kept the gold coins in a bag and fastened its chain around her neck.

The sky opened and dropped a hard rain, long, cool needles stabbing the deck. The *Condor* crept noiselessly toward New Inlet, taking the northern approach to Cape Fear, a strong wind at its back. The waves thrashed and churned. Two lines of Federal vessels skulked in the dark, patrolling just out of range of the Confederate guns of Fort Fisher. Rose was in her quarters, seasick again, her stomach feeling like a separate and defiant thing inside her, disobeying orders to stay still. They were almost there, only three hundred yards from shore.

There was a hard, long jolt, the whole ship shuddering. She was heaved from her bunk. The floor quivered beneath her knees, and then silence, stillness. She scuttled to her wardrobe and dressed in haste, panicked. She gripped her leather pouch, full of dispatches.

The bag of gold swung low, pulling against her skin; she wore it even to bed.

Up on the deck the rain soaked her skin and veiled her eyes. The wind howled and whipped the surf into great white billows that broke across her feet. The Newfoundland puppy ran in frantic circles. She demanded to know what was happening. She strained to listen to Captain Hewett, his words coming in staccato bursts: he'd realized he was being trailed by a Yankee ship, the *Niphon*. Rockets flared and shots cracked, tearing up the water around the *Condor*. Momentary relief as he evaded them. Then he saw what he thought was another Union ship looming ahead and he swerved to starboard. A mistake: the ship was actually the wreck of the *Night Hawk*, another blockade runner that had been driven onto the bar two nights before. Now the *Condor* was stuck on the shoals, immobile and helpless.

She heard enough to understand. The Yankees were out there, and they would get her. They would toss her back into the Old Capitol Prison, where those emissaries of Lincoln would force her to live in filthy quarters with Negroes, taking away her sun and air and dignity. This time they might leave her there until she died.

She grasped Captain Hewett's arm and pulled him close, yelling in his ear over the crush of the waves. She had to go ashore, she said. She had no choice.

No, the captain said. It was far safer to remain on board. The shoal water and the Confederate guns would keep the Yankees at bay. If the *Condor* could withstand the pounding of the waves, she should be freed with the morning tide.

Rose insisted and the captain refused. She insisted again. Two sailors volunteered to row, and at last he agreed.

She climbed into a lifeboat on the leeward side, squeezing next to Lieutenant Wilson. Judge Holcombe, the *Condor*'s pilot, and the pilot's puppy joined them. She clutched her leather satchel of dispatches and stroked the bag of gold at her neck, making sure the cord was fastened. The boat dropped, inch by inch, the water rising to meet them. As they touched down a great swell gathered and lashed at her. Rose sensed herself tilting, tipping. The boat flipped and let her go. The other passengers swam away from her, struggling back to the capsized boat and clinging to the keel. Even the puppy made it.

She was sideways, upside down, somersaulting inside the wet darkness. She screamed noiselessly, the water rushing in. She tried to hold her breath—thirty seconds, sixty, ninety—before her mouth gave way and water filled it again. Tiny streams of bubbles escaped from her nostrils. A burning scythed through her chest. That bag of gold yanked like a noose around her neck. Her hair unspooled and leeched to her skin, twining around her neck. She tried to aim her arms up and her legs down, to push and pull, but every direction seemed the same. No moonlight skimmed along the surface, showing her the way; there was no light at all.

At dawn, a sentry named J. J. Prosper For Me D. Doctor Duval Connor—at three feet eleven inches, said to be the shortest man with the longest name in the Confederate army—was patrolling along the beach at Fort Fisher. Something shiny caught his eye along the water's edge: the metal buckle of a leather pouch. Opening it, he couldn't believe his luck—gold coins, fistfuls of them, more money than he had ever seen. He glanced around, made sure he was alone, and buried it in the sand under a piece of burned log, planning to come back for it later.

As the morning sun slanted across the water, Thomas Taylor, the captain of the grounded blockade runner *Night Hawk*, found Rose's body washed up on the beach. She wore heavy black silk, as if in mourning for herself. Dark ribbons of wet hair reached below her knees. "A remarkably handsome woman she was," Taylor wrote, "with features that showed much character. Although one cannot altogether admire the profession of a spy, still there was no doubt that she imagined herself in following such a profession to be serving her country in the only way open to her. Surely in war the feelings of both men and women become blunted as to the niceties of what is right or wrong."

Taylor ordered a group of slaves to carry Rose to Colonel William Lamb, commander of Fort Fisher, who admired her "lovely face, that graceful form in pure development of womanhood." Lamb's wife, Daisy, wiped the sand and silt from Rose's body and detangled her hair, preparing her for one final journey: a ride aboard the steamer *Cape Fear*, heading upriver to Wilmington. Hundreds of female admirers lined the wharf, awaiting its arrival. One, Eliza Jane De Rosset, president of the Soldier's Aid Society, had the body brought back to her mansion, where she clipped a lock of Rose's hair to send to her daughters. "She was an elegant woman," Eliza wrote, "not at all changed by death." When Doc Connor discovered that the gold had belonged to a Confederate hero, he turned it in.

The wake was held in the chapel of Hospital Number 4. Rose's body lay on a bier, draped with a Confederate flag, surrounded by a phalanx of candles and flowers arranged in crosses. A stream of women, children, and soldiers approached with bent heads and hushed steps. They followed her to the Roman Catholic church of St. Thomas the Apostle, where the presiding priest spoke of "the uncertainty of all human projects and ambition," and followed her again in a procession to Oakdale Cemetery. It rained all the way, stopping only as the pallbearers lowered her into the ground beneath a wilting canopy of magnolia blossoms. A local reporter noticed a rainbow streaking the horizon and took it as a sign: "Let us accept the omen not only for her, the quiet sleeper, who after many storms and a tumultuous and checkered life came to peace and rest at last, but also for our beloved country, over which we trust the rainbow of hope will ere long shine with the brightest dyes."

They called Rose a heroine and compared her to Joan of Arc, but they were still mostly strangers, mourning a symbol more than a person.

THE SWEET LITTLE MAN

LONDON AND THE SHENANDOAH VALLEY

On the night before Samuel departed for America, Belle took his hand and slid onto one finger a small diamond-cluster ring, telling him it had once been the property of an African princess and that it carried a curious power: if it dropped or was taken off, it meant the wearer was in danger. She tried to focus on writing her memoir but was distracted by reports of Rose's death—the tragic news reached Europe by the end of October—and by thoughts of her husband visiting her childhood home.

The entire Shenandoah Valley was now under Union control, and in Martinsburg a tentative serenity had settled over the streets. The provost marshal checked Samuel's pass and bag before waving him through, and by the time he reached the Boyd home the sky was black and gilded with stars. A servant—it had to be "Mauma Eliza," whom Belle spoke of often—greeted him with, "You's Miss Belle's husband, isn't you?" Belle's mother was visiting a friend in Kennysville ten miles away, and so he spent the evening talking with Belle's grandmother, who wept and welcomed him like a son.

Another servant, Jim, showed him to Belle's room. Samuel removed his hat before stepping across the threshold and surveying her things—hair combs and books and a polished palmetto pin—in reverent silence. Sitting before the fire, he opened his journal and composed a passage intended only for Belle's eyes: "This was your room; here you had been held a prisoner, and had suffered the torture of an agonizing doubt as to your fate. Here lay your books just as you had left them. Writings, quotations, every thing to remind me of you, were here; and I do not know how long a time I should have stood gazing about me in silence. . . . When I retired to bed that night, and Jim had been dismissed from further attendance upon me, I lay for a long time thinking, looking into the fire that glimmered and glared about the room, picturing you here, there, and everywhere about the chamber, and thinking of you sadly, far away from me in England—the exile, lonely and sad."

Belle was so charmed by her husband's writings that she included them in her memoir, calling them an "after-piece" that offered another facet of their unlikely love story—albeit one that Samuel, at least, never intended to convey. "Of the two characters," mused an early reviewer of *Belle Boyd in Camp and Prison*, "he is—we say it without meaning disrespect—the truly feminine one."

In the morning, after a servant prodded the fire back to life, Samuel sat near its heat and turned Belle's princess ring on his finger, absentmindedly twisting it off. He recalled her warning and moved through the rest of his day feeling like a "marked man," a premonition that clung to him as he set off for Baltimore in the afternoon. He got as far as Monocacy Station just outside Frederick, Maryland, where Union detectives approached. It was clear to Samuel that the men recognized him. They asked if his wife, Belle Boyd, was "lurking somewhere in the vicinity" and arrested him for desertion—a pretext to take him into custody, since he had already been dismissed from the Union navy. After transporting him to Harpers Ferry, they brought him before General John Stevenson.

The general's face was dominated by a coarse gray mustache, the tips wilting toward his collar like unwatered ferns. He stared at Samuel with a peculiar intensity.

"Is there anything remarkable about me, or that you admire?" Samuel asked.

"Yes, sir," the general replied. "Your duplicity . . . you are a spy. Where are your papers, passes, dispatches?"

"I have none," Samuel insisted.

They were joined by one of the general's aides. "You're the husband of Miss Belle Boyd," he said, delivering the words as an accusation, "and you ought to be hung."

Union officials conveyed him to Washington and conducted another search, confiscating his tall beaver hat and cane and the $14 hidden in his pocket. At Carroll Prison, where Belle had spent four months the previous year, he was tossed into a cell with a blockade runner and a rebel spy. With a crude piece of charred wood he scrawled all three of their names on the wall and sketched a Confederate flag beneath.

Superintendent Wood stopped by to greet him: "Ho, ho, here we are! So you're the husband of the famous Belle Boyd, are you? Well, we haven't got her, but we've got her husband." Samuel received a letter from Belle's mother, who was refused permission to visit him, and despaired because he was unable to hear news of Belle at all. "I have not smiled today," he wrote, "but two or three times my eyes have been filled with tears; for I have been thinking of you, Belle, a stranger in a strange land, waiting sad and lonely for my return."

Belle, meanwhile, had read all about her husband's predicament: he was in "daily danger," one London paper reported, "of meeting with the greatest outrages" at the Yankees' hands. She knew it would be up to her to save him.

In Harpers Ferry, not far from where Samuel Hardinge had been held for questioning, another former Union soldier was reclaiming a piece of her old life. Emma never became a missionary, as she once told Jerome Robbins she might, and instead returned to nursing, working at a hospital run by the US Sanitary Commission, a relief agency that supported sick and wounded soldiers. Not one of her female colleagues knew of Frank Thompson—she would keep that secret for a while longer—and it had taken time to relearn how to move and speak like them, to see herself as they did. Her own eye cataloged curious distinctions: the most highly cultivated and refined women were the least bothered by the hardships of the job, and invariably made better nurses than those from the lower classes. She had no time for the foolish, sentimental girls who expected the hospital to look like a drawing room where, Emma wrote, they might "sit and fan handsome young mustached heroes in shoulder straps and read poetry" instead of combing matted hair and washing dirty skin. Even the hardiest of this sort lasted only days.

She thought often of her own war heroes, General Poe (currently assisting with Sherman's devastating March to the Sea), General McClellan (who just lost the presidential election to Lincoln), and all the men of the 2nd Michigan. She never saw Jerome again, but he wrote and described the siege of Vicksburg with such brutal specificity that she felt as if she'd been there. James Reid disappeared for good, most likely back to Scotland with his family. Now there was someone new, but from an old part of her life: Linus Seelye, also a native of New Brunswick, who had come to Harpers Ferry seeking work as a carpenter and randomly made her acquaintance. "Her whole life was an interesting conundrum," Linus said of her, "for every week something would come up—something she could accomplish, overcome, move or manage, that would eclipse the last." Like Jerome, he was smart and soft-spoken; like James, he was tall and blond and married; his wife remained back home in Canada. He was thirty-two, nine years older than Emma, and listened to all of her stories, even the ones she only pretended were true.

She began writing those stories, beginning with her divinely inspired decision to join the army and including everything she wished to be remembered for: her valor during some of the war's deadliest battles, her inventive disguises and risks behind enemy lines, her care for the wounded and the dead. She titled the book *Unsexed;*

or, The Female Soldier—a risqué choice, she knew—and would later rename it *Memoirs of a Soldier, Nurse and Spy: A Woman's Adventures in the Union Army.* Lest she appear *too* feminine, she chastised all the able-bodied men of the North who failed to enlist, even quoting a famous poem written by Oliver Wendell Holmes, then a young soldier, called "The Sweet Little Man":

> *We send you the buttonless garments of woman!*
> *Cover your face lest it freckle or tan;*
> *Muster the apron-string guards on the common,*
> *That is the corps for the sweet little man.*

At the same time she emphasized that her every violent and "masculine" act—shooting a female rebel through the hand, unloading her pistol in a rebel soldier's face—stemmed from self-sacrifice, that most feminine of qualities; she saw now that her two selves had worked in perfect concert. She mentioned Jerome Robbins only in passing, offering no hint that he knew her secret, and never mentioned Frank Thompson at all, remaining deliberately vague about her name, her background, and exactly what her comrades saw when they looked at her. It was a meticulous balancing act, one that Emma masterfully transferred from her life to the page, leaving the story open to interpretation while its hero kept parts of herself hidden.

LIKE MOST OF HER SEX

RICHMOND

Detectives found Elizabeth's sister-in-law living in a boardinghouse on Canal Street, told her they were conducting
an investigation, and asked if she would testify against the Van
Lews. Mary was happy to oblige, telling them that she had lived
with the family for several years, that she'd heard them express
hope that the Confederacy would fail, that they were strong abolitionists and sent a Negro woman north to be educated, that her
estranged husband had deserted and fled to Union lines, and,
most important, that she didn't want her children growing up in
their home.

The detectives pushed for more, asking Mary if she knew or
had seen or heard anything else, any evidence that the Van Lews
were involved with the spy network that was undermining the Confederacy. She didn't, and their official report concluded with, "And
further this deponent saith not."

After examining all of the testimony against Elizabeth, and
taking into account her family's "wealth and position," Confederate
officials reached a split decision. While they believed that she was
"very unfriendly in her sentiments" toward the Confederacy, and
that, "like most of her sex, she seems to have talked freely," they
didn't believe she had committed any significant acts of disloyalty.
None of their searches of her home yielded any incriminating papers

or escaped Union soldiers. None of their officers had been able to entrap her by offering to smuggle information through the lines. No one could prove she had anything to do with the spy ring assisting the enemy. For now, the adjutant general's office concluded, there was "no action to be taken."

Elizabeth knew how quickly that could change if, at any time, she stopped overestimating them and they stopped underestimating her. She dared not alter her routine. The wheels of the Underground spun on schedule, sending one dispatch after another to Grant's headquarters at City Point. Every evening, from seven to midnight, the general, his officers, and two dozen soldiers gathered around the campfire, smoking cigars and discussing the latest information from their spies in Richmond.

They learned that everyone was preoccupied with Sherman's ruthless march and that Jefferson Davis—publicly, at least—had struck a pose of confidence and optimism. "The Government, from time to time," read one report, "claims to have dispatches of a favorable kind, but this is not believed by the community." They even received updates about North Carolina, learning the exact number of troops General Lee sent to Wilmington to reinforce the forts below the town. They discovered that the weakest point of the enemy's line was between the Nine Mile Road and the Mechanicsville Pike, and debated how to exploit it. On December 21, after Sherman had captured Savannah and presented it as a "Christmas gift" to Lincoln, they heard that the city's "plain classes" were anticipating the evacuation of Richmond. They learned, from information Elizabeth gleaned through a Confederate deserter, that "the enemy are planting torpedoes on all roads leading to the city."

As smoothly as her enterprise ran, Elizabeth still lived in a quiet and pervasive state of terror. She hadn't forgotten that note from the "White Caps," threatening to burn down her house and write with her blood. She knew strange men still spied through her windows, followed her on the street, took careful measure of her words.

Late one night she was awakened by a scratching at the garden entry door. A cold fear sank into her chest. She closed her robe around her and lit a candle with shaking fingers. A shadow at the door came to shape in the light, the features sharpening one by one: Mary Jane Bowser, looking as if she had just run for her life.

[PART FIVE]

1865

The Way a Child Loves Its Mother

LONDON

On the twenty-fourth of January, Belle sat at her desk in the Brunswick Hotel and considered what she wanted to say to the president of the United States. She would approach Lincoln as if she expected him to do her bidding, as she would any man, but she knew better than to cajole or flirt or pretend she was a defenseless woman; his appetites seemed confined to the political, and she would address him in his language, and as his equal. She did not like Lincoln personally, believing him to be a hypocrite and "destroyer," but she appreciated his wit and dry humor, and admired all those, men or women, who were able to create themselves from nothing.

She felt it then: a short, shallow hiccup in her belly, reminding her that she would never again act solely on her own behalf—or at least that she shouldn't. Before the end of the year, if not the end of the war, she would be a mother.

She dipped her nib in the ink and began to write:

Honble Abraham Lincoln

President of the U.S. America
*   I have heard from good authority that if I suppress the Book*
I have now ready for publication, you may be induced to consider
leniently the case of my husband, S. Wylde Hardinge, now a prisoner

in Fort Delaware, I think it would be well for you & me to come to some definite understanding. My Book was originally not intended to be more than a personal narrative, but since my husband's unjust arrest I had intended making it political, & had introduced many atrocious circumstances respecting your government with which I am so well acquainted & which would open the eyes of Europe to many things of which the world on this side of the water little dreams. If you will release my husband & set him free, so that he may join me here in England by the beginning of March—I pledge you my word that my Book shall be suppressed. Should my husband not be with me by the 25 of March I shall at once place my Book in the hands of a publisher.

Trusting an immediate reply,
I am Sir, Yr. Obdt. Sevt.
Belle Boyd Hardinge

The letter traveled on the fastest westbound crossing, Liverpool–New York, reaching America in eight days, fifteen hours, and forty-five minutes. Lincoln was most likely familiar with its author and her involvement in the Valley Campaign of 1862, and the ensuing panic in Washington after Stonewall Jackson's surprise victory. In the spring of 1863 he had suspended the sentence of her first cousin, William Boyd Compton, a Confederate spy in West Virginia who'd been captured and sentenced to be hanged. "Let the execution of William B. Compton be respited or suspended till further orders from me," Lincoln wrote, "holding him in safe custody meanwhile." His act saved the lives of five Union prisoners, whom Jefferson Davis had threatened to hang in retaliation.

Outside his office window the still unfinished Washington Monument, which Mark Twain likened to "a factory chimney with the top broken off," lurched toward a bleached and dismal sky. Encampments of Union soldiers dotted the Potomac River, so frozen that an escaped Confederate prisoner had recently skated across its surface on rag-bound feet, heading toward freedom in Virginia.

Aided by information from Grant's spy ring in Richmond, the Union army and navy had just captured Fort Fisher, guardian of Wilmington, North Carolina, where rebel blockade runners had come and gone and even died. After considerable political wrangling and furious debate, the House of Representatives passed the Thirteenth Amendment, officially abolishing slavery. Lincoln and Jefferson Davis, through emissaries, were engaging in secret negotiations, but reached an impasse when the Confederate president mentioned peace for "the two countries"; Lincoln wanted nothing less than restoration of the Union. But the downfall of arms drew nearer every week.

Lincoln made no notation on Belle's letter, nor did he indicate any knowledge of the "atrocious circumstances" to which she referred. Perhaps because the war was almost over, perhaps because Samuel Hardinge's only crime was being Belle's husband, perhaps because the president admired the twenty-year-old girl's audacity, the prisoner was released on February 3, exactly ten days after Belle made her demand. After stopping at his family home in Brooklyn to say good-bye to his mother, Samuel boarded the steamship *Cuba* and was reunited with Belle a few weeks later.

Her circumstances had grown dire since she had last seen him four months before. She'd spent all of her funds and begun pawning her jewelry and wedding gifts, one by one. She suspected the Federal government was intercepting funds sent by relatives in Virginia. When she had nothing left, the owner of the Brunswick Hotel evicted her from her room, forcing her to move to a boardinghouse in an "unfashionable" part of town. She wept to the London press, insisting that her current state of suffering was worse than what she'd endured in Washington's Old Capitol Prison. "I'd rather be there as I was," she declared, "than here as I am." She took her frustrations out on Samuel, who responded by drinking until he passed out, but even in this condition he couldn't escape Belle's wrath. She stood over him, shaking her fist at his closed eyes, deriding him as "the fool who had married" her. Soon she would no longer have even the war, those four ferocious, thrilling years that had raised her and given her a name.

At night, the strange flutterings growing more persistent, tickling the underside of her skin, she lay awake and imagined what came next. She would publish *Belle Boyd in Camp and Prison* but withhold—at least for now—her knowledge of Union misdeeds. She would start training for the stage, maybe starring in comedies by Edward Bulwer-Lytton and classics by Shakespeare (as Juliet, of course), receiving raves in every city she played. And when she was ready, and the world was ready to hear them, she would dress up in her Confederate costume and reenact scenes from her life, beginning with that afternoon long ago, on the Fourth of July, when the Yankees came to town and threatened to fly their flag over her home. She would tell her fans that she loved the South the way "a child loves its mother," and that she never wanted to fail it again.

As This Mighty Work Was Done

RICHMOND

Elizabeth pulled Mary Jane inside, checking for anyone lurking behind her in the dark. The story came out in hoarse gasps: They suspected her. Not only Varina and Jefferson Davis, but their other, loyal servants. She sensed she was being followed throughout the house. Her manner of clearing plates and polishing wood invited a sudden and intense scrutiny. Voices dropped or fell silent in her presence. In rare unguarded moments the president wondered aloud how the Union knew Lee's plans as soon as the general decided them. She began awakening to a persistent sense of something gone wrong. She grew intently aware of every step behind her back, every shadow beneath her door. She needed to leave before leaving was no longer possible.

In the morning Elizabeth helped Mary Jane into a farm wagon, telling her to lie low and flat. Two other servants shoveled horse manure on top of her from head to foot, several layers deep. She would pass through the lines as if she herself were a secret dispatch, riding to City Point, and from there Elizabeth's contacts would send her to Philadelphia, where brother John would be waiting. Elizabeth had been in touch with John since he deserted the army and fled north, writing to him under the alias "Emma G. Plane," using couriers to deliver messages and money. She watched the wagon rattle off and prayed that her servant would make it.

She knew Mary Jane was right to be worried: rebel authorities had recently arrested nine members of her spy ring, including Confederate railroad superintendent Samuel Ruth, and imprisoned them in Castle Thunder. As with Elizabeth, Ruth's status and wealth gave him a sheen of innocence; the press lauded him as "a most efficient railroad officer" and a "respectable, prudent and cautious man." While the others languished behind bars, Ruth was honorably discharged. He immediately resumed his undercover work, even asking the Federal government to reimburse the $2,500 he'd spent on his defense—money he could use to bribe rebel sources.

The spy ring now spanned the city and three surrounding counties. It was impossible to keep track of every name and face and background. Richmond authorities might try to entrap Elizabeth again at any time, especially in the wake of Mary Jane's sudden disappearance, since Varina would certainly remember hiring the servant on Elizabeth's recommendation. Any new agent was immediately suspect, even if introduced through legitimate channels, as when a Federal scout named Judson Knight knocked on her door in early February.

She led Knight into the darkened library. Behind her, alongside the fireplace, the lion figurines concealed the latest intelligence, waiting for pickup and delivery. Knight told her that an Englishman named R. W. Pole would come through the lines, posing as a Southern sympathizer while working for the Underground. He insisted Pole was a loyal Unionist, but Elizabeth had her doubts.

"My heart sank," she wrote after Knight left, "for here was another avenue of danger."

She waited, for days and then weeks, dealing only with those who had long since proved themselves, and on February 27 heard the news: the Englishman had gone to the provost marshal and reported two of her spies, Lemuel Babcock and William White, both of whom were sent to Castle Thunder. A Confederate detective touched a pistol to White's temple and demanded, "Tell all you know or I'll blow your brains out. Tell all who are concerned in this thing."

White replied, "Blow away."

Elizabeth only increased her efforts. Her spy ring divulged a Confederate plan to exchange $380,000 worth of tobacco for bacon, prompting Grant to organize a raid on Fredericksburg, during which his troops seized the illicit goods, destroyed railroad bridges, and took four hundred rebels prisoner. The spy ring placed ads in newspapers promising payment to "engine-runners, machinists, blacksmiths, molders, and other mechanics" who deserted to work for the Union; large numbers accepted the offer, severely restricting Confederate use of the railroads. The ring wrote of preparations to move the sick and wounded from hospitals, of an extraordinary excitement pervading the city, of alarm bells pealing forth upon the startled air and the home guards seizing their muskets, scrambling directionless toward an unspecified threat. "Everybody turned out upon the streets," read the dispatch. "The cause of the alarm could not be distinguished, except in the unsettled state of the public mind." They reported "an entire prohibition" of all news; people were forbidden to discuss the state of the Confederacy even on the streets.

They reported that the James River Canal was frozen, allowing no boats to pass through and limiting the delivery of provisions. They provided updates on public morale, detailed surging prices (more than $1,000 for a barrel of flour, more than $1,200 for a suit of clothes), and revealed that counterfeiting was so rampant that people no longer discerned the difference between good and bad money. Elizabeth stood on her parapet, counting troops and defenses, listening to the sidewalk hubbub and the wails of the bereaved in their homes. "May God bless and bring you soon to deliver us," she wrote to Grant. "We are in an awful situation here."

She sent him the information he needed to win the last crucial battle of the war. On March 14, Grant told his staff that he had read "a letter from a lady in Richmond" revealing that Confederate troops had been ordered down the Danville Road, which connected Richmond to the sources of supply for the Confederate army; privately owned warehouses of tobacco and cotton had been turned over to the government; and citizens were "ordered to be organized, no doubt to prevent plundering in the city when it is evacuated." The general understood that the enemy intended to fall back southwest to Lynchburg, and that he had less than a week to position his men.

On the final day of March, under a light rain, fifty thousand Union troops commanded by General Philip Sheridan drove beyond the right end of Lee's line, where some ten thousand Confederates tried to stave them off. For a while they did, pushing the Yankees back and settling around the crossroads of Five Forks. The following morning Lee told his forces to "hold Five Forks at all hazards," and the battered Confederates dug in, singing "Dixie" and unleashing a stream of rebel yells: *Woh-who—ey! Who—ey! Who—ey! Woh-who—ey!* Half of them were captured. Among the Southern dead left behind were shoeless old men and boys as young as fourteen. Grant saw a chance to "end matters right there." He ordered an immediate assault all along the lines, and dispatched a correspondent to Lincoln bearing a package of rebel flags collected on the battlefield.

The morning of Sunday, April 2, dawned bright and cloudless, the sun easing into position and turning up its heat, a merciless spotlight on the city below. For the moment church bells replaced alarm bells, their chimes trilling sweetly through the streets, beckoning everyone to the pews. At St. Paul's the usual congregation was in attendance, including Jefferson Davis; he had given Varina a pistol and told her to take the children to safety farther south. Halfway through the service a messenger hurried up the aisle and pressed a scrap of paper into the president's hands. A woman sitting nearby noticed a gray pallor coming over his face as he read Lee's words: "My lines are broken in three places. Richmond must be evacuated this evening." He arose, walked unsteadily from the church, and ordered his government to move to Danville, 140 miles south-west. All of his cabinet boarded a series of freight cars, each gamely labeled—"Treasury Department," "Quarter Masters Department," "War Department"—a veritable government on wheels, rumbling slowly from sight.

Elizabeth stood on Grace Street and watched her city empty itself. Young soldiers on horseback and foot bade frantic farewells, all saying they wished they could stay. She walked to a nearby home and found her neighbor sitting on the steps in shocked acquiescence.

"The war will end now," Elizabeth said. "The young men's lives will be saved."

The woman lifted her head and gazed dully. "I have a son in the army about Petersburg."

Elizabeth lowered herself and met the woman's eyes. She spoke in hushed and dulcet tones. Just think: she might now hope for her son's life. They might never again hear the awful saying, "The last man must die," uttered and acted upon so often during the past four years.

The woman replied, "It would be better, anything would be better, than to fall under the United States Government."

Elizabeth rose and backed away without another word.

Hours had passed, daylight easing from the sky. A Confederate naval officer blew up all that was left of the fleet idling in the James, the shock shattering windows throughout the city. Series of bursting shells formed violent constellations, shaking the air around her. Everywhere everyone was rushing to escape. Wagons rattled furiously down the streets. People bartered for vehicles, offering any price in any currency. Porters toppled under heaps of baggage. Nurses carried pale men on stretchers, unsure of where to go. Lines snaked around the open banks. Rebel soldiers lugged wheelbarrows of paper money, both Federal and Confederate, to Capitol Square for burial. Retreating soldiers burned everything in their path—tobacco warehouses and flour mills, arsenals and ironclads, bridges and depots—the wind kicking up twisted tangles of flames. Vandals set the Confederate White House afire four times; each conflagration was extinguished by the gardener. City Council called a hasty meeting and ordered the destruction of all liquor. Whiskey ran in the gutters ankle-deep. Half-drunk men and women and children dipped cups and pans into the streams. Some lay facedown on the streets and lapped it up like dogs. Mobs plundered shops and broke into abandoned houses. Terrified women tore at their hair and screamed, "The Negro troops will be turned loose on us!" They gathered up silverware in their skirts.

Some of them, both neighbors and strangers, banged on Elizabeth's door, begging her to secrete their valuables, knowing they would be safe in a Unionist's home. She let all of them in, even the ones who had testified against her. During the bedlam three members of her spy ring had escaped from Castle Thunder, and now they got to work, hauling and hiding the neighbors' vases and jewelry and gleaming sets of knives.

One spy, returning through the front door, told Elizabeth that her house was to be burned, that groups of rebel soldiers were right then on the way. She sprinted outside and watched the men charge toward her, jabbing torches at the sky. She took a few steps out onto the lawn, meeting them.

"If this house goes," she said, "every house in the neighborhood will follow." Union troops, she warned, would take vengeance against the homes of Confederate loyalists.

The rebels retreated down Grace Street and left her alone.

The night advanced into dawn, but Elizabeth couldn't sleep. Every now and then a magazine exploded and the ground rocked and trembled beneath her feet, tickling her through the soles of her shoes. There was a terrible roaring and hissing and cracking of flames. The doorbell rang until dawn and more escaped prisoners crowded in, collapsing on cots she and her mother had strewn across the parlor floor. Her nieces, Annie and Eliza, came downstairs, terrified that they would be blown up in their beds. The whole night, Annie said, seemed a dream she would never forget. The girls had packed up their clothes in carpetbags and sheets, ready to run from the rebels. Elizabeth told them that there was no need, that they were all safe now. She walked with them to the window, all three transfixed by the sight of the smoldering city; to Elizabeth the destruction was "the consummation of the wrongs of years."

The ruins of Richmond, April 1865.

Beneath it all she heard a sweet sound—the quick, sure notes of "Yankee Doodle Dandy." They were there, finally, all those boys in their blue wool coats, taking in wild bursts of welcome from former slaves. "No wonder that the walls of our houses were swaying," she would write in her diary. "The heart of our city a flaming altar, as this mighty work was done." She thought of Mary Jane, safe up North, a whole new life opening up to her.

There was a knock at the door. She opened it to find a Union guard, sent by General Grant to protect her. "I want nothing now," Elizabeth told the soldier. "I would scorn to have a guard now that my friends are here." (Grant himself would soon send a personal note, telling Elizabeth, "You have sent me the most valuable information received from Richmond during the war.") She insisted the soldier stay for dinner and the night; he could meet some of her contacts and sources, men who once occupied prominent positions in the Confederate government. Now if he would excuse her for a moment, there was something she needed to do.

She climbed the stairs to her secret room, finding the corner where she'd hidden her American flag, unseen and untouched since Virginia left the Union; it would be the first to fly over the city in four years. She unfurled all eleven by twenty feet of it and counted its thirty-four stars, all of them united again, back where they belonged.

She crossed the hall to a room with a gabled window, the glass reverberating against her palm as she pushed it open, letting the smoke and music—now the first notes of "The Star-Spangled Banner"—rush in. Clutching the flag close, she climbed out and steadied herself, trying each corner to the roof railing and letting it fall, rejoicing at its soft weight in her hands.

Elizabeth Van Lew's family flag.

When searchers retrieved Rose's body from the Cape Fear River, they found a note meant for her youngest daughter:

You have shared the hardships and indignity of my prison life, my darling; and suffered all that evil which a vulgar despotism could inflict. Let the memory of that period never pass from your mind; Else you may be inclined to forget how merciful Providence has been in seizing us from such a people.

Rose O'N Greenhow

Little Rose was devastated by her mother's death. "Les larmes de la pauvre enfant ne tarissaient pas [The poor child's tears never stop]," wrote the mother superior of the Convent du Sacré Coeur. The girl, now eleven and a half years old, asked to be baptized so that she might be better equipped to pray for her mother's soul. After the war Little Rose returned to the states, married William Penn Duvall, a lieutenant with the US Army, and had a daughter of her own.

In 1888, the Ladies' Memorial Association—locally organized groups of Southern white women dedicated to reburying the scattered remains of Confederate soldiers—marked Rose's grave with a marble cross. It was engraved with the words MRS. ROSE O'NEAL GREENHOW. A BEARER OF DISPATCHES TO THE CONFEDERATE GOVERNMENT, a tribute that remains to this day.

Emma's memoir was an unqualified success, selling 175,000 copies; in comparison, *Uncle Tom's Cabin*, the greatest publishing phenomenon of the nineteenth century, sold 300,000 when it came out in 1852. Before the war Emma had spent her Bible salesman's salary on an elegant horse and buggy, but now, uninterested in her newfound wealth, she instructed her publisher to donate all proceeds to the sick and wounded soldiers of the Army of the Potomac.

When the war ended she was thrilled by the ubiquitous sight of the "dear old Stars and Stripes," but, like everyone else in her adopted country, found herself unsure of what to do next. Linus Seelye, her fellow Canadian and friend, had left Harpers Ferry to attend to his pregnant wife back home in Canada, and she was restless for something new. She headed to Oberlin, Ohio, and briefly took classes at Oberlin College, noted for being the first fully coeducational and racially integrated American institution of higher learning. But after the excitement of war she found the academic life "too monotonous," and decided to go back to Canada herself, the first time she'd been there in seven years.

Much had changed since her last visit home, that cold fall day when she had appeared as Frank Thompson, surprising her mother and praying her father wouldn't walk through the door. Both Betsy and Isaac had died during the war without ever knowing what had become of their youngest child, and family legend has it that Emma's father spent his final years sitting and staring out the window, waiting for her return.

She spent that summer in New Brunswick, hunting with her brother Thomas and taking long, solitary walks through the woods, and when the weather turned she decided to go back to Oberlin. On the way she passed through Saint John, where Linus was living and working at his brother's carpenter shop. He was now a widower; his daughter had died in infancy, and his wife six months later. He followed Emma to Ohio and asked her to marry him. She did, on April 27, 1867, at the Wendell House in Cleveland, a fashionable hotel where Lincoln had stayed en route to Washington for his inauguration six years earlier. As happy as she was with Linus—a gentle and genteel man who behaved nothing like her father—she couldn't help viewing marriage, on some level, as a personal defeat. "Well," she said, "you know how the census takers sum up all our employments with the too easily written words 'married woman.' That is what I became; and of course that tells the entire story."

The couple had two boys, Linus and Henry, both of whom died in infancy, and a girl, Alice, who survived; Emma was thirty-two now, with faltering health, and she knew her daughter would be the last child she bore. When she learned through her church of two recently orphaned brothers, Freddy and Charles, an infant and a toddler, she and Linus took them in as their own. Through her church she also learned of an orphanage in St. Mary Parish, Louisiana, home to sixty-seven children, including many whose fathers had died while serving in the "colored" regiments of the Union army. They moved south so that Emma could manage the orphanage, but the muggy Louisiana climate stirred up old ailments. She suffered another bout of the malaria she had contracted during General Mc-

Clellan's Peninsula Campaign, so many years ago. She recovered, but soon afterward her daughter came down with measles. Six-year-old Alice died on Christmas morning of 1880, the third child Emma had lost in eleven years. She hid for weeks in her darkened room, unable to get out of bed.

Hoping for a fresh start, the family moved to Fort Scott, Kansas, a booming military outpost and home to numerous Civil War veterans—all of whom, Emma realized, were collecting pensions for their service. She thought of her recurring malaria, her left leg that still ached and throbbed, her foot that was so deformed she had trouble fitting it into a shoe. She was, she decided, entitled to the same compensation. No matter that the War Department still officially denied that women had served in the Union Army dis-guised as men. Or that even if Emma proved she was Private Frank Thompson, he was listed on the service rolls as a deserter and there-fore ineligible for a pension. She would both clear Frank's name and make sure he got what he deserved.

First thing Monday morning, February 20, 1882, Emma sent a notarized letter to the Michigan adjutant general, John Robertson, swearing that she and Frank Thompson were one and the same, and that she had used that alias to enlist.

Robertson, coincidentally, had recently finished his tome *Michigan in the War*, which included a passage about the mysterious brigade postmaster who went AWOL:

> In Company F, 2nd Michigan, there enlisted at Flint Franklin Thompson (or Frank, as usually called) aged twenty, ascertained afterward and about the time he left the regiment to have been a female, and a good looking one at that. She succeeded in con-cealing her sex most admirably, serving in various campaigns and battles of the regiment as a soldier; often employed as a spy, going within the enemy's lines, sometimes absent for weeks, and is said to have furnished much valuable information. She remained with the regiment until April 1863, when it was supposed she

apprehended a disclosure of her sex and deserted at Lebanon, Kentucky, but where she went remains a mystery.

The adjutant general duly sent Emma her certificate of service.

Next, she had to solicit support from her former comrades. She decided against contacting Jerome Robbins; she needed the testimony of men who had served alongside her without knowing her secret. She thought of Damon Stewart, her bunkmate at the start of the war.

She took a train to Flint, Michigan, where, more than twenty years earlier, she had lived and worked as Frank Thompson and dated many of the belles in town, all of them also now married and middle-aged. She found Stewart's dry goods store, the tinkle of the bell announcing her entry. Adrenaline scoured her mouth dry and muted the pain in her foot. She hadn't seen Stewart since the Battle of Williamsburg in May 1862, when she hauled him off the field and tended to his wounds. He had been sent home and later reenlisted with another Michigan regiment, and she wondered what he'd heard about Frank Thompson's strange disappearance in the spring of the following year.

Now he rose from his desk, approached, and asked how he might be of service.

She found her voice—Emma's, not Frank's. Stewart had never heard it before. "Can you by chance," she said, "give me the present address of Franklin Thompson?"

Stewart was silent, letting his eyes drift up her body and halting at her face, shadowed by the veil of her hat. Behind the veil she bit her lip, forbidding a smile.

"Are you his mother?" he asked.

The past two decades had settled hard on her, carving lines around her eyes and giving her a heavy, seesaw gait. She forgave her old friend his mistake.

"No," Emma said, "I am not his mother."

Stewart tried again, perplexed. "His sister, perhaps?"

The bell chimed behind them, and they were no longer alone.

Emma plucked the pencil from Stewart's hand and took a card from the counter. Hastily she scribbled a few words and watched as he read: "Be quiet! I am Frank Thompson."

He wilted into a chair, looking at Emma as though seeing her for the very first time, and in that she found a small victory. "She was as tranquil and self-possessed," he said, "as ever my little friend Frank had been."

Stewart and other old comrades rallied behind her cause, writing affidavits attesting to her identity, good character, and valiant service during the war. They enlisted the support of two members of Congress: E. B. Winans of the Sixth Michigan district, who served on the Committee for Invalid Pensions; and Byron M. Cutcheon of the Ninth district, a former major with the Twentieth Michigan who now sat on the Committee of Military Affairs. The latter man remembered "Frank Thompson" well, especially his bravery under fire during the Battle of Fredericksburg. He offered to sponsor legislation to remove the deserter charge from Emma's records, while Winans sponsored a bill to grant her a soldier's pension.

President Chester A. Arthur signed the pension bill on July 5, 1884, granting Emma $12 per month, but Cutcheon's desertion bill languished before passing, finally, in July 1886. In the meantime Emma had become both a celebrity and a curiosity back home in Fort Scott, Kansas. Neighborhood children whispered that Mrs. Seelye had been a soldier and spy during the war and made a game of spying on her, peeping around trees to watch as she wielded an ax and chopped wood in her yard, wearing men's trousers and army boots. They whispered about her short hair and worried she might cut off their ears. A neighbor and fellow veteran spoke to the kids on Emma's behalf so they would no longer be afraid of her. "She may have been the means of saving the lives of many soldiers," he said. "A spy does save many lives."

Emma died on September 5, 1898, at the age of fifty-six, finally succumbing to malaria. She was buried in La Porte, Texas, where she and Linus had moved to be near their adopted son Freddy, but in 1901 her fellow members of the Grand Army of the Republic—the nationwide Civil War veterans group—had her remains moved to the Grand Army of the Republic burial ground in Houston's Washington Cemetery following a proper military funeral. Back in Michigan, Jerome Robbins clipped and saved an obituary of his old friend. A "most remarkable character," it read, "has passed away."

Belle Boyd gave birth to a daughter, Grace, in mid-1865, a few months after the South surrendered. To supplement the sales from her memoir (and to keep herself in the public eye), she pursued her dream of stardom on the stage. Coached by the American actress Avonia Jones and the English Shakespearean actor Walter

Belle Boyd on the lecture circuit, circa 1890.

Montgomery, she made her debut on Friday evening, June 1, 1866, at Manchester's Theatre Royal, playing Pauline in Edward Bulwer-Lytton's romantic comedy *The Lady of Lyons*. The reviewers were largely favorable but noticed an odd timidity in her performance—surprising, one mused, for a woman who "has faced shot and shell" and "led by the nose some very astute Federal officers."

Samuel Hardinge seldom attended his wife's performances, instead spending his evenings with a courtesan named "Fannie Sinclair," and Belle decided to leave her "utterly worthless" husband. Taking advantage of President Andrew Johnson's proclamation of amnesty to most former rebels, she and her baby daughter returned to the United States, where she filed for divorce in a New York courtroom. She told reporters she wanted no alimony, and was only anxious to be rid of both her husband and his name. On the day the divorce was finalized, she wore a strand of bells around her neck, and from that point forward she told anyone who asked that Samuel Hardinge had died.

She continued her theatrical career, billing herself "Belle Boyd, of Virginia," sometimes riding onto the stage on horseback and improvising scenes from her memoir. In 1868 she gave a reading in Washington, DC, and invited Lincoln's former bodyguard, Ward Hill Lamon, to attend. Early the following year, after a performance in New Orleans, a fan approached and asked her to dinner. A month later they married in a French Quarter church. This husband, like her first, was a Yankee: John Swainston Hammond, an Englishman by birth, immigrated to America and served with the 17th Massachusetts Infantry. Their son, whom Belle named Arthur Davis Lee Jackson in honor of her favorite Confederate heroes, was born in January 1870 at an insane asylum in Stockton, California, where she had checked in after her mind "gave way." He died in infancy and was buried there. "Her feet and brain had no rest," one reporter sympathized. "We cannot admire, but we must pity this strange soul, and be astonished at its wild, romantic career."

By the fall, Belle's doctor declared her "recovered" and she, John, and Grace moved from city to city, never staying in one place long enough to call it home. They had three more children: a son, John, and two daughters, Byrd and Maria Isabelle, nicknamed "Belle" after her mother. For fifteen years the elder Belle did not appear on-stage, but still the spotlight followed her. Like Mark Twain she was reported dead. After it became clear that such reports were greatly exaggerated, she began hearing of a host of impersonators.

The Masonic lodge in her hometown, Martinsburg, issued a warning that a stout brunette, calling herself Belle Boyd, had been "imposing herself on the fraternity throughout the South." In Philadelphia a female swindler using Belle's name devised a primitive version of a check-cashing scam. Belle and her family had scarcely arrived in Texas when she learned that someone was appearing at opera houses in tiny towns nearby, telling stories of her exploits as the Spy of the Shenandoah. "Belle Boyd, the Confederate Spy, who died recently," wrote one Austin periodical, "is living in Corsicana, Texas, in easy circumstances. She is also living in a garret in Baltimore, where she makes a scanty living by needlework, so the papers say. Belle is beating her Confederate record of being in two places at once." Another impostor—or was it the genuine Belle?—confronted an Atlanta newspaper editor for calling her a fraud, carrying a Smith and Wesson in one pocket and a derringer in the other.

The real Belle Boyd, meanwhile, felt her mind giving way again, betraying her and rousing terrible old ghosts. She told a reporter she'd been subjected to "private sorrows which the world should not demand." In October 1884, when she was forty, she shot and wounded one of nineteen-year-old Grace's suitors for "ruining" her daughter, and then disowned Grace, calling her "dead to the family." That same month, she discovered that her husband had a second wife. She promptly divorced him until he divorced his first wife, and then remarried him. John, in turn, accused Belle of having her own affair. They headed back to divorce court, both of them claiming to be the plaintiff. Grace, despite their differences, took Belle's side, telling the court that her father "boxed Mamma's ears" and barraged her with insults, at least one of them "reflecting on her chastity." Their divorce was final on November 1, 1884. Less than six weeks later, Belle took a third husband. Nathaniel Rue High was an actor and only twenty-four, sixteen years her junior.

Nathaniel encouraged Belle to go back to work, as he alone couldn't support three children. She briefly gave elocution lessons before returning to the stage, reenacting her wartime adventures. She debuted in Toledo, Ohio, on Sunday evening, February 28, 1886, her costume a nod to both genders: a rebel soldier's belt, an elaborate plumed hat, and a specially designed brooch featuring an enameled Confederate battle flag. The debut was, Belle said, an "unqualified success," although that didn't stop her from seeking other means of revenue—namely an unsuccessful $50,000 lawsuit against the *Chicago Tribune* for accusing her of failing to pay her bills.

Belle took her show on the road, incorporating Byrd and Belle Jr. into the act, giving readings as far west as Iowa, and renaming her performance *The Perils of a Spy*. At veteran reunions, she led thousands of rebels in reenactments of her favorite battles, recounting the time she leaped through the fields with the agility of a deer, rushing to tell Stonewall everything she knew. She was again reported dead, this time confused with Belle Starr, the notorious western outlaw who was fatally shot in February 1889. She took to carrying affidavits from the Grand Army of the Republic and Confederate Association attesting that she was "authentic" wherever she performed.

In the coming decade her crowds dwindled, even in the South; the people who had lived through the war were dead or dying off. She was destitute, often unable to pay for housing on the road, but somehow found a way to buy Eliza, her long-devoted servant, a high chair and plate for the birth of her first grandson. As they had during the war, both admirers and detractors commented on her appearance and skillful manipulation of her femininity; she was a relic from the near past who deftly played both lady and killer. "She stood there looking at the dead Union soldier and in that moment she was no longer a child," wrote the *Washington Daily News*. "Life had been crinoline and the smell of roses . . . now . . . . Belle hardened into a strange frightening being."

In a late interview, Belle herself characterized her life as if it were one long sin that needed to be confessed—with one pointed exception. "I have lied, sworn, killed (I guess) and I have stolen," she said. "But . . . I thank God that I can say on my death bed that I am a virtuous woman. . . . Fortune has played me a sad trick by letting me live on and on." On June 11, 1900, at the age of fifty-six, she died of a heart attack in Kilbourn, Wisconsin (now Wisconsin Dells), after giving a reading at the local Methodist church. The Women's Relief Corps of Kilbourn, an auxiliary of the Grand Army of the Republic, chipped in to pay for her burial, and her pallbearers included four veterans of the Union army. On that day, at least, there were no impostors taking her name, wearing military dresses with gilded buttons and trailing trains, caricaturing her character. She was Belle Boyd, the one and only, and fortune's sad trick was done.

A few days after the fall of Richmond, Elizabeth rode in her carriage—pulled by her last remaining white horse—through the wreckage of her city to the state capitol, now headquarters of Union general Godfrey Weitzel. There she found several papers once belonging to abolitionist John Brown, including his constitution, which denounced slavery as "a most barbarous, unprovoked, and unjustifiable War," and which she would keep for the rest of her life. After Lincoln's assassination, she took some small comfort in watching the Union army repair bridges and roads and clear away the rubble, taking with it all evidence of the rebel government, including ninety-one boxes of archives.

In a letter to the War Department, Benjamin Butler praised Elizabeth for furnishing "valuable information during the whole campaign" and requested that John Van Lew be allowed to come home from Philadelphia. She had joyous reunions with her brother and numerous members of the Richmond Underground who had fled in the final months and weeks of the war. Mary Jane Bowser returned too, taking a job teaching two hundred black children at a newly established school in the Ebenezer Baptist Church; she would go on to teach in Florida and Georgia. In November, General Ulysses S. Grant, accompanied by his wife, Julia, visited Richmond during a brief tour through the South. He barely stopped to shake hands with local officials but spent an afternoon drinking tea with Elizabeth on her veranda, in full view of the neighbors, who would never forgive her for what she had done. "She had no moral right to speak of the people of the South as 'our people' and as 'we,'" wrote one. "She had separated herself from us. And, the women of the South being so loyal, so self-sacrificing, so devoted, she made herself not only notorious but offensive. She was vain enough to imagine that she was called upon to make herself a vicarious sacrifice."

Thirteen days after being inaugurated as president, Grant nominated Elizabeth as postmaster of Richmond, one of the highest federal offices a woman could hold in the nineteenth century. It was also one of the most lucrative, paying up to $4,000 per year, and she desperately needed the money. She had depleted a large portion of her estate during the war, spending thousands on food and supplies for Union soldiers and bribes for Confederate officials, and hinted to her Northern contacts that compensation would be welcome. The head of the Bureau of Military Intelligence, George Sharpe, campaigned on her behalf, writing a letter to Congress in which he declared that "for a long, long time, she represented all that was left of the power of the US government in the city of Richmond." Nevertheless Elizabeth received only $5,000, one third of what she wanted. Her brother's hardware business was in such dire straits that she wrote to Grant begging him to employ John, as well: "I earnestly entreat you will give him a position which will enable him to make his living, something which our community has <u>refused to</u>

permit him to do—He is an earnest & faithful Republican." And unlike Rose, Emma, and Belle, she refused to write a memoir, believing that to do so would be in "coarse taste."

Southern newspapers vehemently opposed Elizabeth's appointment—one opined that Grant had chosen a "dried up maid" who planned to start a "gossiping, tea-drinking quilting party of her own sex"—but she began her work unfazed, moving into the Custom House building downtown, where Jefferson Davis once kept an office on the third floor. Post offices had long been bastions of male privilege ("respectable ladies" were encouraged to avoid them altogether and send servants to pick up the mail) and she set about changing the culture, slowly and methodically, in her own way and time. She requested that Virginia newspapers refer to her as "postmaster" instead of "postmistress" and hired numerous female postal clerks, including her friend Eliza Carrington, whose seamstress had been an integral part of the Richmond Underground. She also employed black postal workers, among them her family's former slaves—a practice that, in the tumultuous atmosphere of Reconstruction, made her unpopular even among some Unionists.

President Grant remained loyal and retained her during his second term. As she had during the war Elizabeth refused to wilt under the increased scrutiny, and flung her behavior in the faces of her critics. She spoke publicly about "the thirst for knowledge among our colored citizens" and sponsored a library for them. Before the presidential election of 1876 she sent an impassioned plea to Northern Democrats, printed in both Washington and Richmond newspapers, urging them to repudiate their counterparts in the former Confederacy. She spoke of votes rigged against Republicans, of Southern whites who still wielded whips, of the "gross personal insults" she constantly endured. She signed off with a lament about her own disenfranchisement. "As a woman," she wrote, "I have no power but through your vote." When her beloved mother, Eliza, died, she couldn't find enough pallbearers to carry the casket. Neighbors ridiculed her service, calling it the "nigger funeral."

After the election of Rutherford B. Hayes, a moderate Republican who promised to end Radical Reconstructionism—the idea that blacks were entitled to the same rights and opportunities as whites—Elizabeth had to fight to keep her job. Her opponents called her erratic, eccentric, mentally unstable, and masculine; any vigorous attempts to defend herself only seemed to prove them right. In May 1877, Hayes replaced her with the moderate William W. Forbes, who at least kept many of her black and female hires.

Elizabeth mourned the loss of not only her position and influence but also her income. Her efforts to sell off various family properties proved fruitless. No one in Richmond would even give her a fair mortgage on the mansion. She put it up for sale, thinking she would never again sit by the fireplace that hid her dispatches or peer into the secret room, but each offer was too insulting to accept. She sought out her old friend Ulysses S. Grant. "I tell you truly and solemnly I have suffered," she wrote. "I have not one cent in the world. . . . I am a woman and what is there open for a woman to do?" She asked Grant if he might persuade current president James Garfield to make her postmaster again. Grant agreed, but his efforts were in vain.

At last Elizabeth retreated, withdrawing entirely from public life. She had no target for her ferocious will. Her one political act was to attach a note of "solemn protest" to her annual tax payment, declaring it unjust to tax someone who was denied the vote. In 1890 she watched as her city erected a sixty-one-foot-tall statue of Robert E. Lee on horseback before a roaring crowd of 100,000. She felt there was no place and no one left for her. Her brother John had moved and was living with his second wife and their children on a farm in Louisa County. He died in 1895 at age seventy-five. His older daughter—her niece Annie—lived with her husband in Massachusetts. Eliza, the younger niece, was all she had, and a strange distance had come between them—the sort of fraught, crackling distance that can come only from being too close. Eliza had never married and still lived with her in the mansion. The same neighbors who had shunned Elizabeth, who had warned their children away from the parched, wizened old lady with the sharp blue eyes and curled, clawlike hands, also grew wary of her niece. The daily rejection and isolation wore at Eliza; they were no longer welcome even at St. John's Church, since Elizabeth had made a habit of barging in late and disrupting the service. The pastor had taken to locking the doors once all of his regulars were inside. She told her aunt that if she had a child and it became a Republican she would kill it rather than watch it suffer her same fate. She cleaned the mansion with a grim, manic energy, flitting like a hummingbird among its fourteen rooms. If Elizabeth tried to stop her Eliza turned on her in fury, ordering her to leave.

Elizabeth Van Lew (bottom left) *in the garden of her mansion, circa 1895.*

Terrified, Elizabeth obeyed, spending hours wandering the city, aware of every person who crossed the street to avoid catching her eye. She began gathering relics of her life, piece by piece, and donated them to the poor: the exquisite china and fine silverware, the antique furniture, the family photos in gilded frames. The famous backyard gardens grew tangled and overgrown, years of planning and care undone. Eliza cleaned and cleaned. Elizabeth asked for one favor: that she leave up the decorations from their last Christmas together, in 1899. In the spring of the following year Eliza became unexpectedly and severely ill, and died on the tenth of May. Elizabeth had hovered by her bedside, weeping, "Save her! Save her! I love her better than anything in the world."

Richmond waited for Elizabeth to follow. In August she read her own obituary, complete with pictures of herself and Ulysses S. Grant and tales of her espionage during the war. "They say I am dead?" she asked a reporter. "Well, I am not, but I am very feeble and sick. My heart is heavy and I am sad. My hours are lonely and long." She died on September 25, 1900, at age eighty-two, and was buried in Shockhoe Cemetery, across from the graves of her parents. Because the family plot had insufficient space, her casket was positioned vertically in the ground. A group of abolitionist admirers in Boston, including Colonel Paul Revere—the grandson and namesake of the Revolutionary War patriot—raised money for a memorial stone:

SHE RISKED EVERYTHING THAT IS DEAR TO MAN— FRIENDS—FORTUNE—COMFORT—HEALTH—LIFE ITSELF—ALL FOR THE ONE ABSORBING DESIRE OF HER HEART—THAT SLAVERY MIGHT BE ABOLISHED AND THE UNION PRESERVED.

Soon after her death, the people of Richmond—adults and children alike—began reporting sightings of the ghost of "Crazy Bet." She haunted her own home, now owned by a civic organization, casting the outline of her figure against the basement walls, scaring the servants to death: "I done hear Miss Lizzie walkin' 'bout," one said. "I knowed all 'long she was here." The city condemned the mansion in 1911 and had it torn down the following year, but Elizabeth's ghost still stalked the streets. Parents warned their children, "Crazy Bet will get you" if they misbehaved.

The sightings continued as late as World War II. In 1943, one Richmond woman, out for a walk with her young son, felt a cat brush against her leg. Suddenly the cats were everywhere—Elizabeth was rumored to have dozens—and then the woman heard the soft rustle of taffeta. When she turned she saw her: Elizabeth, in a black Victorian dress, a beribboned hat perched upon her head, her face like dried fruit beneath its brim. The ghost waved her hands and spoke in crisp, urgent tones. "We must get these flowers through the lines at once," she said, "for General Grant's breakfast table in the morning," and with the push of the wind she was gone.

ACKNOWLEDGMENTS

I hadn't given much thought to the Civil War until one summer day in 2002 when I found myself stuck in traffic on Route 400 outside Atlanta, idling for hours behind a pickup truck emblazoned with a bumper sticker: DON'T BLAME ME—I VOTED FOR JEFF DAVIS. As a native Philadelphian newly transplanted to the Deep South, I was struck by the idea that Civil War personalities and politics lived on, in ways both frivolous and sincere, nearly a century and a half after the last body was buried and the final sacrifice made.

During the long process of researching and writing this book I depended on the assistance, generosity, and kindness of numerous people. First and foremost, I am deeply grateful to Bart Hall, the great-grandson of Elizabeth Van Lew's niece, who graciously shared family lore and insights about Elizabeth's extraordinary life. The next white clam pie (or two) at Pepe's is on me.

Cathy Wright, the curator at the American Civil War Museum (formerly the Museum of the Confederacy) gave me a personal tour of the museum, read an early draft, and shared numerous anecdotes that made this a much richer book. David Gaddy, a former CIA code breaker, answered dozens of questions and vetted the manuscript with patience, humor, and grace. Karen Needles, intrepid director of the Lincoln Archives Digital Project, tracked down countless documents and helped me navigate the National Archives. Matthew Boylan, of the ASK NYPL division of the New York Public Library, lent his formidable research skills and wise counsel. Keith

Hammersla of the Martinsburg Public Library helped me understand, and bring to life, the strange and indomitable Belle Boyd. Ray Swick, historian with the Blennerhassett Museum, located archives that hadn't been seen in decades. Dorothy King cheerfully tackled several queries about all four spies.

Special thanks to my former editor Julia Cheiffetz, whose passion for this story (and for strong female characters in general) inspired me from the very beginning, and to my former editor Susanna Porter, for her friendship and ongoing support.

My editor, David Hirshey, enthusiastically took the reins on this book and guided it expertly to completion. I am eternally grateful for his skills, friendship, humor, and penchant for happy hours. Ditto for Michael Morrison, who has gone above and beyond the call of duty to support me and this book, and who made Harper-Collins feel like home. Barry Harbaugh's sage comments were invaluable during the rewrites. The indefatigable Sydney Pierce is a godsend and a saint.

I owe my deepest gratitude to the entire HarperCollins team: Jonathan Burnham (who offered his spirited support from the beginning); the brilliant and tireless publicity/marketing team of Jane Beirn, Katie O'Callaghan, Tina Andreadis, and Rebecca Welbourn; production editors Emily Walters and David Koral; designer Michael Correy; digital media producer Marisa Benedetto; and all of the reps who worked to get this book into readers' hands.

I am incredibly lucky to have Simon Lipskar as my longtime agent and friend, even though he is the worst sore winner in history. Just wait till next time, babe. Thanks, also, to Joseph Volpe, for keeping all the wheels turning.

Early versions of my manuscript were turned over to a number of trusted friends who were generous enough to offer valuable advice and encouragement. Joshilyn Jackson and Sara Gruen cheered me on from the beginning and talked me off the ledge at regular intervals; I am thankful every day for their friendship. Emma Garman offered her sharp, critical eye and unflinching support in all things.

Gilbert King, who knows a thing or two about telling a good story, made savvy suggestions that improved this book immeasurably. Anna Schachner is a brilliant line editor and a devil with a red pen. Meredith Hindley and Holly Tucker lent their vast historical knowledge and well-honed narrative instincts, and Matthew Goodman his astute judgment and big-picture sensibilities. Lydia Netzer talked me through each character and asked exactly the right questions. Alexandra Shelley's structural suggestions and margin notes proved indispensable. The Writers Room provided me with a peaceful place to write and ample storage for my books and files.

Thanks also to the numerous friends who dragged me out for cocktails, spurred me on, or otherwise helped out along the way: Kathy Abbott, Mary Agnew Turley, Nick Barose, Brooke Berry, Jason Buck and Scott Sjue, Patti Callahan Henry, the Coven (aka Bess Lovejoy, Michelle Legro, and Angela Serratore), Katie Decker, Laura Dittmar Kutina, Jennifer Fales, Chip and Susan Fisher, Tom Frail, Brian Wolly, Beth France and Susan Ciccarone, Jess Graves Bianco, Ramona Huegel, Sandy Kahler, Denise Kiernan, Susan Keyock, Rick Kogan, Erik Larson, Alison Law, Elisa Ludwig, Meenoo Mishra, Melisa Monastero Steinberg, Laura Neilson, Gesha-Marie Bland-Sebrien, Maud Newton, Jack Perry, Renee Rosen, Anne Scarborough, Rachel Shteir, Martin Starger, Neal Thompson, Steven Wallace, the "Warriors" (Vanessa Beasley, Christine Jones, Eric Laursen, Miranda Nessler, Lynn Ramey), Philip Weiss, and Chuck Wilson. An extra huzzah to the marvelous, big-mouthed booksellers who have hawked my wares over the years, and Book Club Queen Kathy Murphy.

And finally to Chuck Kahler: You're the best and I'm the luckiest, and that is the last word.

CREDITS

Grateful acknowledgment is made for permission to reprint the following:

Page xiv: Courtesy of the Florida Center for Instructional Technology, fcit.usf.edu.

Page 13: Library and Archives, Canada.

Page 25: National Archives.

Page 31: Courtesy of the Florida Center for Instructional Technology, fcit.usf.edu.

Page 37: Library of Virginia.

Page 78: Courtesy of Bart Hall.

Page 83: Valentine Richmond History Center.

Page 91: Bentley Historical Library, University of Michigan.

Page 101: American Civil War Museum (formerly Museum of the Confederacy).

Page 301: Library and Archives, Canada.

Page 331: New York Public Library.

Page 332: New York Public Library.

Page 343: Courtesy of Bart Hall.

Page 384: Library of Virginia.

Page 419: South Carolina Confederate Relic Room and Military Museum.

Page 427: Valentime Richmond History Center.

Note: Other images are courtesy of the Library of Congress.

NOTES

Splendid and Silent Suns

xi   "Extry—a *Herald*!": Linden, *Voices from the Gathering Storm*, 197.

xi   "hates the Yankees": Chesnut, *Diary from Dixie*, 48.

xi   "domestic insurrection": Jones, *Rebel War Clerk's Diary*, 12.

xi   "Down with the Old Flag!": Brock Putnam, *Richmond During the War*, 18.

xii   "many a bitter experience": Sandburg, *Lincoln: The War Years*, 3:253.

xii   backing up orders with violence: Harper, *Women During the Civil War*, 341.

xii   "It merely looks unbecoming": Heidler, Heidler, and Coles, *Encyclopedia of the American Civil War*, 2143.

xii   "Wear these, or volunteer": Ibid.

The Fastest Girl in Virginia (or Anywhere Else for That Matter)

3   "the fastest girl in Virginia": Sperry Papers, 72.

4   fifteen thousand of them: There were 14,344 men in the Department of Pennsylvania, but only 3,500 made it across the river to fight at Falling Waters before the Confederates withdrew. US War Department, *The War of the Rebellion: Official Records* (hereafter cited as *OR*): ser. 1, 2:187; Gary Gimbel, "The End of Innocence: The Battle of Falling Waters," *Blue & Gray*, Fall 2005.

4   a few as young as thirteen: Wiley, *Life of Billy Yank*, 299.

4   "romantic spot": Boyd, *In Camp and Prison*, 63.

4   four cannon and 380 boys of his own: *Harper's Weekly*, July 20, 1861; *OR*, ser. 1, 2:187. Colonel Thomas Jackson (he was not yet a general nor nicknamed "Stonewall") reported that 380 men engaged at the battle. In most instances, for the sake of simplicity, I use "General" regardless of a character's specific rank at the time.

4   twenty-one wounded and three Yankee dead: Gary Gimbel, president of the Falling Waters Battlefield Association, e-mail to author, October 2011. These numbers differ from both the Union and the Confederate official reports. "The object of North and South," Gimbel explains, "was to both raise the morale at home by exaggerating the damage to the enemy *and*

by overstating the size of the opposing forces to hopefully receive more support in men and material."

4 enlisted as a private in Company D: Muster roll record for Benjamin Reed Boyd, 2nd Regiment Virginia Infantry, May 11, 1861.

4 a general sadness and depression: Boyd, *In Camp and Prison*, 59.

4 voted three to one against secession: Phillips, "Transfer," 16–17; Keith Hammersla, director of information services at Martinsburg Public Library, e-mail to author, November 2011.

4 five for the Confederacy and two for the Union: *West Virginia Public Affairs Reporter* (Bureau for Government Research, West Virginia University) 19, no. 4, Fall 2002, 7.

4 Citizens formed a volunteer Home Guard: Berkeley County Historical Society, "Martinsburg," 2.

5 "too tame and monotonous": Boyd, *In Camp and Prison*, 60.

5 "fond vows": Ibid., 61.

5 "War will exact its victims": Ibid., 62

5 avoid the "sin of being surprised": Rable, *God's Almost Chosen Peoples*, 133.

5 "Be very careful what you say": Louis Sigaud to Colonel John Bakeless, January 11, 1962, Sigaud Papers.

6 Yankee Doodle: Boyd, *In Camp and Prison*, 69.

6 John O'Neal: Wood, *History of Martinsburg*, 28.

6 rebel troops had destroyed: *New York Times*, June 26, 1861; Berkeley County Historical Society. "Martinsburg, West Virginia," 5.

6 quarrying native limestone: *Martinsburg Journal*, February 14, 2010.

6 Colt 1849 pocket pistol: Simens, historical gun expert and dealer (historical arms.net), e-mail to author, March 2011.

6 "an uprising of the Negroes" and "Northerners were coming down to murder us": Coffin, *Stories of Our Soldiers*, 43.

7 "I am tall": Belle Boyd to her cousin, Willie Boyd, July 22 (no year), Boyd Papers.

7 she carved her name: Scarborough, *Siren of the South*, 7.

7 "my horse is old enough": Sigaud, *Belle Boyd*, 2.

7 "Surely so high a spirit": Ibid.

8 "Mauma Eliza": Coffin, *Stories of Our Soldiers*, 44.

8 she defied the law: Hammersla, e-mail; *Martinsburg News*, June 22, 1951.

8 "Slavery, like all imperfect forms": Boyd, *In Camp and Prison*, 54.

8 five other slaves: 1850 US Census slave schedule for Benjamin R. Boyd.

8 four of her eight children had died: Benjamin Reed Boyd Jr. died in April 1846 at thirteen months; Anna Boyd died in April 1849 at age three; Fannie Boyd died in December 1849 at fourteen months; and Annie Boyd died in March 1851 at seven months. All of the children are buried in Green Hill Cemetery in Martinsburg, West Virginia.

8 "saucy": Coffin, *Stories of Our Soldiers*, 44.

9 "a great big Dutchman": Ibid.

9 "Are you one of those damned rebels?": Ibid.

9   "every member of my household will die": Boyd, *In Camp and Prison*, 73.

9   "very handsome woman": Coffin, *Stories of Our Soldiers*, 44.

10 "prettiest girl in Baltimore": *Augusta (GA) Daily Constitutionalist*, July 19, 1861.

10 "too inhuman and revolting to dwell upon": *Carolina Observer*, June 10, 1861.

10 "Let go my mother!": Coffin, *Stories of Our Soldiers*, 44.

10 looked up at her and grinned: Ibid.

10 "roused beyond control"; "literally boiling": Boyd, *In Camp and Prison*, 73.

10 the force of her shot: Word of the shooting spread quickly among residents of the Shenandoah Valley. Elizabeth Lindsay Lomax, the mother of Confederate general Lunsford Lindsay Lomax, mentioned it in her diary on July 12, 1861, as did Letitia Blakemore of Front Royal, Virginia, on July 24, 1861. The July 7, 1861, issue of the *American Union* (a newspaper briefly published in Martinsburg by the Federal army) reported the death of Private Frederick Martin of Company K, Seventh Pennsylvania Volunteers; I speculate that he was Belle's victim. Elizabeth Lindsay Lomax and Letitita Blakemore, diary entries, Sigaud Papers.

10 "Only those who are cowards": Coffin, *Stories of Our Soldiers*, 47.

Our Woman

11 he posed the question to God: Edmonds, *Nurse and Spy*, 18.

12 suffering from diphtheria: Ward, Burns, and Burns, *Civil War*, 39.

12 "You have pretty good health": Wiley, *Life of Billy Yank*, 23.

12 "two or three little sort of 'love taps'": Stillwell, *Story of a Common Soldier*, 5.

12 "what sort of living": Dannett, *She Rode with the Generals*, 52.

12 "Up to the present": Ibid.

13 "for the defense of the right": *Fort Scott Monitor*, January 17, 1884.

13 Emma was one of fifty thousand Union soldiers: Furgurson, *Freedom Rising*, 116.

13 some came with pieces of rope: Klingaman, *Lincoln and the Road to Emancipation*, 108.

13 Soldiers lounged on the cushioned seats: Edmonds, *Nurse and Spy*, 28.

14 Runaway slaves from Virginia and Maryland: Klingaman, *Lincoln and the Road to Emancipation*, 109.

14 "dress the line": Wright, *Language of the Civil War*, 95.

14 "The first thing in the morning is drill": Davis and Pritchard, *Fighting Men of the Civil War*, 38.

14 Members of the Seventh New York Infantry: Ward, Burns, and Burns, *Civil War*, 39.

14 Many of the immigrant regiments: Ibid., 40.

15 "silver mounted harness": *Fort Scott Monitor*, January 17, 1884.

15 "our woman": Dannett, *She Rode with the Generals*, 247.

15 Emma could be arrested or jailed: Dempsey, *Michigan and the Civil War*, 49.

15 "this great drama": Edmonds, *Nurse and Spy*, 18.

15 Some even turned in wearing coats and boots: Blanton and Cook, *They Fought Like Demons*, 47.

15 took care of "the necessaries": Ibid., 46.

16 stop her menstrual cycle: Ibid.

16 scoured it with dirt: Wiley, *Life of Billy Yank*, 47.

16 As many as four hundred women: Tendrich Frank, *Encyclopedia of Women at War*, 146.

16 one couple even enlisted together: Eggleston, *Women in the Civil War*, 77.

16 twelve-year-old girl who joined as a drummer boy: Blanton and Cook, *They Fought Like Demons*, 35.

16 $13 per month for Union soldiers, $11 for Confederates: Varhola, *Life in Civil War America*, 127.

16 "slavery was an awful thing": Blanton and Cook, *They Fought Like Demons*, 41.

16 "shoulder my pistol and shoot some Yankees": Ibid., 25.

16 "magnetic power": *Fort Scott Monitor*, January 17, 1884.

16 illnesses that would ultimately kill twice as many: Cathy Wright, curator of the American Civil War Museum (formerly Museum of the Confederacy), e-mail to author, December 2011.

16 "Bowels are of more consequence than brains": Davis and Pritchard, *Fighting Men of the Civil War*, 188.

17 "a little eau de vie to wash down the bitter drugs": Edmonds, *Nurse and Spy*, 282.

17 "the stranger"; scene of Emma at home: Dannett, *She Rode with the Generals*, 37–41.

19 "FORWARD TO RICHMOND!": *New York Tribune*, June 25, 1861.

19 didn't even possess a map of Virginia: White, *Lincoln*, 431.

19 "This is not an army": Detzer, *Donnybrook*, 85.

19 "You are green, it is true": McPherson, *Battle Cry of Freedom*, 336.

19 thirty-seven thousand Union recruits: Ward, Burns, and Burns, *Civil War*, 53.

19 "many, very many": Edmonds, *Memoirs of a Soldier*, 32.

A Shaft in Her Quiver

20 "within easy rifle range of the White House": Furgurson, *Freedom Rising*, 126.

21 Bettie Duvall background and mission: Duvall, *Recollections of the War by Grandmama*.

22 "McDowell has certainly been ordered": The piece of paper featuring Beauregard's name is located in the National Archives, RG 59, but the message accompanying it is lost. For a discussion of the value of Rose Greenhow's contributions to the First Battle of Bull Run, see Fishel, *Secret War for the Union*, 59; Blackman, *Wild Rose*, 305–7.

22 "reliable source": Greenhow, *My Imprisonment*, 15.

22 Another source provided her with a map: Ibid., 233.

22 Rose had been the head: Tidwell, *April '65*, 60.

22 "the same kind of intimacy": McKay, *Henry Wilson*, 152; Nevins, *Hamilton Fish*, 609.

23 "indispensable to the peace and happiness": Calhoun, *Speeches*, 630.

23 "positive good": Reynolds, *John Brown, Abolitionist*, 439.

23 "the best and wisest man of this century": Greenhow, *My Imprisonment*, 59.

23 "glorious as a diamond": *New York Times*, April 12, 1858.

23 "is no more": Blackman, *Wild Rose*, 165.

23 "confidential relations": Frémont, *Letters*, 504.

23 money she lost speculating in stocks: Blackman, *Wild Rose*, 12.

23 compared Rose to the notorious Peggy O'Neale Eaton: Ross, *Rebel Rose*, 7.

24 "hunted man with that resistless zeal": Tidwell, *April '65*, 58.

24 "one of the most persuasive women": Keyes, unpublished autobiography, 330–31.

24 "Believe me, my dear": Joseph Lane to Rose O'Neal Greenhow, undated, Greenhow, seized correspondence.

24 "You know that I do love you": Henry Wilson to Rose O'Neal Greenhow, Greenhow, seized correspondence. Historians have long debated the exact nature of Rose Greenhow's relationship with Wilson. Ernest McKay (*Henry Wilson*, 154) wrote that the official clerk of the Military Affairs Committee was a young man with the initials "H. W."—Horace White—and that his penmanship was similar to that in the love letters. On the other hand, Hamilton Fish, who served with Wilson in the senate and was Ulysses S. Grant's secretary of state while Wilson was his vice president, was among those who later claimed that Wilson and Rose had a romantic relationship; Leech, *Reveille in Washington*, 137.

24 cipher of the type used in Edgar Allan Poe's "The Gold-Bug": David Gaddy, retired CIA code breaker, e-mail to author, December 2011.

25 Thomas J. Rayford: Thomas Jordan to Judah P. Benjamin, October 29, 1861, *OR*, ser. 1, 5:928.

25 "Infantry" was two parallel lines: Greenhow Papers, North Carolina State Archives.

25 Jordan pointed out that the upper windows: Burger, *Confederate Spy*, 66.

26 forced to take seamstress jobs: Blackman, *Wild Rose*, 7.

26 "I am so much worried about the latest news": Florence Moore to Rose O'Neal Greenhow, June 23, 1861, Greenhow, seized correspondence.

26 "beasts of the field": Greenhow, "European Diary," 9.

26 "every capacity with which God has endowed me": Greenhow, *My Imprisonment*, 193.

26 less than 3 percent: Mitchell, *Maryland Voices of the Civil War*, 9.

26 barely 1 percent: Holzer, *Lincoln President-Elect*, 42.

26 "smack of sympathy": Leech, *Reveille in Washington*, 136.

27 in the habit of eavesdropping outside the Cabinet room doors: Bakeless, *Spies of the Confederacy*, 5; Morrow, *Mary Todd Lincoln*, 51.

27 The Confederate government had no fund set aside: Gaddy, e-mail, February 2013.

27   "a railroad man willing to undertake": Bakeless, *Spies of the Confederacy*, 16.
27   If a scout was captured: Gaddy, "Lee's Use of Intelligence."
27   Her group included a lawyer: Ross, *Rebel Rose*, 154–55.
28   Battery Martin Scott: Cooling and Owen, *Mr. Lincoln's Forts*, 142.
28   "beautiful young lady": Davis, *Battle at Bull Run*, xi.
30   "Trust Bearer": Williams, *P. G. T. Beauregard*, 76.
30   "order issued for McDowell": Ibid.
30   "This must go thro' ": Tidwell, *April '65*, 64.
30   relay system designed for maximum efficiency: Ibid., 65.
30   "Let them come": Greenhow, *My Imprisonment*, 16.

As If They Were Chased by Demons

31   "a bristling monster lifting himself": Eicher, *Longest Night*, 92.
32   considered the best cure for diarrhea: Flannery, *Civil War Pharmacy*, 128; Smith, "Polite War."
32   wrestle beehives from stands: Haydon, *For Country, Cause and Leader*, 52.
32   boots that didn't distinguish right from left: Ward, Burns, and Burns, *Civil War*, 220.
32   "By the left flank, *march!*": Detzer, *Donnybrook*, 286.
32   As she cupped and lifted his head: Edmonds, *Nurse and Spy*, 40.
33   "We've whipped them!": Wheeler, *Voices of the Civil War*, 36.
33   currently detailed as a clerk: Muster roll record for Benjamin Reed Boyd, 2nd Regiment Virginia Infantry, May 11, 1861.
33   she would later claim he took a bullet: Coffin, *Stories of Our Soldiers*, 47.
33   "Look, there is Jackson": The origin of Jackson's nickname was widely reported, including in Beauregard's recollection of First Bull Run. Jackson's hometown paper, the *Lexington Gazette*, did not print stories of his earning the nickname "Stonewall" until August 25, more than a month after the battle.
33   One private from the 4th South Carolina: This Bible is on display at the American Civil War Museum.
33   Union shells tore through the wall: Davis, *Battle at Bull Run*, 37.
33   "There is nothing like it": Kagan and Hyslop, *Eyewitness to the Civil War*, 152.
34   "Hoo-ray! Hoo-ray!": Dew, *Yankee and Rebel Yells*, 955.
34   "In reckless disorder the enemy fled": Brock Putnam, *Richmond During the War*, 62.
34   Every angle, every viewpoint, offered a fresh horror: *New York Times*, July 24, 1861; *Harper's Weekly*, August 17, 1861; Wilson, *Sufferings Endured*, 24–25. Throughout the war both the North and the South exaggerated the atrocities committed by the enemy, and it's difficult to determine which incidents were real and which were apocryphal (although there is strong documentation that Confederates unearthed Yankee graves and devised creative uses for the bones). Such atrocities may have been more common early in the war because soldiers feared it would be over soon, and that they would

not have many chances to acquire a macabre souvenir; Wright, e-mail, May 2012.

35 "great skedaddle": Bulla and Borchard, *Journalism in the Civil War Era*, 45.

35 "Many that day who turned their backs": Edmonds, *Nurse and Spy*, 41.

35 "Our President and our General": Greenhow, *My Imprisonment*, 18.

36 "got shorn": Ward, *Slaves' War*, 59.

Never as Pretty as Her Portrait Shows

37 "palpable state of war": Elizabeth Van Lew, "Occasional Journal," Van Lew Papers, New York Public Library. The original journal papers are out of order and lack page numbers.

38 "hooted at and insulted him": Speer, *Portals to Hell*, 23.

38 "stirring up the animals": *Norwich* (CT) *Morning Bulletin*, August 29, 1861.

38 "What did you come here for?": Ibid.

38 "never as pretty as her portrait shows": Anne B. Hyde to India Thomas, November 6, 1957, courtesy of Cathy Wright.

39 a pedigree that prevented her: For a discussion of Richmond society during wartime, see Furgurson, *Ashes of Glory*, 74–77.

39 "It was my sad privilege": Van Lew, "Occasional Journal."

39 "Left the Tredegar Iron Works": *Richmond Dispatch*, July 24, 1861.

39 "is arrogant—is jealous and intrusive": Van Lew, "Occasional Journal."

39 "Good Miss Van L. could not refrain": Bremer, *Homes of the New World*, 509–10.

40 standing in the Van Lew family pew: For details about Elizabeth Van Lew's family background and espionage operation I consulted Bart Hall, the great-grandson of Elizabeth's niece, Annie Van Lew Hall.

40 the stipulation that she was not to sell or free: Will of John Newton Van Lew, signed October 2, 1843, Local Records Collection, Library of Virginia.

40 system of "hiring out": For a discussion of this practice, see Varon, *Southern Lady, Yankee Spy*, 27–28.

40 "From what I have seen of the management": Testimony of George Watt, August 1, 1864, Letters Received by the Confederate Adjutant and Inspector General, RG 109, Entry 12, Box 9, Folder 3, National Archives.

41 "Loyalty was called treason": Van Lew, "Occasional Journal."

41 "butchery, rape, theft, and arson": *Richmond Dispatch*, September 13, 1861.

41 "damned rascally": Coulter, *Confederate States of America*, 90.

41 "kill as many Yankees as you can for me": Van Lew, "Occasional Journal."

41 "I longed to say to them": Ibid.

41 "Mr. Lincoln's head or a piece of his ear": Ibid.

41 "offended and disgusted": Watt, testimony.

42 sewed a representative star: Bart Hall, e-mail to author, February 2012.

42 "calm determination and high resolve": Van Lew, "Occasional Journal."

42 "Keep your mouth shut": Elizabeth's childhood notebook is on display in the "Sisterhood of Spies" exhibit in the International Spy Museum, Washington, DC.

42 When the war broke out, Lieutenant Todd: Speer, *Portals to Hell*, 162–63.

42 "violent appearance": Van Lew, "Occasional Journal."

42 "I would like to be made hospital nurse": Ibid.

43 "You are the first and only lady": Ibid.

43 "Trial of the Ploughs": *Alexandria Gazette*, November 2, 1854.

43 succumbing to the yellow fever epidemic: Hall, e-mail, September 2013.

43 "Let me see the prisoners": Van Lew, "Occasional Journal."

44 "I could not think of such a thing": Ibid.

44 "Once I heard you at a convention": Ibid.

44 when a projectile struck a nearby soldier: Welsh, *Medical Histories*, 237.

44 great regret: Blakey, *General John H. Winder*, 8.

44 Winder's oldest son, William: Ibid.

45 "alien plug-uglies": Ibid., 51.

46 "Your hair would adorn the temple of Janus": Van Lew, "Occasional Journal."

46 "I should be glad to visit the prisoners": Ibid.

46 "I can flatter almost anything": Blakey, *General John H. Winder*, 53.

46 "Two ladies, a mother and a daughter": *Richmond Examiner*, July 29, 1861.

46 "They are Yankee offshoots": *Richmond Dispatch*, July 31, 1861.

47 Alien Enemies Act: *OR*, ser. 2, 2:1370.

47 "These ladies were my mother and myself": Clipping, Van Lew Papers, NYPL.

47 hoping to be made governor of Virginia: Ely, *Journal*, 236; *Carolina Observer*, December 2, 1861.

47 "Poor Calvin Huson": *Charleston Daily Courier*, October 19, 1861.

48 hanging boldly on her entry parlor wall: Photo in the Eleanor S. Brockenbrough Archives, American Civil War Museum.

Little Rebel Heart on Fire

49 Private Frederick Martin: The official Pennsylvania records confirm his existence but not his death or burial (Bates, *History of Pennsylvania Volunteers*, 69). War Department records confirm his death (cause unknown) but omit the place of burial; Sigaud, *Belle Boyd*, 15. As previously mentioned, the *American Union* of July 7, 1861, reported Martin's death and named Martinsburg as the place of his burial.

49 a violation that would degrade and declass: Harper, *Women During the Civil War*, 220.

49 "one shadow of remorse": Boyd, *In Camp and Prison*, 72.

50 "left no stain": Ibid.

50 "insult and outrage": Ibid., 73.

50 Washington was still practicing appeasement: Sigaud, *Belle Boyd*, 16.

50 "plucky girl" and "do it again": Coffin, *Stories of Our Soldiers*, 47.

50 any soldier who insulted a lady: *Boston Post*, July 15, 1861.

50 "The old ass thinks he can starve us out": *Richmond Dispatch*, May 18, 1861.

50 Merchants reopened for business: Berkeley County Historical Society. "Martinsburg, West Virginia," 10–11. Gold and silver were scarce by 1862.

50   "thronged the streets in perfect security": *New York Times*, July 6, 1861.
51   They operated on varying levels of importance and authenticity: Gaddy, e-mail, June 2013.
51   "did a little spying": Bakeless, *Spies of the Confederacy*, 130.
51   slipping questions between the pauses: Belle left no record of the exact questions she posed to her guards, but on July 5 a Martinsburg correspondent reported a rumor that McClellan's column was "two days' march from us"; *New York Times*, July 6, 1861.
51   although she kept the pistol that killed him: *New York Times*, July 20, 1862.
52   *lettres de cachet*: Boyd, *In Camp and Prison*, 78.
52   "lovely girl" named Sophia B.: Ibid.
52   "a loud, coarse laugh": Sigaud, *Belle Boyd*, 100; Sandburg, *Lincoln: The Prairie Years and the War Years*, 504.
53   She summoned the bravery: Boyd, *In Camp and Prison*, 77.
53   Ward Hill Lamon . . . had intervened on her behalf: Sigaud, *Belle Boyd*, 19. If President Lincoln intervened, it would have been at Lamon's behest.
53   "Thank you, gentlemen of the jury": Boyd, *In Camp and Prison*, 78.
53   "My little 'rebel' heart": Ibid.
54   "Miss D. was a lovely": Boyd, *In Camp and Prison*, 91.
54   "intrepidity and devotion": Ibid., 92.
54   Her first cousin . . . was a spy himself: Louis Sigaud, "William Boyd Compton, Belle Boyd's Cousin," *Lincoln Herald 67* (Spring 1963): 22–23.
54   "at home on a horse's back"; "a good deal of boy myself": Coffin, *Stories of Our Soldiers*, 44.
54   where Confederate soldiers engraved their names: Miller, *Kartcher Caverns*, 24.
54   She even trained her beloved horse: Sigaud, *Belle Boyd*, 25.
54   she began riding as a courier: Boyd, *In Camp and Prison*, 83.
55   expected to deliver it orally: Gaddy, e-mail, March 2013.
55   the challenge being "Stonewall": Ibid.
55   "We have the same old signal": Reid Hanger, *Diary*, October 20, 1861.
55   one, a boy exactly her age: *Clarksville* (TX) *Standard*, July 27, 1861.
55   "Where are you going?": Coffin, *Stories of Our Soldiers*, 49.
55   exchanges detailing a coordinated effort: *OR*, September 7, 1861, ser. 1, 5:587.
56   "I have no papers": Dialogue from Coffin, *Stories of Our Soldiers*, 49.

Admirable Self-Denial
57   "One case I can never forget": Edmonds, *Nurse and Spy*, 46.
57   "simply eyes, ears, hands": Ibid., 58.
57   "I was not in the habit": Ibid., 60.
57   its most depressing form: Ibid., 55.
58   Greeley removed the "Forward to Richmond!" banner: Foote, *Fort Sumter to Perryville*, 85.
58   "On every brow sits sullen": Ward, Burns, and Burns, *Civil War*, 60–61.

58   "grace and dignity": Edmonds, *Nurse and Spy*, 61.

58   "hatred of male tyranny": *Fort Scott Monitor*, January 17, 1884.

59   able to bend a quarter: Ward, Burns, and Burns, *Civil War*, 61.

59   "By some strange operation of magic": Eicher, *Longest Night*, 101.

59   "The army under McClellan began to assume": Edmonds, *Nurse and Spy*, 61.

59   "perfect pandemonium": *McClellan's Own Story*, 67.

59   all work would be suspended on the Sabbath: *Vermont Journal* (Windsor, VT), September 14, 1861; *New York Times*, September 8, 1861.

59   "utmost decorum and quiety": *Vermont Journal*, September 14, 1861.

60   splendid: Edmonds, *Nurse and Spy*, 61.

60   watching the festivities from a carriage: Ward, Burns, and Burns, *Civil War*, 61.

60   "They received him with loud shouts": *Works of Nathaniel Hawthorne*, 8:133.

60   "Under the auspices of the 'Young General' ": Greenhow, *My Imprisonment*, 35.

61   infested by maggots and weevils: Billings, *Hardtack and Coffee*, 115.

61   "the President is nothing more than a well-meaning baboon": quoted in McPherson, *Battle Cry of Freedom*, 364; Sears, *Young Napoleon*, 132.

61   "If the men pursue the enemy": Haydon, *For Country, Cause and Leader*, 4.

62   she could be arrested for prostitution: Blanton and Cook, *They Fought Like Demons*, 124. Commanding officers were likely to have feared censure from superiors for allowing a woman to serve.

62   "one of the inmates of a disreputable house": Ibid.

62   "implacable enemy" of her sex: *Fort Scott Monitor*, January 17, 1884.

62   "going down the line": Wright, *City Under Siege*, 129.

62   names of brothels: Lowry, *Story the Soldiers Wouldn't Tell*, 63–64.

62   An estimated fifteen thousand black, white, and mulatto streetwalkers: Furgurson, *Freedom Rising*, 207.

62   "handle a gun": Lowry, *Story the Soldiers Wouldn't Tell*, 24.

63   "truly wife-like in their tented seclusion": Wiley, *Life of Johnny Reb*, 52.

63   "Almost all the women are given to whoredom": Clinton, *Public Women and the Confederacy*, 18.

63   "Oh, how my heart has ached": Edmonds, *Nurse and Spy*, 64.

63   "was in the habit of eating rebels": Dannett, *She Rode with the Generals*, 209.

64   "cow-bell dodge": *New York Times*, September 25, 1861.

64   trading tobacco and newspapers and buttons from their coats: *Philadelphia Inquirer*, October 21, 1861.

64   Munson's Hill: *Charleston Daily Courier*, September 23, 1861.

64   With Emma leading the way: Edmonds, *Nurse and Spy*, 62.

The Birds of the Air

66   offal rotting three feet deep: Leech, *Reveille in Washington*, 102.

66   whole situation amusing: Greenhow, *My Imprisonment*, 39.

66   every movement and act: "H" to Rose Greenhow, Greenhow seized corre-

spondence.

66   "The Southern women of Washington": Greenhow, *My Imprisonment*, 39.

67   "Ape" Lincoln: Tagg, *Unpopular Mr. Lincoln*, 73.

67   "touching simplicity": Greenhow, *My Imprisonment*, 49.

67   "a short, broad, flat figure": Ibid., 201.

67   gaudy flowers: Ibid., 202.

67   "I don't think": Ibid., 202.

68   "It gives a quaint look": Chesnut, *Diary from Dixie*, 167.

68   $40,000, likely in donations from private citizens: Tidwell, *April '65*, 60.

68   three gunshots from the provost marshal's office: Greenhow, *My Imprisonment*, 37.

68   "thereby creating still greater confusion": Ibid., 38.

68   "expects to surprise you": These letters were pieced together from fragments after Rose's arrest. Proceedings of the Commission Relating to State Prisoners, Records of the Department of State, RG 59, E962, National Archives.

69   "Tonight, unless *Providence* has put its foot": "H" to Rose O'Neal Greenhow, Greenhow, seized correspondence.

69   "information on the movements": Mortimer, *Double Death*, 36.

69   "Chief of the United States Secret Service": Fishel, *Secret War for the Union*, 54.

69   "In operating with my detective force": Pinkerton, *Spy of the Rebellion*, 247.

69   "a sort of synonym for 'detective' ": Ibid., 156.

70   "bland gentleman with distinguished black whiskers": Taft Bayne, *Tad Lincoln's Father*, 139–40.

70   "What's wrong with Mrs. Greenhow?": Ibid.

71   "Very well": Ibid.

71   "That is treason": Greenhow, *My Imprisonment*, 29.

71   "My remarks were addressed": Ibid.

71   "Madam, if he insults": Ibid., 30.

72   "Oh, never mind": Ibid.

72   "beware": Ibid.

72   "McClellan is vigilant": Proceedings of the Commission Relating to State Prisoners.

72   "his manner of polishing": Waugh and Greenberg, *Women's War in the South*, 37.

72   "Lt. Col. Jordan's compliments": Greenhow, *My Imprisonment*, 55.

72   she showed it to Senator Wilson: Ibid.

72   "McClellan is very active": Proceedings of the Commission Relating to State Prisoners.

72   "an occasional and *useful* visitor to my house": Greenhow, *My Imprisonment*, 38.

73   The Confederate States of America had established a Post Office Department: Wright, e-mail, December 2011.

73   "McClellan's excessive vigilance": Greenhow, *My Imprisonment*, 38.

73 "always very sorry that no opportunity": Ibid.

73 "Do not talk with anyone about news": Proceedings of the Commission Relating to State Prisoners.

74 a customary closing among the betrothed: Gaskell, *Compendium of Forms*, 233.

74 "I was slow to credit": Greenhow, *My Imprisonment*, 40.

74 She began urging secessionist friends: Ibid., 57.

74 fastening a pearl-and-ivory tablet: Martha Elizabeth Wright Morris, "Memories," diary, Wright Morris Papers.

The Secret Room

75 background on Calvin Huson: *Albany Evening Journal*, October 19, 1861; *Philadelphia Inquirer*, October 19, 1861; US Census, 1860.

75 "aid and comfort" to a "Black Republican enemy": Ely, *Journal*, 166.

75 enemas consisting of oil of turpentine: Beach, *American Practice of Medicine*, 402.

75 In tears, she told Congressman Alfred Ely: Ely, *Journal*, 166.

76 "You dare to show sympathy": Van Lew, "Occasional Journal."

77 moved into her home: Ibid.

77 Salt, which used to sell: Coulter, *Confederate States of America*, 220.

77 placed ads seeking hunting dogs: Ibid.

77 stamped with figures of lions: John Albree notes, Box 1, Folder 6, Elizabeth Van Lew Papers, Special Collections Research Center, Swem Library, College of William & Mary.

77 "not a man of much intellect": Van Lew, "Occasional Journal."

77 "I would give my right arm": Ibid.

77 background on Mary Carter West Van Lew: Hall, e-mail, October 2011.

78 "niggers": Hall, e-mail, March 2013.

79 "The Negroes have black faces": Albree notes, Elizabeth Van Lew Papers, College of William & Mary.

79 which had been built: Hall, e-mail, March 2013.

79 Mary Jane had married another Van Lew servant: Varon, *Southern Lady, Yankee Spy*, 166.

79 They began sleeping in separate chambers: Hall, e-mail, March 2013.

80 the escape of eleven prisoners: Gibbs to Winder, September 7, 1861, *OR*, ser. 2, 3:718.

80 usually horse or mule meat: Speer, *Portals to Hell*, 125.

80 "I should have perished for want": Testimony of Lewis Francis, US Congress, *Report of the Joint Committee on the Conduct of the War*, 477–78.

80 "The custard was very nice": Van Lew, "Occasional Journal"; Wheelan, *Libby Prison Breakout*, 87.

80 "God help us": Van Lew, "Occasional Journal."

80 somehow slipped past their guards: *Norwich* (CT) *Morning Bulletin*, August 29, 1861.

81 several soldiers from New York: *New London* (CT) *Daily Chronicle*, October 12, 1861.

81 distinguished by a bit of red ribbon: *San Francisco Bulletin*, October 11, 1861.

81 the secret room: Albree notes, Elizabeth Van Lew Papers, College of William & Mary.

82 "dread and fear the Yankee": Brock Putnam, *Richmond During the War*, 102.

82 he could leave Church Hill and work: Hall, e-mail, March 2013.

82 an ad in the *Richmond Dispatch*: *Richmond Dispatch*, August 14, 1861.

82 "get suitable servants": Silber, *Landmarks of the Civil War*, 38.

83 Mary Jane was highly educated: Historians have long debated the role Mary Jane Richards/Mary Jane Bowser (often called Mary Elizabeth Bower) played in Elizabeth Van Lew's espionage ring. In 1910 Elizabeth's niece, Annie Randolph Van Lew Hall, gave an interview in which she named Bowser and described her as a "favorite" servant of the Van Lews. During the war, she said, "Mary sort of disappeared" from the Van Lews' home, a disappearance that coincided with her employment at the Confederate White House. Bart Hall, Annie's great-grandson, told me that he heard about Mary Bowser from his grandfather (Annie's son), an army captain during World War I: "Because of my grandfather's interest in intelligence, his mother [Annie, then in her early seventies] conveyed to him the information [about the ring] she had learned from her aunt when visiting her in Richmond in the 1880s and early 1890s." In 1867 Reverend Crammond Kennedy, the secretary for the American Freedmen Union Commission, wrote in the *American Freedman* that Bowser, "while appearing as a slave, was in the secret service of the U.S. She could write a romance from her experience in that employment." In an 1867 letter to a Freedman's Bureau official, Bowser herself wrote of "having been in the service . . . as a detective"; John Reynolds interview with Annie Hall, William Gilmore Beyer Papers, UTA; Hall, e-mail, November 2011; Mary Richards to G. L. Eberhardt, April 4, 1867, Frank Wuttge Jr. Research Files, Schomburg Center, New York Public Library. For a discussion of Mary Bowser's identity and various aliases, see Varon, *Southern Lady, Yankee Spy*, 165–68. I use "Mary Jane" in the text to distinguish her from Mary Van Lew, Elizabeth's sister-in-law.

Stakeout

84 three of his best detectives: Mortimer, *Double Death*, 103.

84 Pinkerton strolled its perimeter: Scene of the stakeout based on Pinkerton, *Spy of the Rebellion*, 253–66.

85 he flipped the slats of the blinds until the parlor: Ibid., 256. Rose's home probably had operable louver bifold shutters, which were popular with the middle class by the mid-nineteenth century. Wright, e-mail, September 2011.

86 Captain John Elwood: Rose's visitor was most likely Captain Elwood, Fifth Infantry. Pinkerton uses a pseudonym, "Captain Ellison," in his memoir *Spy of the Rebellion*. Bakeless, *Spies of the Confederacy*, 39; Van Doren Stern, *Secret Missions*, 61; Louis A. Sigaud, "Mrs. Greenhow and the Rebel Spy Ring," *Maryland Historical Magazine* 41–42 (1946): 177.

Hard to Name

89 In reality the Confederate troops numbered: McPherson, *Illustrated Battle Cry of Freedom*, 296.

89 "by the sound of gongs": Sears, *Young Napoleon*, 164.

89 "the panic is great": Proceedings of the Commission Relating to State Prisoners.

89 plum-sized swellings: Lowry, *Story the Soldiers Wouldn't Tell*, 104.

90 "black wash": Bumstead, *Pathology and Treatment of Venereal Diseases*, 418.

90 "Wash as fast as you can": Marten, *Civil War America*, 116–17.

90 "plain-looking" and modestly dressed: Brown, *Dorothea Dix*, 303.

91 "To me, nothing in the whole of human actions": Jerome Robbins, journal, Robbins Papers.

92 "I felt as if an angel had touched me": *Fort Scott Monitor*, January 27, 1884.

92 "had no higher ambition": Ibid.

92 "In our family the women were not sheltered": Ibid.

93 "I had a very pleasant conversation": Robbins, journal.

93 "if there were instances of it in the Bible": Dannett, *She Rode with the Generals*, 20.

93 "I visited my friend Thompson this evening": Robbins, journal.

94 "a pretty little girl": *Fort Scott Monitor*, January 27, 1884.

94 "I arose greatly refreshed": Robbins, journal.

94 "I revere as a blessing": Ibid.

94 "I can do it all": Sears, *Young Napoleon*, 125.

95 "a delicious morsel for our thirsty souls": Robbins, journal.

95 "My time is greatly eased": Ibid.

95 "the only young lady correspondent": Ibid.

95 told Jerome she didn't feel well: Ibid.

Crinoline and Quinine

96 The want-ad columns in newspapers: Van Doren Stern, *Secret Missions*, 318.

96 A Doctor Line: Hartzler, *Marylanders in the Confederacy*, 52.

97 A Postmaster Line: Hastedt, *Spies, Wiretaps*, 193.

97 met at the home of a Mrs. Jack Taylor: Bush, *Louisville and the Civil War*, 77.

97 The proprietor of a nearby boardinghouse: Ibid.

97 "could take hints quickly": Pavlovsky, "Riding in Circles," 272.

97 she reached into her purse and produced a white dog skin: *Atlanta Constitution*, October 10, 1919.

97 "Some of the old and ugly ladies": George Cadman to Esther Cadman, May 24, 1862, Cadman Papers.

97 "Some of the boys met the woman Belle Boyd": Marvin, *Fifth Regiment, Connecticut Volunteers*, 144–45.

98 At other times she wore a rebel soldier's belt: *New York Tribune*, June 4, 1862.

98 she could recite the names of every general: Ibid.

99 seized twenty-one thousand bushels of wheat: Long and Long, *Civil War Day by Day*, entry for October 16, 1861.

99 "My dear Beauty": A copy of this note, in Belle's handwriting, belonged to Sue Stribling Snodgrass, a longtime Martinsburg resident whose sisters were Belle's childhood playmates. Sue's great-granddaughter, Lyn Snodgrass, kindly shared the note with me.

99 One day the 28th Pennsylvania: Bates, *History of Pennsylvania Volunteers*, 419.

99 "I had been confiscating and concealing": Boyd, *In Camp and Prison*, 77.

100 Some merchants accepted sewing pins: Wright, e-mail, December 2013.

100 newspapers printed suggestions for ersatz brews: *Augusta* (GA) *Daily Chronicle & Sentinel*, August 25, 1861.

100 Powers Weightman and Rosengarten Sons: Flannery, *Civil War Pharmacy*, 99–100.

100 A farmer named Thomas A. Jones: Taylor, *Signal and Secret Service*, 20.

101 some of them conducted by nine-year-old Robert Fitzgerald: *Guardian*, September 18, 2007.

102 Some male agents used an acorn-shaped brass contraption: Cathy Wright told me that a visitor brought in this "anal acorn," which had belonged to her Civil War veteran ancestor.

102 One woman managed to conceal: Pryor, *Reminiscences*, 223.

102. Another found a functioning pistol: Royce, *Genteel Spy*, 41–42.

102 Mothers packed quinine in sacks of oiled silk: The American Civil War Museum owns two such dolls.

102 offering to deliver letters to Richmond for $3 each: *Newark Daily Advertiser*, July 30, 1862.

102 "as grand a flirt as ever lived": Hassler, *Colonel John Pelham*, 53.

103 The army was experiencing daily desertions: *Harrisburg Patriot*, November 14, 1861.

103 Alcohol was at times rationed out to the men: Hambucken and Payson, *Confederate Soldier*, 63.

103 Beauregard served special guests: Williams, *P. G. T. Beauregard*, 99.

103 "bloody fracas": *Leavenworth* (KS) *Daily Times*, November 23, 1861.

103 more than twice the number of casualties: *New York Times*, November 24, 1861.

Dark and Gloomy Perils

104 "by no means a person of sharp": *Washington Evening Star*, December 4, 1862.

104 the widow shared Rose's secessionist sympathies: *National Aegis*, December 13, 1862.

105 At least, she thought, she would enjoy even greater access: Greenhow, seized correspondence.

105 "A number of prominent gentlemen": E. J. Allen (Pinkerton) to Brigadier General Porter, November 1861, *OR*, ser. 2, 5:566–69.

105 "Tell me what to send you": Greenhow, seized correspondence.

106 "a distinguished member of the diplomatic corps": Greenhow, *My Imprisonment*, 53.

106   "Those men will probably arrest me": Ibid.

106   "The fate of some of the best and bravest": Ibid., 54–55.

106   "a German Jew": Ibid., 203.

106   "Is this Mrs. Greenhow?": Ibid., 54.

107   "I have no power": Ibid.

107   "That would have been wrong": Ibid.

107   "Mother has been arrested!": Taft Bayne, *Tad Lincoln's Father*, 62.

107   "Take charge of this lady": Mortimer, *Double Death*, 105.

108   unlettered scribblings: Greenhow, *My Imprisonment*, 56.

108   concerning his "feebleness": Greenhow, seized correspondence.

108   "Wants to see Mrs. G very much": Ibid.

108   "a beautiful woman": Mortimer, *Double Death*, 105.

109   "winning way": Ibid.

109   "If I had known who you were": Ibid.

109   "The revolver has to be cocked": Ibid.

109   "one of those India rubber dolls": Greenhow, *My Imprisonment*, 62.

110   "I began to realize": Ibid., 65.

110   "I did not know what they had done": Ibid., 58.

110   "slaves of Lincoln": Mortimer, *Double Death*, 106.

110   "nice times": Greenhow, *My Imprisonment*, 65.

111   house on fire: Ibid., 66.

111   beneath her mattress: Ibid., 68.

111   "allowed the clue to escape them": Ibid., 77.

111   "so noble a lady": Ibid., 70.

112   "I led Pat a dance": Ibid., 71.

112   "sublime fortitude": Ibid.

112   "know that you had forgiven me": Ibid.

113   "Little Bird": Ibid., 110. Family members later suspected the intermediary Rose called "Little Bird" to be Little Rose; Blackman, *Wild Rose*, 194.

113   invariably peering inside: Greenhow, *My Imprisonment*, 72.

113   to send an armada: Ibid., 110.

113   known derisively as the Mosquito Fleet: Wright, e-mail, June 2011.

114   "Tell Aunt Sally that I have some old shoes": Greenhow, *My Imprisonment*, 92.

114   "for a clever woman": Ibid.

114   "artillery is constant and severe": Proceedings of the Commission Relating to State Prisoners.

114   "I have signals": Ibid.

115   corresponding with the enemy: Blackman, *Wild Rose*, 193.

115   "complimented as being equal to": Greenhow, *My Imprisonment*, 81.

115   after taking an oath of allegiance: William H. Seward to Col. Martin Burke, *OR*, ser. 2, 2:597.

116   "fashionable woman spies": McCurry, *Confederate Reckoning*, 100.

116   "The 'heavy business' in the war of spying": *Albany Evening Journal*, October 5, 1861.

116 calls to condemn her to the ducking stool: *Weekly Wisconsin Patriot*, September 7, 1861.

116 "Let it come": Greenhow, *My Imprisonment*, 96.

116 "I felt now that I was alone": Ibid., 101.

117 "she had had too much of my teachings": Ibid., 134.

117 "Your whole bankrupt treasury": Ibid., 83.

118 "beyond a woman's ken": Ibid., 120.

118 "cruel and dastardly tyranny": Wallenstein and Wyatt-Brown, *Virginia's Civil War*, 126.

118 "almost irresistible seductive powers": E. J. Allen (Pinkerton) to Brig. Gen. Andrew Porter, *OR*, ser. 2, 2:567.

118 "I crushed down the impulse": Greenhow, *My Imprisonment*, 166.

118 "a party well known to the government": Ibid., 210.

Unmasked

120 "But the doctor must be consulted": Edmonds, *Nurse and Spy*, 60.

121 "a mixture of awful consternation": *Walt Whitman's Civil War*, 25.

121 "brilliant beyond description": Lesser, *Rebels at the Gate*, 239.

122 "For God's sake heed this": Jordan to Beauregard, *OR*, ser. 1, 5:1038.

122 "No living man ever made such a desperate effort": Ibid.

122 Mary Todd Lincoln had donated all unsolicited gifts: Harvey Baker, *Mary Todd Lincoln*, 186.

The Defenseless Sex

124 "Nothing is so hideous": *Richmond Dispatch*, August 14, 1861.

125 John and Elizabeth refused to entrust their education: Hall, e-mail, April 2013.

125 "If one Confederate soldier kills 90 Yankees": Wiley, *Life of Johnny Reb*, 123.

125 "I will try to do the best I can by her": Varon, *Southern Lady, Yankee Spy*, 30.

125 first-class cabin instead of steerage: Ibid. Elizabeth's request was not honored.

125 "perambulating the streets": Ibid.

127 earned $137 per year: Lebergott, "Wage Trends, 1800–1900," 453.

128 "trumpery": Varina Davis, *Jefferson Davis*, 576.

128 Every evening before bed: Allen, *Jefferson Davis*, 292.

128 dark skin and "tawny" looks: *Richmond Dispatch*, December 11, 1861.

128 calling her a mulatto and a "squaw": Cashin, *First Lady of the Confederacy*, 119.

128 most men "dressed right": Wright, e-mail, June 2013.

128 "half-breed Yankee on one side": Cashin, *First Lady of the Confederacy*, 299.

129 "harbinger of gladness in the future": *Richmond Whig*, December 2, 1861.

129 how fortunate that her injuries weren't grave: *Richmond Enquirer*, October 5, 1861.

130 "an excellent house servant": Dorothy Lewis Grant, "Lady of Refinement," *Torch Magazine*, Spring 1997, 23. Van Lew's descendant Bart Hall

suggests that Elizabeth didn't place Mary Jane in the Confederate White House with the specific intention of conducting espionage, and that her servant's role as a spy evolved over time. Considering that Rose Greenhow and female spies in general were national news at the time Elizabeth visited Varina Davis, I believe Mary Jane's placement was intentional. In 1905, Varina Davis wrote a letter to Isabelle Maury, regent of the Confederate White House Museum, denying that she had ever employed "an educated negro woman whose services were 'given or hired by Miss Van Lew' as a spy in our house during the war. . . . My maid was an ignorant girl born and brought up on our plantation who, if she is living now, I am sure cannot read, and who would not have done anything to injure her master or me if even she had been educated. That Miss Van Lew may have been imposed upon by some educated negro women's tales I am quite prepared to believe." Eleanor S. Brockenbrough Library, American Civil War Museum.

Not Your Ideal of a Beautiful Soldier

133 Martinsburg's law against traveling faster than at a canter: Wood, *History of Martinsburg*, 9.

133 busy building earthworks fortifications along its perimeter: *New York Evening Post*, January 2, 1862.

133 "She is quite a favorite with me": Sigaud, "More About Belle Boyd," 176.

134 "Not what I call a beauty": James Webb Papers, #760, Southern Historical Collection, University of North Carolina.

134 She also had a brief romance with one Dr. Cherry: Susan Earle Glenn (Belle Boyd's aunt) to "Nat," January 10, 1962, Sigaud Papers.

134 "perils and its pleasures": Boyd, *In Camp and Prison*, 215.

134 serenades from a regimental brass band: *Richmond Examiner*, January 4, 1862.

134 "I am in great distress": Hansen, *Civil War*, 174.

135 "I beg your pardon": Boyd, *In Camp and Prison*, 84–86. Most Union forces in the vicinity of Martinsburg were in Hancock, about twenty-five miles away and too far for the purposes of Belle's story. It is possible, however, that a Union cavalry unit had sent out patrols, pickets, and scouts, extending its actions and awareness in that general area, and that Belle had come into contact with some of those officers. Gaddy, e-mail, November 2013.

136 "If this Valley is lost": Tanner, *Stonewall in the Valley*, 36.

137 blankets freezing: Selby, *Stonewall Jackson*, 232; Wright, e-mail, December 2013. Some of the luckier soldiers had waterproof sheets.

137 "when, all of a sudden": Coffin, *Stories of Our Soldiers*, 47.

137 "sorrowfully upon me": Boyd, *In Camp and Prison*, 272.

137 "I was dumb or I should have spoken": Ibid.

137 " 'It is time for us to go' ": Ibid.

138 more scarecrow than human: Commire, *Historic World Leaders*, 376.

138 His horse, Fancy: Robertson, *Stonewall Jackson*, 230.

138 rode him with his feet drawn up: Ibid., 300.

138 "O my god!": Greene, *Wherever You Resolve to Be*, 88.

138 "out of balance": Davis, *Civil War*, 126.

138 A partial deafness in one ear: James I. Robertson Jr., "Stonewall in the Shenandoah: The Valley Campaign of 1862," *Civil War Times Illustrated*, May 1972, 4–49.

138 he self-medicated with a variety of concoctions: Robertson, *Stonewall Jackson*, 85.

138 "My afflictions": Ibid., 90.

138 Twice a day, rain or shine: Addey, *Stonewall Jackson*, 233–34; Corsan, *Two Months in the Confederate States*, 101.

139 "my little dove": Ward, Burns, and Burns, *Civil War*, 115.

139 "He would have a man shot": Ibid.

139 "Very commendable": Ibid., 234.

139 "Really, ladies": Chase, *Story of Stonewall Jackson*, 442; Hall, *Patriots in Disguise*, 101. There is a record of such an incident with Stonewall Jackson at Ramer's Hotel in Martinsburg on September 11, 1862. The diary of Susan Nourse Riddle, of Martinsburg, Virginia, includes the following entry: "September 11 (1862)—I had scarcely gotten home with a dreadful headache when the Rebels began to come in, and before dinner the whole army was here. The girls all went to see General Jackson at Mr. Ramer's and stripped his coat of its buttons." Belle Boyd specified that her first meeting with Jackson occurred before First Manassas, when he was still a colonel and not yet known as "Stonewall." Either there were two separate such incidents, or Belle incorrectly recalled the timing of her encounter. Thanks to Keith Hammersla, director of information services at the Martinsburg Public Library, for sending the Riddle diary excerpt.

139 "won't you give me a button?": Coffin, *Stories of Our Soldiers*, 47.

140 "He did not look old": Ibid.

She Will Fool You out of Your Eyes

141 She yanked the cake from the platter: *Boston Daily Adviser*, January 3, 1862.

142 "I trust that in the future": Greenhow, *My Imprisonment*, 206.

142 real means of communication: Ibid., 216.

142 "same strange fancy of the eye": Ibid., 207.

143 barred with wood latticing: Davis, "'Old Capitol,'" 206–34.

143 a farm called Conclusion: Blackman, *Wild Rose*, 59.

143 the final *e* was dropped from the family name: Ross, *Rebel Rose*, 3.

143 a particular taste for: Blackman, *Wild Rose*, 60.

143 thrown from his horse and lying on the ground: Ibid., 61.

144 the "extra" ones: Ibid., 68.

144 occasional lessons in a classroom: Tidwell, *April '65*, 58.

144 "She was a celebrated belle and beauty": Blackman, *Wild Rose*, 72.

145 "kindest and best friend": Calhoun, *Papers*, 26:366.

145 "short, ugly, and slovenly in his dress": Davis, " 'Old Capitol,' " 211.

145 "You have one of the hardest little rebels here": Greenhow, *My Imprisonment*, 207.

145 "Rose, you must be careful": Ibid.

146 "dirty enough": Ibid., 213.

146 located across the prison yard: James I. Robertson Jr., "Prison for the Capital City," *UDC Magazine*, August 1990, 18–19.

146 "she will fool you out of your eyes": Greenhow, *My Imprisonment*, 214.

146 "not to shoot the damned Secesh woman": Ibid., 218.

146 "Massa Lincoln": Ibid., 220.

146 "The tramping and screaming": Ibid., 222.

147 "indomitable rebel": Ibid., 230.

148 "Oh Mamma, never fear": Ibid., 216.

148 "Oh Mamma, the bed hurts": Ibid., 222.

148 "To-day the dinner for myself and child": Ibid., 217.

148 "round chubby face, radiant with health": Ross, *Rebel Rose*, 171.

148 one of every seven Confederates serving in northern Virginia had contracted measles: McArthur, Burton, and Griffin, *A Gentleman and an Officer*, 54.

149 "unless it be the intention of your Government": Greenhow, *My Imprisonment*, 244.

149 "Cyclops": Ibid., 260.

149 "I command you to go out" to "It was farcical in the extreme": Ibid., 245–46.

150 an involuntary jerking of the head: William Preston Johnson wrote to his wife that Greenhow "looked nervous and careworn. . . . Her eyes have a dim and somewhat faded look . . . and her head has a jerking motion, which the President says was *not* her manner, before her imprisonment." *Papers of Jefferson Davis*, 8:245.

150 "vivacity is considerably reduced": *New York Times*, April 15, 1862.

150 pried loose a plank in the floor: Beymer, *On Hazardous Service*, 158.

Rebel Vixens of the Slave States

151 "as a soldier, Frank Thompson was effeminate looking": Edmonds, pension file.

151 "The mail was even more heartily received": Dannett, *She Rode with the Generals*, 93.

151 "I will bring you now": Sears, *Young Napoleon*, 166.

152 It would take three weeks: Ward, Burns, and Burns, *Civil War*, 95.

152 "stride of a giant": Kagam and Hyslop, *Eyewitness to the Civil War*, 132.

152 "a fair specimen of Virginia mud": Edmonds, *Nurse and Spy*, 75.

152 "black as midnight": Ibid., 74.

153 "Why should blue eyes and golden hair": Ibid., 73.

153 "rebel vixens of the slave states": Ibid., 90.

153 which lasted up to two and a half years: Schroeder-Lein, *Encyclopedia of Civil War Medicine*, 81.

153 "To what fortunate circumstance": The scene between Emma and the "rebel vixen" is depicted in Edmonds, *Nurse and Spy*, 90–94.

155 he had staged a performance of *Othello*: Settles, *John Bankhead Magruder*, 41.

155 a masterly display of special effects: Eicher, *Longest Night*, 216.

156 "making balloon reconnaissances": Edmonds, *Nurse and Spy*, 90.

156 "It seems clear I have the whole force": Sears, *Young Napoleon*, 178.

156 "I think you had better break": *McClellan's Own Story*, 265.

156 "The President very coolly telegraphed me": Sandburg, *Lincoln: The Prairie Years and the War Years*, 287.

156 "*No one* but McClellan": Catton, *This Hallowed Ground*, 138.

157 "I know of a situation I could get": Edmonds, *Nurse and Spy*, 105.

157 "The subject of life and death": Ibid.

Wise as Serpents and Harmless as Doves

158 "I suffer a double death": *Richmond Examiner*, April 30, 1862.

158 "designing men in this City": W. S. Ashe to Jefferson Davis, February 28, 1862, Letters Received by the Confederate Secretary of War, 1861–1865, RG 109, National Archives.

158 the arrest of thirty suspected Northern sympathizers: Speer, *Portals to Hell*, 20.

158 "It is the universal conviction": *Richmond Examiner*, March 4, 1862.

158 "If you are going to imprison": Ibid., March 3, 1862.

159 disloyalty and giving aid: Mortimer, *Double Death*, 179.

159 "die like a man": Pinkerton, *Spy of the Rebellion*, 552.

159 "poor agonized creature": Van Lew, "Occasional Journal."

159 "desperate brigand looking villain": Ibid.

159 winning three matches against full-grown bears: Speer, *Portals to Hell*, 94.

160 "The body of Webster has been brought back": Van Lew, "Occasional Journal."

160 oddly polite tone: Ibid.

160 was even picked as a juror: *Richmond Whig*, April 29, 1862.

160 a washer and ironer and a seamstress: *Richmond Dispatch*, August 7, 1862.

161 In an attempt to secure it further: Wheelan, *Libby Prison Breakout*, 32.

161 The place was overrun with vermin: Ibid., 45.

161 "sporting for Yankees": Speer, *Portals to Hell*, 91.

161 "To 'lose prisoners' was an expression": Van Lew, "Occasional Journal."

162 "utter depravity": Glazier, *The Capture*, 45.

162 kicking dying prisoners: Speer, *Portals to Hell*, 91.

162 threatened to shoot if he didn't get up: Ibid.

162 "He never called the rolls": Parker, *Chatauqua Boy*, 54–64.

164 She stationed two more of her bravest: Ibid.

164 he refused to allow her into Libby Prison: Van Lew, "Occasional Journal."

164 Most escapees had cut their blankets into thirds: Casstevens, *George W. Alexander*, 89.

165 "I think I'll have a look": Horan, *Desperate Women*, 134.

165 "You have been reported several times": Van Lew, "Occasional Journal."

165 "I wish to tell you something": Ibid.

165 "Nothing of that sort": Ibid.

165 "Let me board here": Ibid.

166 "We have to be watchful and circumspect": Ibid.

166 A few days later she saw the same man: Ibid.; Furgurson, *Ashes of Glory*, 246.

A Woman Usually Tells All She Knows

167 "was obscured by clouds as dark": Greenhow, *My Imprisonment*, 268.

167 "ethereal ear inhalator": *Washington Evening Star*, January 15, 1862.

167 the senator and his wife had spent $75,000 annually: Ross, *Proud Kate*, 56.

167 "the most magnificent entertainments of the kind": *Alexandria Gazette*, April 12, 1858.

168 "one of the few Northern politicians": Blackman, *Wild Rose*, 216.

169 New York superior court judge Edwards Pierrepont: Rose was unfamiliar with Pierrepont, and misidentified him in *My Imprisonment* as "Governor Fairfield."

169 "resume your seats": Rose's appearance before the commission—including the dialogue, which is reported here verbatim—is based on the Proceedings of the Commission Relating to State Prisoners. Rose's depiction of the proceedings is remarkably true to the record; *My Imprisonment*, 269–70.

176 "I ask of your clemency": Greenhow, *My Imprisonment*, 282.

176 "in cold blood": Ibid., 286.

176 "I had been treated well and kindly": Ibid., 288.

176 "knew his plans better than Lincoln": Ibid., 289.

176 Another possible reason was that Lincoln: Pinkerton, *Spy of the Rebellion*, 546–48.

176 "you will be refused a pass": Greenhow, *My Imprisonment*, 289.

177 she promised two rebel prisoners: Ibid., 301.

A Slave Called "Ned"

178 "the nation's most feared address": Dunkelman, *War's Relentless Hand*, 236.

178 "engage in so perilous an undertaking": Edmonds, *Nurse and Spy*, 106.

179 phrenological studies on women: Fowler, *Fowler's Practical Phrenology*, 32–33.

179 "largely developed": Edmonds, *Nurse and Spy*, 106.

179 Lafayette C. Baker, the Federal government's chief detective: Fishel, *Secret War for the Union*, 25.

179 A Confederate spy named Benjamin Franklin Stringfellow: Sulick, *Spying in America*, 68.

180 One of Pinkerton's men, Dave Graham: Ibid.

180 John Burke, a rebel scout: Ibid.

180 Another Confederate spy, Wat Bowie: Ibid.

180 plantation-style suit: Edmonds, *Nurse and Spy*, 105 6.

180 "reconnaissance business": Ibid.

180 "I found myself without friends": Ibid., 107.

181 "Who do you belong to": This scene is based on Edmonds, *Nurse and Spy*, 112–20.

183 a third more than even the general had believed: Heidler, Heidler, and Coles, *Encyclopedia of the American Civil War*, 1732.

183 The cautious McClellan was always eager to accept: Waugh, *Lincoln and Mc-Clellan*, 91.

183 including a slave who escaped at the same time: Edge, *Major McClellan*, 80.

183 heading northwest toward Williamsburg: Eicher, *Longest Night*, 182.

184 "You will be surprised to hear of our departure": *New York Times*, May 6, 1862.

184 "spread throughout the Federal army like lightning": Edmonds, *Nurse and Spy*, 122. Lincoln, however, did not view Magruder's retreat as a victory.

184 "unequalled excitement": Robbins, journal.

184 "our success is brilliant": McClellan, *Civil War Papers*, 254.

184 "leave, not to return": Edmonds, *Nurse and Spy*, 123.

184 tripping and tumbling headlong: Haydon, *For Country, Cause and Leader*, 231.

184 "There was plenty of work": Edmonds, *Nurse and Spy*, 124.

185 "Colonel, you are not wounded at all": Ibid., 125.

185 "mentally regretting that the lead": Ibid., 126.

185 "It was his first attempt at carving": Ibid., 373.

186 General McClellan calling it "brilliant": Bonekemper, *McClellan and Failure*, 68.

186 Two men lay face-to-face: Haydon, *For Country, Cause and Leader*, 234.

186 "quite red": *New York Times*, June 8, 1862.

186 "It was indescribably sad": Edmonds, *Nurse and Spy*, 127.

Perfectly Insane on the Subject of Men

187 no "clear evidence of guilt": Dix, *Memoirs*, 2:43–44.

187 "all surface, vain, and hollow"; "captive in the parlor": Buck, *Sad Earth, Sweet Heaven*, 18, 34.

187 "Of all fools I ever saw": Diary of Kate S. Sperry, Sperry Papers, 74.

187 "write the boys by me"; "Poor boys!": Buck, *Sad Earth, Sweet Heaven*, 74.

188 Fishback Hotel: Information about the hotel from an e-mail from Judith Pfeiffer, archivist with the Warren Heritage Society, January 2012.

188 "It is here that some of the most accomplished women": *Philadelphia Inquirer*, July 19, 1862.

188 "disloyal or insane": Nathaniel Page to Sydney H. Gay, January 27, 1863, Gay Papers.

188 "He was an Irishman": Boyd, *In Camp and Prison*, 101.

188 "tender mercies": Ibid., 102.

189 "He was completely off his guard": Ibid.

189 its vocabulary included: *Atlanta Constitution*, October 10, 1919.

189 "looking well and deporting herself": Strother, *Virginia Yankee*, 37.

189 "without being beautiful": Andrews, *North Reports the Civil War*, 257.

189 "closeted four hours": Page to Gay, January 27, 1863, Gay Papers.

189 subsequently wrapped a rebel flag: *Washington Evening Star*, August 4, 1862; Starr, *Bohemian Brigade*, 100.

189 "indebted for some very remarkable effusions": Boyd, *In Camp and Prison*, 102.

190 she shut herself in a closet: I asked Cathy Wright of the American Civil War Museum about the veracity of this scene (specifically Belle's claim that she hid in a closet) and she replied, "There is a myth that old houses don't have closets because they were taxed as a room. Actually, closets could be common, although they were not usually found in bedrooms, but rather might be a china closet, linen closet, etc." Wright, e-mail, November 2011.

190 General McClellan was advancing toward Richmond: Geary, *Politician Goes to War*, 43–44.

190 "so in coming east you will be following him": Sigaud, *Belle Boyd*, 36.

190 "various circumstances": Boyd, *In Camp and Prison*, 104.

190 If she carried a pass from divisional headquarters: Bakeless, *Spies of the Confederacy*, 157.

191 dressed in the garb of a boy: Coffin, *Stories of Our Soldiers*, 48.

191 "Who is there?": The scene between Belle, "Mr. M," and Turner Ashby is from Boyd, *In Camp and Prison*, 105–6.

192 "Who comes there?": Coffin, *Stories of Our Soldiers*, 49.

192 "Let the boy pass": Ibid.

The Other Side of the River

193 McClellan prepared to move his base of operations: *New York Times*, May 13, 1862.

193 The river was generally fordable: Cullen, *Peninsula Campaign*, 51–52.

193 "which must in that event suffer terribly": Sears, *Young Napoleon*, 189.

193 A contraband who left Richmond: *New York Times*, May 13, 1862.

194 The Richmond newspapers quoted Jefferson Davis: *New York Times*, May 21, 1862.

194 Allan Pinkerton delivered estimates: Sears, *Young Napoleon*, 186.

194 nearly three times its actual strength: Sears, *To the Gates of Richmond*, 98.

194 "reconnaissances, frequently under fire": McClellan, *Report on the Organization and Campaigns*, 157.

194 "noble steed": The scene with Emma, the dying soldier, and Major McKee is based on Edmonds, *Nurse and Spy*, 148–70.

195 die upon the scaffold at Richmond: Ibid., 151.

199 some officers began to wonder: *Detroit News*, May 9, 1937.

My Love to All the Dear Boys

200 "I know you have permission to go"; Boyd, *In Camp and Prison*, 108.

200 "Jeff Davis" bonnets: Chase and Lee, *Winchester Divided*, 35.

201 "The women are almost universally bitter": Duncan, *Beleaguered Winchester*, xviii.

201 Winchester changed hands almost daily: Holsworth, *Civil War Winchester*, 95.

201 the female population never failed to harass the enemy: *New York Times*, June 5, 1862.

201 Colonel William Denny of the 31st Virginia Militia Regiment: Louis Sigaud to Harry Lemley, September 26, 1962, Sigaud Papers.

201 "will you take these letters": Boyd, *In Camp and Prison*, 109.

202 "repulsive-looking": The scene between Belle, the Union detective, and Colonel Beal, ibid., 110–16.

203 "with her usual adroitness": *Washington Evening Star*, May 31, 1862.

204 no citizen was allowed to leave: Buck, *Sad Earth, Sweet Heaven*, 77.

204 "buy some pies and pigs": Ibid.

205 "Oh, Miss Belle, I t'inks": Boyd, *In Camp and Prison*, 118.

205 "we are endeavoring to get the ordnance": Ibid., 119.

205 "But what will you do with the stores": Ibid.

205 "insult an innocent young girl": Ibid., 118.

205 "Great heavens!": Ibid., 120.

206 "I'm going to skedaddle": Coffin, *Stories of Our Soldiers*, 48.

206 "No, no. You go!": Boyd, *In Camp and Prison*, 123.

207 "I shall never run again": Ibid., 125.

207 "That was just to my taste": Douglas, *I Rode with Stonewall*, 51.

207 "Good God, Belle, you here!": Boyd, *In Camp and Prison*, 127.

207 "I knew it must be Stonewall": Douglas, *I Rode with Stonewall*, 51. Confederate general Richard Taylor also witnessed Belle's mad dash across the battlefield: "A moment later, there rushed out of the wood to meet us a young, rather well-looking young woman, afterward widely known as Belle Boyd. Breathless with speed and agitation, some time elapsed before she found her voice. Then, with much volubility, she said we were near Front Royal, beyond the wood; that the town was filled with Federals, whose camp was on the west side of the river, where they had guns in position to cover the wagon bridge, but none bearing on the railway bridge below the former; that they believed Jackson to be west of Massanutten, near Harrisonburg; that General Banks, the Federal commander, was at Winchester, twenty miles northwest of Front Royal, where he was slowly concentrating his widely scattered forces to meet Jackson's advance, which was expected some days later. All this she told with the precision of a staff officer making a report." Taylor, *Destruction and Reconstruction*, 51.

208 "I must hurry back": Douglas, *I Rode with Stonewall*, 52.

208 Belle blew him a kiss: Ibid.

208 who that "young lady" was: Ibid.

208 "rosy with excitement and recent exercise": Ibid.

208 "Remember, it is *blood red*": Ibid.

208 "Miss Belle Boyd, I thank you": Boyd, *In Camp and Prison*, 133. Belle claims to have lost this paper and others in a house fire after the war.

209 "All honor to the brave Jackson": Greenhow, *My Imprisonment*, 311.

209 "My child is so nervous": Ibid., 303.
209 "tears of blood": Ibid., 296.
209 "dangerous, skillful spy": *OR*, ser. 2, 2:271.

One Grain of Manhood

210 "hell's artillery": Sears, *To the Gates of Richmond*, 120.
210 "Had it not been for McClellan's faith": Edmonds, *Nurse and Spy*, 175.
211 "silent battles": Heidler, Heidler, and Coles, *Encyclopedia of the American Civil War*, 7–8.
211 Johnston had no idea his men were fighting: Garrison, *Civil War Curiosities*, 291.
212 "a little shaking up": Edmonds, *Nurse and Spy*, 184.
212 "not the possessor of one grain": Dannett, *She Rode with the Generals*, 149.
213 within speaking distance of each other: Guernsey and Aiden, *Harper's Pictorial History*, 352.
213 sight of their commander: Edmonds, *Nurse and Spy*, 187.
213 "The ground around that tree": Ibid., 191.
213 "Soldiers of the Army of the Potomac!": Ibid., 187–88.
214 "I am tired of the sickening sight": Sears, *To the Gates of Richmond*, 145.
214 "most lonely spot": *Fort Scott Monitor*, January 17, 1884.

The Madam Looks Much Changed

215 "The weather was very warm": Van Lew, "Occasional Journal."
215 "return north of the Potomac": *OR*, ser. 2, 2:577.
215 "rendered historic"; "the monsters": *Richmond Whig*, June 6, 1862.
216 The hotel was the city's most prestigious: Ludy, *Historic Hotels*, 248.
216 Rose ignored these and other pleas: Mortimer, *Double Death*, 199.
216 "all was warlike preparation and stern defiance": Greenhow, *My Imprisonment*, 322.
216 "I belong to the country but my heart": Cashin, *First Lady of the Confederacy*, 125.
216 "charming chamber"; "new matting and pretty curtains"; "General McClellan's room": Van Lew, "Occasional Journal."
216 "She must not come handicapped": Chesnut, *Diary from Dixie*, 311.
217 "if there were any Yankees where she was": *Papers of Jefferson Davis*, 8:244.
217 "The Madam looks much changed": Ibid.
217 "But for you, there would have been": Greenhow, *My Imprisonment*, 322.
217 "the proudest moment of my whole life": Ibid.

The Secesh Cleopatra

218 "camp cyprian": *Philadelphia Inquirer*, May 31, 1862.
219 "Lieutenant Colonel, 5th Virginia Regiment": Ibid.
219 "Where do you wish to go?": Boyd, *In Camp and Prison*, 137–38.
219 "with the utmost good-nature and pleasantry": Ibid., 138.
219 "a killing set of whiskers": *Boston Daily Globe*, May 28, 1890.
220 "the sensation of the village": Sigaud, *Belle Boyd*, 64.

220 "I suppose you came to report me again": E. J. Allen, report to Edwin Stanton, June 25, 1862, RG 94, Turner-Baker Papers.

220 "She gets around considerably": Ibid.

220 "I communicate to you a fact": Washington Duffee to Edwin Stanton, July 30, 1862, Turner-Baker Papers.

221 "C. W. D. Smitley": Scene between Belle and Smitley from Reader, *History of the 5th West Virginia Cavalry*, 257–58; Boyd, *In Camp and Prison*, 139–43.

221 "travel the slightest distance"; "These cows have permission to pass": Boyd, *In Camp and Prison*, 111.

The Bright Rush of Life, the Hurry of Death

224 "decisive step": Sears, *Young Napoleon*, 204.

224 "fit subjects for the hospital": Edmonds, *Nurse and Spy*, 203–4.

224 "behaved splendidly": McClellan, *Civil War Papers*, 285.

225 "I regret my great inferiority": Sandburg, *Lincoln: The War Years*, 3:491.

225 In reality Lee had just eighty thousand men: Gansler, *Mysterious Private Thompson*, 115.

225 as many as forty shots per minute: Haydon, *For Country, Cause and Leader*, 255.

225 "The excitement on the Mechanicsville Turnpike": Van Lew, "Occasional Journal."

226 "I almost begin to think": Cullen, *Peninsula Campaign*, 98.

226 "one of the most difficult undertakings": *McClellan's Own Story*, 412.

226 "into the hands of the enemy": Edmonds, *Nurse and Spy*, 212.

227 "hour after hour, the enemy advancing": Ibid., 225.

227 just six miles downstream: Hall, e-mail, March 2012.

227 "supremely ridiculous": Van Lew, "Occasional Journal."

227 forged exemption tickets: Varon, *Southern Lady, Yankee Spy*, 88.

228 Her connection—Mary Jane Bowser's husband, Wilson: Hall, e-mail, April 2013.

228 "damned Yankees": Hall, e-mail, March 2013.

228 "awful sin": Van Lew, "Occasional Journal." The girls never saw their mother again.

228 "Keep it all a secret": Hall, e-mail, May 2013.

229 "Are any of the prisoners related to you?": Statement from George D. Harwood, July 27, 1864, RG 109, Records of the Adjutant and Inspector General's Department, National Archives.

230 "But it does not say we must visit": Ibid.

230 In one far corner: Description of Libby Prison's hospital in McCreery, *My Experience as a Prisoner of War*, 15.

230 "Read the pinpricks": Hall, e-mail, August 2012.

231 Wilson Bowser, message hidden in his shoe: Ibid.

232 Varina a devil: Cashin, *First Lady of the Confederacy*, 126.

232 Nevertheless she strolled across the street: Hall, e-mail, June 2013.

233 "I'm going through the lines": Van Lew, "Occasional Journal." Van Lew's

descendant, Bart Hall, believes that Elizabeth's "muttering" during this scene was the origin of the "Crazy Bet" myth, which is ably discounted in Varon, *Southern Lady, Yankee Spy*, 253–61.

She Breathes, She Burns

234 "Miss Belle, de Provo' ": Boyd, *In Camp and Prison*, 148.

234 After her initial trepidation: Belle doesn't explicitly state that she wished to be mistreated while in the Old Capitol Prison, but her claims that she *was* mistreated directly contradict other reports of her rather mild—and occasionally pleasant—experience behind bars (an experience that sharply contrasted with Rose's genuinely difficult punishment). Belle told George August Sala, the English journalist who helped her write her memoir, of the "insults, sufferings, and persecutions" she endured while in Union custody, and she herself wrote that her situation was "too painful to admit of real, lasting consolation." I concluded that Belle did indeed wish to use her prison experience to cement her status as a Confederate heroine. Later, during her second stint in prison, Belle engaged in more blatant cognitive dissonance when she insisted that she "became a celebrity" through "no desire of my own."

234 "outrageous persecutions": *Daily Constitutionalist* (Augusta, GA), January 9, 1862.

234 "the best-looking in the Confederacy": Unidentified clipping, Laura Virginia Hale Archives.

235 He reminded her of Edgar Allan Poe's raven: Boyd, *In Camp and Prison*, 157.

235 "Major Sherman has come to arrest you": Ibid., 149.

235 "sorrow and sympathy": Ibid., 156.

235 "Belle Boyd was taken prisoner": Buck, *Sad Earth, Sweet Heaven*, 124.

236 "My poor, dear child!": Boyd, *In Camp and Prison*, 168.

236 In the near distance Belle spotted: Sigaud, *Belle Boyd*, 72.

236 Rose's old cell: Boyd, *In Camp and Prison*, 193.

236 "And so this is the celebrated": Ibid., 177.

236 "many happy hours": Ibid., 179.

237 "Ain't you tired of your prison": Ibid., 181.

237 "Sir, if it is a crime to love the South": Ibid., 182.

238 "Bravo!": Ibid., 183.

238 "She was dressed today": *Washington Star*, August 2, 1862.

238 "I can afford to remain here": Doster, *Lincoln*, 102.

239 "The pathos of her voice": Mahony, *Prisoner of State*, 271.

239 "a lump up in my throat": Williamson, *Prison Life in the Old Capitol*, 50.

239 "I shan't do it!": Mahony, *Prisoner of State*, 275.

239 stopped at every floor to announce services: Williamson, *Prison Life in the Old Capitol*, 33–34.

239 "separate the goats from the sheep": Boyd, *In Camp and Prison*, 202.

239 inscribed "Belle Boyd, Old Capitol Prison": Sigaud to Judge Harry Lemley, November 9, 1962, Sigaud Papers.

240 "did them more good than the preaching": Williamson, *Prison Life in the Old Capitol*, 51.

240 one rebel soldier in particular: Mahony, *Prisoner of State*, 278.

240 "I have no objection": Boyd, *In Camp and Prison*, 199.

240 "I'll break every bone in your body": Ibid., 200.

241 She rolled up her sleeve: *Atlanta Constitution*, September 2, 1900.

The Still, Small Voice

242 eaten by hogs: Wright, e-mail, November 2013.

242 Army of the Potomac swore: Edmonds, *Nurse and Spy*, 255.

242 "I think I begin to see his wise purpose": Sears, *Young Napoleon*, 235.

242 rebels were courteous: Robbins, journal.

243 and a $5 note: Ibid.

243 "She slept in the tent with us": Haydon, *For Country, Cause and Leader*, 276.

243 ordered instead by General Samuel P. Heintzelman: It's uncertain whether Emma was being truthful about this order. She claims that during this mission she stole crucial battle plans from the unattended jacket of a Confederate officer, but there is no record of such plans being discovered under such circumstances. Her story, however, is similar to a widely (but not entirely accurately) reported incident in which a group of Union cavalry surprised Confederate general J. E. B. Stuart while he rested at a farmhouse before the main engagement. Stuart fled without his coat, which contained a letter from Robert E. Lee. If she did go on a mission behind Confederate lines, it was before the main engagement on August 29, when she was involved in a well-documented riding accident. Fishel, *Secret War for the Union*, 191–93.

243 one day teach them: Edmonds, *Nurse and Spy*, 73.

244 "preferred to live in bondage": Ibid., 262.

244 "coosh": Fisher and Fisher, *Food in the American Military*, 75.

244 strained to listen: Edmonds, *Nurse and Spy*, 263.

244 same small voice : Ibid., 264.

244 only a half dozen barrels of hard bread: *New York Times*, August 31, 1862.

244 The cars were pierced with rifle shots: Haydon, *For Country, Cause and Leader*, 133.

244 the reflection of the flames lighting the sky: *New York Times*, August 31, 1862.

245 One piece swiped the cap clear: Haydon, *For Country, Cause and Leader*, 279.

245 *The mail!* she thought: *Fort Scott Monitor*, January 17, 1884.

245 The surgeon wanted to examine her: Ibid. After this accident, for the rest of her life, Emma's foot was so deformed that she had trouble wearing shoes.

246 "more like a butcher's shambles": Strother, *Virginia Yankee*, 79.

246 lawmakers had to navigate carefully: Furgurson, *Freedom Rising*, 197.

246 "How long did it take you to come back": Mahony, *Prisoner of State*, 276.

246 "Hush up, you damn bitch": Ibid.

246 "Go meet men, you cowards": Ibid., 277; Leech, *Reveille in Washington*, 157.

246 A steam warship was anchored: Dannett, *She Rode with the Generals*, 187.

247 "Here is a paper with which": Davis and Pritchard, *The Battlefields of the Civil War*, 76.

247 "here are the dreadful details": Samuels, *Facing America*, 71.

247 "Federal buried, Confederate unburied" and "Bloody Lane, Confederate Dead, Antietam": http://www.nps.gov/anti/photosmultimedia/Historic-Photogaphs.htm.

247 "I feel some little pride": Furgurson, *Freedom Rising*, 201.

248 "decided this question in favor": McPherson, *Crossroads of Freedom*, 139.

248 Elizabeth celebrated by arranging for the private purchase: "Receipt for Purchase of Louisa," January 1, 1863, Van Lew Album.

248 "I stooped down": Edmonds, *Nurse and Spy*, 271.

249 "I can trust you": Ibid., 272.

Richmond Underground

250 wrapping them around marbles: Mahony, *Prisoner of State*, 279.

250 as per Southern custom: Church, *Weddings Southern Style*, 9.

250 "All you rebels get ready": Boyd, *In Camp and Prison*, 197.

250 "I forward likewise": *OR*, ser. 2, 4:461.

251 Superintendent Wood . . . shopped for her: Sigaud, *Belle Boyd*, 87.

251 including a young mother who threw kisses: *Washington Star*, August 30, 1862.

251 She had two gold saber knots: Boyd, *In Camp and Prison*, 198.

251 "Three cheers for the Devil!": Ibid., 204.

252 her next-door neighbor, R. J. White, who was hoping to patent a machine: 1860 census; Confederate patent office files, R. J. White and George Lehner, August 25, 1863.

252 There Mary Jane was waiting with the seamstress: Hall, e-mail, June 2013.

252 leaving the roads strewn with the bodies of bloodhounds: Ward, *Slaves' War*, 216.

252 through kind words and sacred hymns: Ibid.

253 Cotton dresses were constructed: Details about dress construction, Wright, e-mail, March 2012.

253 "a friend to be trusted": McNiven, "Recollections." Historians have long debated the veracity of McNiven's "recollections" and his role in Elizabeth Van Lew's spy ring. Van Lew's descendant Bart Hall believes (and I agree) that he was a peripheral player whom Elizabeth occasionally used as a liaison.

254 "General Limpy, the food fop": Lowry, *Story the Soldiers Wouldn't Tell*, 157. In January 2011 Lowry, a psychiatrist turned historian who is the author of several Civil War–related books, confessed to altering the date on a Lincoln document in the National Archives. A month later he recanted, claiming on his website that the National Archives had "coerced an innocent man into a false confession." In his endnotes to *The Story the Soldiers Wouldn't Tell*, Lowry

explains that he himself had never seen the diary but that he obtained these excerpts from the now-deceased Robert K. Waitt, who had served as the executive secretary of the Richmond Civil War Centennial Commission. Lowry told me he doesn't know the current whereabouts of the diary, and I was unable to track down descendants of Waitt and three others who claimed to have seen it.

254 "the bravest of the brave": Elizabeth Van Lew to Ulysses S. Grant, October 2, 1869, Van Lew Papers, NYPL.

255 the adults created a recognition system: McNiven, "Recollections."

255 "trusty Union man": Parker, *Chatauqua Boy*, 54–64.

255 by certain tombstones in the graveyard: McNiven, "Recollections."

256 "to be placed in cipher": Gaddy, e-mail, June 2013.

256 "scattered broadcast over the land": Woodworth, *Davis and Lee at War*, 206–7.

256 "deceiving the guards and evading the scouts": Ibid., 196.

256 "the army will melt away": Davis, *Virginia at War, 1862*, 146.

256 Mary Jane transcribed pages of information: Hall, e-mail, June 2013.

256 she hung a red shirt on the laundry line: Statement from Annie Randolph Van Lew Hall, Box 2K394, "Correspondence, 1910," Beymer Papers.

257 Certain quantities of items corresponded: Hall, e-mail, June 2013.

258 stuffed cotton into her cheeks: John Albree notes on interview with Annie Whitlock, Van Lew Papers, College of William & Mary.

258 at her elbow; strange faces: Van Lew, "Occasional Journal."

258 accompanying her and Little Rose to Richmond's hospitals: *Alexandria Gazette*, October 9, 1862.

259 Stonewall had chosen to camp on the lawn: Chambers, *Stonewall Jackson*, 242.

259 "not altogether assured": Ibid., 246.

259 "share his dangers": *Washington Star*, August 2, 1862.

259 "If I ever catch you in Martinsburg": Sigaud, *Belle Boyd*, 105–6.

260 "God bless you, my child": Belle recounts this alternative version of her meeting with Stonewall Jackson in her memoir, 208. It's unclear whether this encounter occurred after the general's initial refusal to see her, or if it never occurred at all.

Playing Dead

261 in fine condition: Edmonds, *Nurse and Spy*, 289.

261 "hemorrhaging from the lungs": Dannett, *She Rode with the Generals*, 192.

262 "irregular fighters": *New York Times*, August 23, 1862.

262 shoot any Yankees: Stevenson, *History of the 78th Regiment*, 169.

262 "by the heat of his beautiful head": Edmonds, *Nurse and Spy*, 293.

262 a group of Confederate guerrillas: Incident described ibid., 294–96.

263 "unprofitable life": Ibid., 296.

264 "delaying on little pretexts": McPherson, *Battle Cry of Freedom*, 570.

264 "had no confidence in the ability of General Burnside": Fishel, *Secret War for the Union*, 255.

264 "political cabal at Washington": Pinkerton, *Spy of the Rebellion*, 579.

264 "To the Army of the Potomac": *New York Times*, November 14, 1862.

264 "In parting from you": McClellan, *Civil War Papers*, 521; Williams, 93.

265 "Have been somewhat busy": Robbins, journal.

266 the hand's perfect condition: Edmonds, *Nurse and Spy*, 299.

266 "short petticoats and bare knees": *Don Troiani's Regiments*, 42.

266 three-year-old George and one-year-old William: US Census, 1860.

267 "It looks rather dismal": Haydon, *For Country, Cause and Leader*, 294.

267 "If you make the attack as contemplated": Hawkins, in *Battles and Leaders of the Civil War*, 126.

267 "The carrying out of your plan": Ibid.

268 "If it rests his poor feet": Sigaud, *Belle Boyd*, 107.

268 "Truly your friend": Boyd, *In Camp and Prison*, 213. This note, which has never been found, was also possibly destroyed in a postwar fire; Sigaud, *Belle Boyd*, 212.

268 Six thousand citizens were suddenly homeless: Ward, Burns, and Burns, *Civil War*, 137.

268 posing as a Union telegraph operator: Dannett, *She Rode with the Generals*, 209.

268 "zeal and energy": Fishel, *Secret War for the Union*, 257.

269 "The work went steadily on": Edmonds, *Nurse and Spy*, 301.

269 Dead rebels lay strewn: Haydon, *For Country, Cause and Leader*, 297.

270 "I wish my friends could see me": Edmonds, *Nurse and Spy*, 301.

270 "A chicken could not live in that field": Eicher, *Longest Night*, 401.

270 "grass before the scythe": O'Reilly, *Fredericksburg Campaign*, 318.

270 "It is well that war is so terrible": Ward, Burns, and Burns, *Civil War*, 141.

271 "with a fearlessness that attracted the attention": Dannett, *She Rode with the Generals*, 213.

271 an officer who lay writhing in agony: Emma describes this incident in her memoir (p. 306), but does not explicitly mention Reid's name.

When You Think He May Be Killed Tomorrow

275 revealed more of herself: Boston, *Diary*. Emma never explicitly described the nature of her relationship with James Reid.

275 "friend Frank"; "extremely fond": Robbins, journal.

275 "Have not had a very long chat": Ibid.

275 "individuals who repose": Ibid.

275 "one of nature's noblemen": Ibid.

276 "Dear Jerome": Robbins Papers.

277 "Charles Norton": Blanton and Cook, *They Fought Like Demons*, 108.

277 "While out on the foraging expedition": Ibid., 108–9.

277 Two women serving with the 95th Illinois Infantry: Ibid., 109.

277 "a quick jerk of her head": Ibid.

277 A recruit in Rochester: Ibid.

277 "complained of feeling unwell": Ibid., 103.

278 "of barroom and brothel": Gold, *Studies in Etymology*, 113.

278 "Beware of rashness": Burlingame, *Inner World of Abraham Lincoln*, 81.

278 "The weather department is in perfect keeping": Edmonds, *Nurse and Spy*, 311.

279 Confederate forces under General John Pegram: *OR*, ser. 1, 23:168.

279 young widow: Edmonds, *Nurse and Spy*, 313.

279 "Having made up my mind": *Southern Literary Messenger* 38, no. 4 (April 1864): 124.

280 "the blockade don't keep out babies": *Diary of Miss Emma Holmes*, 266.

280 "One looks at a man so differently": Wayne, *Women's Roles*, 8.

280 "See here, my lad": Edmonds, *Nurse and Spy*, 313.

281 rebel forces at Danville: *OR*, ser. 1, 23:168.

281 "Well, you will not have to wait long": Edmonds, *Nurse and Spy*, 314.

282 "to cancel all obligations"; "graceful curve of his mustache": Ibid., 316–17.

Bread or Blood

283 On March 27 Jefferson Davis declared: *Portland Daily Advertiser*, March 9, 1863.

283 better dinner than usual: Van Lew, "Occasional Journal."

284 In Atlanta a dozen women: Fowler and Parker, *Breaking the Heartland*, 161.

284 his wife and children were emaciated: Jones, *Rebel War Clerk's Diary*, 381.

284 "the luxury of mule meat and fricasseed kitten": John Wright, *Language of the Civil War*, iv.

284 remains of the girl's pet bird: Loughborough, *My Cave Life*, 137.

284 "Alas for the suffering of the very poor!": Van Lew, "Occasional Journal."

284 "That's all there's left of me!": Furgurson, *Ashes of Glory*, 193.

284 "catch something from them poor white folks": Ibid.

284 "Bread or blood!": Davis, *Virginia at War*, 89.

285 "abstain from their lawless acts": Varina Davis, *Jefferson Davis*, 375.

285 "You say you are hungry": Ibid.

285 "The Union!": *New York Times*, April 5, 2013.

285 "We do not desire to injure anyone": Varina Davis, *Jefferson Davis*, 375.

285 "order your men to load": *Richmond Dispatch*, December 16, 1888.

285 "if this street is not cleared": Ibid.

286 "This demonstration was made use of": Brock Putnam, *Richmond During the War*, 209.

286 "devilish and secret motives": Varon, *Southern Lady, Yankee Spy*, 104.

286 "glaringly incongruous"; $500 cash: *Richmond Examiner*, April 13, 1863. Thomas McNiven boasted that "a lot of American dollars went into organizing the riots," but Elizabeth Van Lew remained discreet, never referring to any of the defendants by name; McNiven, "Recollections."

286 "drearily": Varina Davis, *Jefferson Davis*, 369.

286 "secession had made havoc of all female charms and graces": *Cincinnati Daily Enquirer*, June 27, 1862.

287 "respond to the call of patriotism": *New York Times*, April 17, 1863.

287 lost to convalescent camps: Bill Lewis, "Horses and Mules: Gap in Civil War Research," *Civil War Talk*, May 1959.

287 "the present immobility of the army": *OR*, ser. 1, 2:725.

287 mounted diminutive slaves: *New York Times*, May 8, 1864.

288 sending her nieces with baskets of fresh apples: Hall, e-mail, April 2012.

288 "thoroughly understood matters": Van Lew, "Occasional Journal."

A Wean That's Born to Be Hung

289 "rejoice in the beauty": Edmonds, *Nurse and Spy*, 317.

289 "I would not be permitted to go out again": Ibid., 318. The conclusion of this mission is dubious; if a Union officer did recognize Emma, his first thought would probably have been that she was a Confederate double agent, not a Union spy.

290 "It is an auld saying": Ibid., 129.

290 "far worse than death": Hall, *Patriots in Disguise*, 92.

290 A shell had exploded randomly: Edmonds, *Nurse and Spy*, 358.

291 "soldierly qualities": *Fort Scott Monitor*, January 17, 1884.

291 "a poor, cowardly, nervous": Ibid.

291 "I passed most of the day with Frank": Robbins, journal.

292 "A young woman wearing soldier's apparel": *Louisville Daily Democrat*, April 16, 1863.

292 around the night of April 17: By April 19, Frank Thompson was officially listed as a deserter on the regimental muster rolls.

A Dreadful Blow

293 "All my wounds are by my own men": *Cleveland Plain Dealer*, May 19, 1863.

293 "If it was in my power to replace": Chase, *Story of Stonewall Jackson*, 520.

293 A week later pneumonia had set in: Some historians argue that Jackson already had pneumonia before he was shot; there are accounts of him shivering and huddling near the fire, despite the reasonably mild weather.

294 "unable to think of anything": Woodworth, *Davis and Lee at War*, 225.

294 "How is he?": Ibid.

294 just as he'd always wanted: Selby, *Stonewall Jackson*, 209.

294 "You must excuse me": Woodworth, *Davis and Lee at War*, 225.

294 striking across the Potomac: Cooper, *Jefferson Davis, American*, 468.

294 "It is not for me to trace": Boyd, *In Camp and Prison*, 217.

294 "She was a brilliant talker": Avary, *Virginia Girl in the Civil War*, 52.

295 "Like General Johnston, I can fight": *New York Herald*, February 24, 1863.

295 the wound was minor: Boyd, *In Camp and Prison*, 216.

295 "Please telegh if Gen. Jackson": Robertson, *Civil War Virginia*, 757.

296 "The last scene in a stupid farce": *Richmond Whig*, July 3, 1863.

296 "Was Property of Belle Boyd": *Martinsburg Evening Journal*, December 26,

1916.

296   "'Tis said Belle Boyd is in town": Buck, *Sad Earth, Sweet Heaven*, 209.

297   "Oh, here comes de Yankees": Boyd, *In Camp and Prison*, 225.

297   "General Kelly commanded me": Ibid., 226.

297   "Good morning": Ibid., 227.

297   bothered again: Ibid., 227.

298   "abdominal binding": Selin and Stone, *Childbirth Across Cultures*, 210.

298   "Doctors are gentlemen": Hager, *Demon Under the Microscope*, 119.

298   "Know who that is?": Coffin, *Stories of Our Soldiers*, 47.

298   "send a dagger": Boyd, *In Camp and Prison*, 228.

299   "room for distinguished guests": Ibid., 231.

299   she loosened a brick from her windowsill: Diary of Henry T. Bahnson, Bahnson Papers.

300   "I never for a moment considered myself": Hall, *Patriots in Disguise*, 92.

300   "Do you know I have learned": Robbins, journal.

301   "We are having quite a time at the expense": Boston, *Diary*.

302   female friends: Emma Edmonds to Jerome Robbins, Robbins Papers.

302   "Dear Jerome": Ibid.

302   Anna had told him she intended: Gansler, *Mysterious Private Thompson*, 185.

302   "I received a letter from my friend": Robbins, journal.

No One Ignorant of the Danger

303   "the instincts of cows": Davis, *The Man and His Hour*, 428.

303   casualties stretching fourteen miles: Blackman, *Wild Rose*, 254.

303   "foolish and disastrous"; "utter want of generalship": Thomas, *Robert E. Lee*, cxvi.

304   The wives of US officials: Blackman, *Wild Rose*, 255.

304   "a clever woman . . . equally at home": Tidwell, *April '65*, 58.

304   an example of Yankee vulgarity: Blackman, *Wild Rose*, 255.

304   "affords me every facility": Rose Greenhow to Alexander Boteler, June 19, 1863, Greenhow Papers.

305   explaining why his services were no longer necessary: Evans, *Judah P. Benjamin*, 236. Judah P. Benjamin wrote a letter to Mason dated August 4, 1863, the day before Rose left for England: "Perusal of the recent debates in the British Parliament satisfies the President that the government of her Majesty has determined to decline the overtures made through you for the establishing by treaty, friendly relations between the two governments, and entertains no intention of receiving you as the accredited minister of this government and the President therefore requests that you consider your mission at an end, and that you withdraw with your secretary from London."

305   the masses had held hundreds of meetings: Du Bois, *Black Reconstruction in America*, 80.

305   the Confederate government gave her: Gaddy, e-mail, November 2013.

305   calm, confident tone: Rose Greenhow to Jefferson Davis, June 16, 1863, Greenhow Papers, Duke University.

305 regiments composed entirely of Negroes: *Daily National Intelligencer* (Washington, DC), August 14, 1863; *Charleston Mercury*, August 7, 1863.

306 "Be assured that every body": Rose Greenhow to Jefferson Davis, July 16, 1863, Greenhow Papers, Duke University.

306 "no printed paper could be kept secret": *OR*, ser. 1., 32:487.

306 the *Phantom*: Description in Horner, *Blockade-Runners*, 44.

307 "There they go ahead of me": Greenhow, "European Diary," 16.

307 "Oh never mind": Ibid.

307 At half past midnight a Union blockader spotted her: *Official Records of the Union and Confederate Navies*, ser. 1, 9:150–51.

307 bacon and turpentine-soaked cotton: Gragg, *Confederate Goliath*, 11.

307 "How my blood boils": Greenhow, "European Diary," 5.

308 "I can't stand to have": Ibid., 18.

308 "The negroes are lazy, vicious, and insubordinate": Ibid., 8.

309 discussed how he might send her coded dispatches: Rose Greenhow to Alexander Boteler, December 10, 1863, Greenhow Papers, Duke University.

309 admit a pair of Confederate officers: Hall, e-mail, May 2013.

309 "These good men are tired of the war": Ibid.

310 bearing gifts for Winder: Hall, e-mail, March 2013.

311 fifteen to twenty-five men died there: Ransom, *Andersonville Diary*, 15.

311 A servant would cover them with a tarp and pile horse manure: Hall, e-mail, May 2013.

311 a "most exciting" rumor: *Wisconsin Daily Patriot*, September 19, 1863.

La Belle Rebelle

312 "Take Me Back": Boyd, *In Camp and Prison*, 238.

312 "Poor girl! You have the deepest sympathy": Ibid., 239-40.

313 yielded to curiosity: Ibid., 240.

313 "The discomfited Yankee"; "These are the kind of men": Graham and Torrance, "Running the Blockade," np.

314 "She is a damn rebel": Boyd, *In Camp and Prison*, 244.

314 "several official lovers": *Frank Leslie's Illustrated Newspaper*, September 12, 1863.

315 "Had I been a queen": Boyd, *In Camp and Prison*, 254.

315 She encouraged rumors: Coffin, *Stories of Our Soldiers*, 49.

315 a pair of field glasses she claimed: Boyd, *In Camp and Prison*, 261.

316 "annoying and ungentlemanly": Ibid., 256.

316 Merchants sold chamber pots: Wilkie, *Dixie*, 289.

317 "Make like Butler": Chesnut, *Diary from Dixie*, 177.

317 "I expect to be killed": *Aberdeen Journal*, December 23, 1863.

318 "Ah, so this is Miss Boyd": Boyd, *In Camp and Prison*, 257.

318 "rage depicted on every lineament": Ibid., 258.

319 she dropped her face in her hands and wept: Ibid., 261.

319 a drunken rebel soldier once rode his horse: Dabney, *Richmond*, 164.

319 "being torn": Boyd, *In Camp and Prison*, 273.

Women Make War upon Each Other

320 "a smoky chimney which had taken fire": Froude, *Thomas Carlyle*, 246.

320 dressing gown and slippers: Blackman, *Wild Rose*, 269.

320 "What sort of looking animal?": Greenhow, "European Diary," 48.

320 "The flat nosed Negro of Haiti": Ibid.

321 "I see him": Ibid.

321 Carlyle promised to consider it: Rose O'Neal Greenhow to Alexander Boteler, February 17, 1864, Greenhow Papers, Duke University.

321 split the profits for all sales: Blackman, *Wild Rose*, 271.

321 "Men may not war on women": *London Standard*, December 2, 1863.

321 "She boasts that she made herself as obnoxious": *London Morning Post*, November 27, 1863.

322 "as bitter as a woman's hate can make it": *New York Times*, December 5, 1863.

322 "All the accounts come through the Yankee press": Rose O'Neal Greenhow to Alexander Boteler, December 10, 1863, Greenhow Papers, Duke University.

322 "the greatest evil": Greenhow, "European Diary," 36–37.

323 "O, Mama": Ibid., 38.

323 "She fell 4 times": Ibid., 42.

323 "My little one is very shy": Greenhow, "European Diary," 44. The convent is now the site of the Rodin Museum.

324 "directly or by implication": Long and Long, *Civil War Day by Day*, 444.

324 "the crisis which threatened to divide": Ibid.

324 "Why is the attitude of European Powers": *Richmond Whig*, December 19, 1863.

Please Give Us Some of Your Blood

325 thought the whole episode great fun: Hall, e-mail, May 2013.

326 "Would you be free?": Varon, *Southern Lady, Yankee Spy*, 109.

326 He wouldn't be the first prisoner: Speer, *Portals to Hell*, 229; Chambers McKean, *Blood and War*, 1166.

326 "Follow me at a distance": H. S. Howard, Deposition, Secret Service Accounts, RG 110, National Archives.

327 gray uniforms made of rebel blankets: *Connecticut Courant*, January 2, 1864.

327 abandoning their ranks in growing numbers: *OR*, ser. 4, 2:674.

327 including revoking draft exemptions: Woodworth, *Davis and Lee at War*, 302.

328 "The 'dead' Yankee has arrived": *Richmond Examiner*, January 23, 1864.

329 fourteen locally organized groups of citizen spies: Fishel, *Secret War for the Union*, 1.

329 "My Dear Boutelle": Butler to Charles O. Boutelle, December 19, 1863, *Private and Official Correspondence of General Benjamin F. Butler*, 3:228–29. Elizabeth's original note to Butler has not survived.

330 "My Dear Aunt": Butler to Elizabeth Van Lew, January 18, 1864, *Private and Official Correspondence*, 3:319.

330 "S. S. Fluid": Fishel, *Secret War for the Union*, 423.

331 "They are coming at night": Van Lew Papers, NYPL. The note is undated.

That Unhappy Country

335 improbably grand spaces: De Vooght, *Royal Taste*, 150.

335 "His majesty will receive you": Greenhow, "European Diary," 47.

335 "final adieu": Cooper, *Jefferson Davis, American*, 8.

335 "Entrez, Madame, dans la cabinet de l'Empereur": Greenhow, "European Diary," 47. I corrected any grammatical or spelling errors in Rose's French.

335 wearing a large diamond eagle: Baguley, *Napoleon III and His Regime*, 132.

336 It was rumored that he had: Ibid., 26.

336 finding the whole business "disgusting": Ibid., 220.

336 "the terror of ugly husbands": Hartje, *Van Dorn*, 308.

336 "Vous parlez français, madame?": Greenhow, "European Diary," 47.

336 "Yes sir, from that unhappy country": Ibid.

336 He was a nationalist who sympathized: Blackburn, *French Newspaper Opinion*, 6.

337 "large, hairy hands": *New York Times*, August 2, 2011.

337 "was dressed in the French mode": Ibid.

337 "evaded or rejected": Greenhow, "European Diary," 48.

337 "Tell the President that I have thoughts": Ibid.

337 "The President is fully convinced": Ibid.

338 "I wish to God I could": Ibid., 49.

338 "So much for my interview": Ibid., 50.

339 "like so many petrified steel clad warriors": Ibid., 51.

339 "very like the emperor": Ibid., 52.

339 "fat and vulgar looking"; "thin with sharp features"; "not at all pretty": Ibid., 51.

340 "no desire to enter further": Ibid., 52.

340 "I think things are looking better": William B. (unintelligible surname) to Rose O'Neal Greenhow, February 7, 1864, Greenhow Papers, Duke University.

340 every rat, mouse, and pigeon: Wixson, *From Civility to Survival*, xxxix.

340 A single hotel meal cost: Chesnut, *Diary from Dixie*, xiv.

340 One woman sold her hair for $100: Duvall, *Recollections of the War by Grandmama*.

340 then he had best resign: *Hartfort Daily Courant*, January 13, 1864.

340 he urged passage of a law: Thomas, *The Confederate Nation*, 260.

340 "As soon as General Butler has a sufficient force": *Milwaukee Sentinel*, January 11, 1864.

340 "hope deferred": Greenhow, "European Diary," 151.

Despicable Remedies

341   "The condition of our servants": Ross, *First Lady of the South*, 193.

341   "loud-talking, violent": Furgurson, *Ashes of Glory*, 241.

341   who used the code name "Quaker": Stuart, "Of Spies and Borrowed Names"; Stuart, "Colonel Ulric Dahlgren and Richmond's Union Underground."

342   requiring 1,304 laborious strokes: Fishel, *Secret War for the Union*, 553.

342   "They do a great deal of harm": OR, ser. 1, 33:519–21.

344   "make a dash": Ibid.

344   "To arms! To arms!": *Hartford Daily Courant*, February 10, 1864.

345   "corruption and faithlessness": Butler, *Private and Official Correspondence*, 3:408.

345   there was to be an exit: Van Lew, "Occasional Journal."

345   "We were so ready for them": Ibid.

345   "damned Yankees": Moran, "Colonel Rose's Tunnel at Libby Prison."

346   two or three at a time: Wheelan, *Libby Prison Breakout*, 163. (Of the 109 men who escaped, 59 reached safety, 2 drowned, and 48 were recaptured.)

346   imprinted with the word *coward*, or were branded on the cheek: Arnold and Weiner, *American Civil War*, 66.

346   Jefferson Davis begged the women: Coulter, *Confederate States of America*, 468.

347   "this one horse barefooted naked": Ibid., 469.

347   "greatly distressed": Van Lew, "Occasional Journal."

347   Desperate situations sometimes require despicable remedies: Ibid.

347   he was especially pleased when a fourteen-year-old boy: Blakey, *General John H. Winder*, 139.

348   "of a delicate appearance": Varon, *Southern Lady, Yankee Spy*, 285.

348   "Bring him to me tomorrow": Furgurson, *Ashes of Glory*, 246.

348   he could board temporarily at her home: *National Tribune*, April 20, 1893.

348   to live with relatives in Hillsborough: Blakey, *General John H. Winder*, 48.

The Sanctuary of a Modest Girl

349   "You knew my Father": There is a copy of this note in the Sigaud Papers.

349   "great attention as one of the heroines": *Alexandria Gazette*, February 9, 1864.

350   bearing $500 in gold: Foreman, *World on Fire*, 627; Sigaud, "More About Belle Boyd," 174–81.

350   "Captain Henry": OR, ser. 1, 10:42.

351   "E. A. Parkinson": Sigaud, *Belle Boyd*, 152.

351   "Steady"; "Cut off your smoke!": Pollard, *Observations in the North*, 10.

351   "Unless Providence interposes": Boyd, *In Camp and Prison*, 281.

352   "There's another they shall not get": Ibid., 283.

352   "More steam!": Ibid.

352   "I declare to you": Ibid.

352   "act without reference to me": Ibid.

353   "It is too late to burn her": Ibid., 284.

353 "Good day to yourself, sir": Ibid., 287.

353 He wore lavender kid gloves: Horner, *Blockade-Runners*, 156.

353 "An officer as unfit for authority": Boyd, *In Camp and Prison*, 287.

354 "That man, whoever he may be": Ibid., 291.

354 "what a good fellow": Ibid., 293.

355 "I am now in command": Ibid., 296.

You Are Very Poor Company

356 "would not by itself remove the blockade": Fitzmaurice, *Life of Granville George Leveson Gower*, 443.

356 "Our people earnestly desire recognition"; Greenhow, "European Diary," 119.

357 "Took my child back to the convent": Ibid.

357 "showed that he was utterly ignorant of everything": Ibid., 58.

357 "the best diplomatist": Ibid., 61.

357 "The most interesting object": Ibid., 112.

357 "They are all stupid": Ibid., 107.

357 "The place assigned for ladies": Ibid., 57.

358 "no conversation going on": Ibid., 66.

358 "What trifles color life": Ibid., 65.

358 "Sir, if you accept the scientific weight": Ibid., 80–81.

359 "What about the massacre": Ibid., 81.

359 one man nailed to the boards: *Massachusetts Spy*, May 11, 1864.

359 Negro children as young as seven: *New York Herald Tribune*, May 6, 1864.

359 "If I were a Negro": Greenhow, "European Diary," 82.

359 "Never! Either extermination": Ibid., 82.

Be Prudent and Never Come Again

360 "Some I know to be Unionists": Van Lew, "Occasional Journal."

361 "irritate and inflame": Ibid. Preeminent Civil War historians, including Stephen Sears, have concluded that the Dahlgren papers were authentic.

361 he seemed to have a soft spot: Hall, e-mail, March 2013.

361 the pantaloons of a Confederate: *Memoir of Ulric Dahlgren*, 227.

361 his prosthetic wooden leg had been sent: Chesnut, *Diary from Dixie*, 389.

361 nor treated with carbolic acid: Dammann, *Images of Civil War Medicine*, 45.

361 "fair, fine, and firm"; "The comeliness of the young face": Van Lew, "Occasional Journal."

362 the "good death": Faust, *This Republic of Suffering*, 6.

362 "Every true Union heart": Van Lew, "Occasional Journal."

362 affection of his company: Ibid.

362 "clear case of personal animosity": Furgurson, *Ashes of Glory*, 247.

363 he recognized her as a true patriot: Hall, e-mail, July 2013.

363 from Eliza's home: Ibid.

363 "nothing in Richmond except Home Guards": Varon, *Southern Lady, Yankee Spy*, 149.

363   "My Dear Niece": Butler, *Private and Official Correspondence*, 3:485.

364   "muff of the latest style": Feis, *Grant's Secret Service*, 238.

364   "only those you know to be faithful": Butler, *Private and Official Correspondence*, 3:564–65.

364   chief aide and "pet": Blakey, *General John H. Winder*, 144.

364   copies of orders issued from Winder's office: *Baltimore Sun*, April 19, 1864.

364   importance as a Confederate official: Varon, *Southern Lady, Yankee Spy*, 150.

364   "a contemptible ass": Blakey, *General John H. Winder*, 144.

365   "deadly pale"; "Be prudent": Van Lew, "Occasional Journal."

365   three Union soldiers, just escaped from Libby: *Richmond Sentinel*, May 2, 1864.

366   There they hid in dank, dark corners: Bart Hall, interview with author, July 2013.

366   whose costume of men's pants: *Richmond Sentinel*, May 2, 1864.

366   "I am a lady, gentlemen": Buhk, *True Crime in the Civil War*, 15.

366   "this disgusting production of Yankeeland": *Richmond Whig*, May 2, 1864.

366   the transportation of a hundred Libby prisoners: *Richmond Examiner*, May 7, 1864.

367   "your ordering your underlying officers": John Albree notes, Box 1, Folder 7, Van Lew Papers, College of William & Mary.

367   a cup and a half of flour: Northrop, *Diary of a War Prisoner in Andersonville*, 179.

367   "talk Southern Confederacy": Van Lew, "Occasional Journal."

Good-Bye, Mrs. Greenhow

368   "sick with anxiety": Greenhow, "European Diary," 83.

368   stuffing their nostrils with leaves: Ward, Burns, and Burns, *Civil War*, 252.

368   "handsomely driven back": *OR*, ser. 1, 36:974.

368   "Thank God it is good": Greenhow, "European Diary," 83.

368   "I am going to propose": Ibid., 84.

369   "Thank you all": Ibid.

369   "Yes Missis, them deep ones": Van Lew, "Occasional Journal."

369   "Confederate loss is heavy": Greenhow, "European Diary," 95.

369   "Grant intends to *stink*": Foote, *Red River to Appomattox*, 295.

369   "God grant that those vandals": Greenhow, "European Diary," 111.

370   "The very person we want": *New York Times*, September 9, 1864.

370   "holy saint and martyr": Greenhow, *My Imprisonment*, 190.

370   "He was a traitor": Ibid.

371   "Poor fellow": Ibid., 114.

371   She spent many of the rest of her days with Fighting Joe: Commentary from John O'Neal, descendant of Rose O'Neal Greenhow, in Greenhow, *European Diary*, 55–56.

371   hurt the South: Ibid., 124.

371   "Does it never occur": Ibid.

372   "O how sad has been": Ibid., 96.

372   "Alas, inexorable destiny": Ibid., 123.

372 "A sad sick feeling": Ibid., 128.
372 capturing or destroying nearly 92 percent: Fonvielle, *Wilmington Campaign*, 13.
372 using deadly force: *Daily Constitutionalist* (Augusta, GA), June 29, 1864.

This Verdict of Lunacy

373 "endeavored to paint the home": Boyd, *In Camp and Prison*, 302.
373 "If he felt all that he professed": Ibid.
374 "They must be in the lower cabin": Ibid., 319.
374 "It is impossible!": Ibid., 321.
374 "the ill-used South": Ibid., 316.
375 "sufficiently persuasive": *Daily National Republican* (Washington DC), June 21, 1864.
375 "She was proving a troublesome customer": Keyes, unpublished autobiography, Curtis Carroll Davis Papers.
375 "her ladyship": Ibid.
376 "For this verdict of lunacy": Boyd, *In Camp and Prison*, 327.
376 "in possession of a vast amount": Ibid., 27.
376 "My dear Miss Belle": Ibid., 327.
377 "deeply touched at this new proof": Ibid., 336.
377 "neglect of duty": Sigaud, *Belle Boyd*, 235.
377 "has made a fool of him": *Boston Post*, September 8, 1864.
378 "Hon. Jefferson Davis": Belle Boyd Hardinge to Jefferson Davis, RG 109, Unfiled Papers and Slips, National Archives.
379 there were ugly rumors: http://jeffersondavis.rice.edu/JosephEvanDavis.aspx.
379 "I have said all the prayers": Chesnut, *Diary from Dixie*, 306.
379 Grant's Circular No. 31: Weitz, *A Higher Duty*, 57.
380 he had to cancel the entire operation: Davis, *The Man and His Hour*, 574.
380 "I say to the servant": Van Lew, "Occasional Journal."

The Delicacy of the Situation

381 "Van Lew rode out from his friend's": *Richmond Whig*, July 29, 1864.
381 "puts me to a great disadvantage": Fishel, *Secret War for the Union*, 553.
382 With her back to the mantel: John P. Reynolds Jr., "Biographical Sketch of Elizabeth Van Lew," Van Lew Papers, NYPL.
382 bacon was now $20 per pound: Robertson, *Civil War Virginia*, 109.
382 high-laced shoes cost $100: *Richmond Enquirer*, January 28, 1864.
382 never wearing out of Richmond the same shoes: *Richmond Daily Dispatch*, July 17, 1883.
383 "Do you think I am being watched?": W. W. New to Miss King, Van Lew Papers, NYPL.
385 "good Yankees": Testimony of John Liggon, September 26, 1864, RG 109, National Archives.
385 "I wish the Yankees would whip": Testimony of Watt.
385 "violent and outrageous": Testimony of Reverend Philip B. Price, Septem-

ber 17, 1864, RG 109, National Archives.

385 "go to the Yankees and take": Testimony of A. B. Mountcastle, November 8, 1864, RG 109, National Archives.

Not at All Changed by Death

386 "Lock your doors. Keep inside": Kaemmerlen, *Sherman and the Georgia Belles*, 67.

386 "complete and efficient": *Richmond Enquirer*, November 28, 1864.

386 but not one cheer: Cooper, *Jefferson Davis, American*, 526.

386 "cold isolation": Greenhow, "European Diary," 128.

387 "elusive white": Wise, *Lifeline of the Confederacy*, 150.

387 she had no superior: McNeil, *Masters of the Shoals*, 59.

387 had brought along his Newfoundland puppy: Ibid., 62.

389 Even the puppy made it: Ibid.

389 "A remarkably handsome woman": Taylor, *Running the Blockade*, 128–29.

390 "lovely face, that graceful form": Diary of William Lamb, Lamb Papers, College of William & Mary.

390 "She was an elegant woman": Blackman, *Wild Rose*, 300.

390 "the uncertainty of all human projects": *Wilmington Sentinel*, October 1, 1864.

390 "Let us accept the omen": Ibid.

The Sweet Little Man

391 the property of an African princess: Boyd, *In Camp and Prison*, 346.

391 a tentative serenity: Wood, *History of Martinsburg*, 46.

391 "You's Miss Belle's husband": Boyd, *In Camp and Prison*, 344.

391 "This was your room": Ibid., 345.

392 "Of the two characters": *Pall Mall Gazette*, June 9, 1865.

392 "lurking somewhere in the vicinity": *Daily Constitutionalist* (Atlanta, GA), December 16, 1864.

393 "Is there anything remarkable about me": Boyd, *In Camp and Prison*, 350.

393 "daily danger": *London Miner and Workman Advocate*, December 31, 1864.

394 "sit and fan handsome young": Edmonds, *Nurse and Spy*, 371.

394 "Her whole life was an interesting conundrum": *State Republican*, June 26, 1900.

395 "We send you the buttonless garments": Edmonds, *Nurse and Spy*, 333.

Like Most of Her Sex

396 "And further this deponent saith not": Testimony of Mary C. Van Lew, October 15, 1864, RG 109, National Archives.

396 "wealth and position"; "no action to be taken": Order of Charles Blackford, October 18, 1864, RG 109, National Archives.

397 Every evening, from seven to midnight: Badeau, *Military History of Ulysses S. Grant*, 243.

397 "The Government, from time to time": *OR*, ser. 1, 42:710.

397 "plain classes" were anticipating: *OR*, ser. 1, 42:1108.

397 "the enemy are planting torpedoes": Butler, *Private and Official Correspondence*, 5:355.

398 Mary Jane Bowser, looking as if: Hall, e-mail, March 2013.

The Way a Child Loves Its Mother

401 "destroyer": Boyd, *In Camp and Prison*, 460.

401 "Honble Abraham Lincoln": Sigaud, "When Belle Boyd Wrote Lincoln," 15–22.

402 reaching America in eight days: Ibid.

402 "Let the execution of William B. Compton": *OR*, ser. 2, 5:711.

402 "a factory chimney with the top broken off": Brown and Dickey, *Rough Guide to Washington*, 46.

402 an escaped Confederate prisoner had recently skated: *Washington Post*, December 4, 2008.

403 "the two countries": Long and Long, *Civil War Day by Day*, 623.

403 She suspected the Federal government: *Harrisburg Patriot*, March 30, 1865.

403 an "unfashionable" part of town: *London Standard*, February 3, 1865.

403 "I'd rather be there as I was": Ibid.

403 "the fool who had married": Keyes, unpublished autobiography, Curtis Carroll Davis papers.

404 "a child loves its mother": *Boston Journal*, March 31, 1893.

As This Mighty Work Was Done

405 shoveled horse manure on top of her: Hall, e-mail, March 2013.

405 "Emma G. Plane": Fay Papers, Library of Virginia.

406 "a most efficient railroad officer": *Richmond Sentinel*, January 26, 1865.

406 honorably discharged: Varon, *Southern Lady, Yankee Spy*, 186.

406 "My heart sank": Van Lew, "Occasional Journal."

406 "Tell all you know": Ibid.

407 "Blow away": Ibid. The detective sighed, lowered his pistol, and threw his prisoner into solitary confinement.

407 a Confederate plan: Stuart, "Samuel Ruth and General R. E. Lee: Disloyalty and the Line of Supply to Fredericksburg." *New York Herald Tribune*, March 10, 1865.

407 "engine-runners, machinists, blacksmiths": Varon, *Southern Lady, Yankee Spy*, 188.

407 "Everybody turned out upon the streets": *OR*, ser. 1, 46:525.

407 "May God bless you": *OR*, ser. 1, 46:79–80.

408 "a letter from a lady in Richmond": *OR*, ser. 1, 46:963.

408 "hold Five Forks at all hazards": Pickett, *Pickett and His Men*, 266.

408 "end matters right there": Furgurson, *Ashes of Glory*, 298.

408 he had given Varina a pistol: Wright, e-mail, November 2013.

408 "My lines are broken in three places": Lossing, *History of the Civil War*, 404.

409 "The war will end now": Van Lew, "Occasional Journal."

409 "It would be better": Ibid.

410 "The Negro troops will be turned loose on us!": Duvall, *Recollections of the War by Grandmama*.

410 "If this house goes": Annie Randolph Van Lew to Anna Van Lew Klapp, courtesy of Bart Hall.

411 "the consummation of the wrongs of years": Van Lew, "Occasional Journal."

411 "No wonder that the walls of our houses": Ibid.

411 "I want nothing now": Parker, *Chatauqua Boy*, 56.

412 it would be the first to fly: Eggleston, *Women in the Civil War*, 84; Wheelan, *Libby Prison Breakout*, 220; Hall, e-mail, January 2013.

412 a room with a gabled window: Hall, e-mail, January 2013.

Epilogue

413 "You have shared the hardships": David Rankin Barbee Papers.

414 "The poor child's tears never stop": Blackman, *Wild Rose*, 303.

414 asked to be baptized: Ibid.

414 selling 175,000 copies: Gansler, *The Mysterious Private Thompson*, 185.

414 "dear old Stars and Stripes": Edmonds, *Nurse and Spy*, 354.

414 "too monotonous": Gansler, *The Mysterious Private Thompson*, 187.

415 "you know how the census takers": Dannett, *She Rode with the Generals*, 239.

416 War Department still officially denied: Blanton and Cook, *They Fought Like Demons*, 190.

416 "In Company F": Robertson, *Michigan in the War*, 205.

417 "Can you by chance": This scene between Emma and Damon Stewart is from Dannett, *She Rode with the Generals*, 246–47.

418 "She may have been the means": Gansler, *The Mysterious Private Thompson*, 214.

419 "most remarkable character": Ibid., 219.

420 "has faced shot and shell": *New Orleans Times-Picayune*, August 12, 1866.

420 "utterly worthless": *Boston Traveler*, April 28, 1866.

420 anxious to be rid of: *Memphis Daily Avalanche*, January 24, 1868.

420 Arthur Davis Lee Jackson: Letter from Belle Boyd to Jefferson Davis, May 10, 1882, Eleanor S. Brockenbrough Archives, American Civil War Museum.

420 her mind "gave way": *New York World*, February 11, 1889.

420 "Her feet and brain": *Jackson City* (MO) *Patriot*, November 18, 1869.

421 "imposing herself on the fraternity": Curtis Carroll Davis, introduction, *Belle Boyd in Camp and Prison*, 45.

421 confronted an Atlanta newspaper editor: *Atlanta Constitution*, November 10, 1874.

421 "private sorrows": Curtis Carroll Davis, introduction, *Belle Boyd in Camp and Prison*, 25.

421 "dead to the family": *New York World*, February 11, 1889.

421 accused Belle of having her own affair: Undated clipping, Laura Virginia Hale Archives.

421 "boxed Mamma's ears": *Dallas Morning News,* January 1, 1964.

422 elocution lessons: *New York Evangelist,* June 4, 1885.

422 unsuccessful $50,000 lawsuit: *Canton Ohio Repository,* June 12, 1886.

422 she led thousands of rebels: *Philadelphia Inquirer,* August 13, 1886.

422 "She stood there looking": *Washington Daily News,* September 19, 1941.

423 In a late interview: Abbott, "Belle Boyd, Teenage Spy for the Confederacy."

423 "I have lied, sworn, killed": *Atlanta Constitution,* September 2, 1900.

423 now headquarters of Union general Godfrey Weitzel: *Alexandria Gazette,* April 7, 1865.

423 "a most barbarous, unprovoked": Elizabeth Van Lew Papers, College of William & Mary.

423 including ninety-one boxes of archives: *Salem Register,* May 15, 1865.

423 "valuable information": Benjamin Butler, letter to the War Department, April 5, 1865, Elizabeth Van Lew Papers, NYPL.

424 taking a job teaching two hundred black children: Varon, *Southern Lady, Yankee Spy,* 197–98.

424 "She had no moral right": J. Staunton Moore to William Gilmore Beymer, Box 2K394, Correspondence 1910, William Gilmore Beymer Papers.

424 "for a long, long time": Sharpe to Comstock, Janaury 1867, Elizabeth Van Lew Papers, NYPL.

424 "I earnestly entreat you will": Elizabeth Van Lew to Ulysses S. Grant, April 6, 1869, *Papers of Ulysses S. Grant,* November 1, 1876–September 30, 1878, 161.

425 "coarse taste": Elizabeth Van Lew to "Gentlemen": July 11, 1887, Elizabeth Van Lew Papers, NYPL.

425 "dried up maid": Varon, *Southern Lady, Yankee Spy,* 216.

425 "the thirst for knowledge": Ibid., 229.

425 "As a woman": Ibid., 230–31.

425 "nigger funeral": Ibid., 230.

426 "I tell you truly": Elizabeth Van Lew to Ulysses S. Grant, February 3, 1881, *Papers of Ulysses S. Grant,* October 1, 1880–December 31, 1882, 132.

426 a sixty-one-foot-tall: *Cleveland Plain Dealer,* May 39, 1890.

428 "Save her! Save her!": Varon, *Southern Lady, Yankee Spy,* 250.

428 "They say I am dead?": *Washington Post,* August 5, 1900.

429 "I done hear Miss Lizzie": Varon, *Southern Lady, Yankee Spy,* 257.

429 "We must get these flowers": Roberts, *Civil War Ghost Stories and Legends,* 129–30.

BIBLIOGRAPHY

Abbott, Karen. "Belle Boyd, Teenage Spy for the Confederacy." *New York Times,* May 23, 2012.

Addey, Markinfield. *Stonewall Jackson: The Life and Military Career of Thomas Jonathan Jackson.* New York: C. T. Evans, 1863.

Albree, John. "A Woman Who Would Not Tell." Elizabeth Van Lew Papers. Special Collections Research Center, Swem Library, College of William & Mary.

Allen, Felicity. *Jefferson Davis, Unconquerable Heart.* Columbia: University of Missouri Press, 1999.

Andrews, J. Cutler. *The North Reports the Civil War.* Pittsburgh: University of Pittsburgh Press, 1955.

——. *The South Reports the Civil War.* Princeton, NJ: Princeton University Press, 1970.

Arnold, James R., and Roberta Weiner. *American Civil War: The Essential Reference Guide.* Santa Barbara, CA: ABC-CLIO, 2011.

Avary, Myrta Lockett. *A Virginia Girl in the Civil War.* New York: D. Appleton, 1903.

Badeau, Adam. *Military History of Ulysses S. Grant: From April 1861 to April 1865.* New York: Houghton, 1887.

Baguley, David. *Napoleon III and His Regime: An Extravaganza.* Baton Rouge: Louisiana State University Press, 2000.

Bahnson, Henry T., Papers. Southern Historical Collection, University of North Carolina at Chapel Hill.

Bakeless, John. *Spies of the Confederacy.* Philadelphia: Lippincott, 1970.

Barbee, David Rankin, Papers. Series 6. Georgetown University.

Bates, Samuel P. *History of Pennsylvania Volunteers, 1861–1865.* Harrisburg, PA: B. Singerly, 1869.

Beach, Wooster. *The American Practice of Medicine.* New York: Charles Scribner, 1852.

Berkeley County Historical Society. "Martinsburg, West Virginia, During the Civil War." *Berkeley Journal* 27 (2001): 1–51.

Beymer, William Gilmore. *On Hazardous Service: Scouts and Spies of the North and South.* New York: Harper, 1912.

————. Papers. Briscoe Center for American History, University of Texas at Austin.

Billings, John Davis. *Hardtack and Coffee.* 1887. Reprint, Chicago: Lakeside Press, 1960.

Blackburn, George M. *French Newspaper Opinion on the American Civil War.* Westport, CT: Greenwood Press, 1997.

Blackman, Ann. *Wild Rose: Rose O'Neale Greenhow, Civil War Spy.* New York: Random House, 2005.

Blakey, Arch Fredric. *General John H. Winder, C.S.A.* Gainesville: University of Florida Press, 1990.

Blanton, DeAnne, and Lauren M. Cook. *They Fought Like Demons: Women Soldiers in the American Civil War.* Baton Rouge: Louisiana State University Press, 2002.

Bonekemper, Edward H. *McClellan and Failure.* Jefferson, NC: McFarland, 2007.

Boston, William. *Diary.* William L. Clements Library, University of Michigan.

Boyd, Belle. *In Camp and Prison.* New York: Bleluck, 1865.

————. Papers. Berkeley County Historical Society, Martinsburg, WV.

Bremer, Fredrika. *The Homes of the New World: Impressions of America.* New York: Harper, 1853.

Brock Putnam, Sallie. *Richmond During the War: Four Years of Personal Observation.* New York: G. W. Carleton, 1867.

Brown, Jules, and Jeff Dickey. *The Rough Guide to Washington, DC.* London: Rough Guides, 2011.

Brown, Thomas J. *Dorothea Dix: New England Reformer.* Cambridge, MA: Harvard University Press, 1998.

Buck, Lucy Rebecca. *Sad Earth, Sweet Heaven: The Diary of Lucy Rebecca Buck During the War Between the States.* Birmingham, AL: Cornerstone, 1973.

Buhk, Tobin T. *True Crime in the Civil War.* Mechanicsburg, PA: Stackpole, 2012.

Bulla, David W., and Gregory A. Borchard. *Journalism in the Civil War Era.* New York: Peter Lang, 2010.

Bumstead, Josiah Freeman. *The Pathology and Treatment of Venereal Diseases.* Philadelphia: H. C. Lea, 1870.

Burger, Nash K. *Confederate Spy: Rose O'Neale Greenhow.* New York: Franklin Watts, 1967.

Burlingame, Michael. *The Inner World of Abraham Lincoln.* Urbana: University of Illinois Press, 1994.

Bush, Bryan S. *Louisville and the Civil War.* Charleston, SC: History Press, 2008.

Butler, Benjamin F. *Private and Official Correspondence of Benjamin F. Butler,* Vols. 1–5. Norwood, MA: Plimpton Press, 1917.

Cable, G. W., ed. *Famous Adventures and Prison Escapes of the Civil War.* New York: Century, 1893.

Cadman, George Hovey, Papers. Southern Historical Collection, University of North Carolina at Chapel Hill.

Calhoun, John C. *The Papers of John C. Calhoun.* Vol. 26, *1848–1849.* Edited by Clyde N. Wilson and Shirley Bright Cook. Columbia: University of South Carolina Press, 2001.

————. *Speeches of John C. Calhoun.* New York: Harper & Brothers, 1843.

Cashin, Joan E. *First Lady of the Confederacy.* Cambridge, MA: Harvard University Press, 2006.

Casstevens, Frances Harding. *George W. Alexander and Castle Thunder.* Jefferson, NC: McFarland, 2004.

Catton, Bruce. *This Hallowed Ground: The Story of the Union Side of the Civil War.* Garden City, NY: Doubleday, 1956.

Chambers, Lenoir. *Stonewall Jackson.* New York: William Morrow, 1959.

Chambers McKean, Brenda. *Blood and War at My Doorstep.* Bloomington, IN: Xlibris, 2001.

Chase, Julia, and Laura Lee. *Winchester Divided: The Civil War Diaries of Julia Chase and Laura Lee.* Edited by Michael G. Mahon. Mechanicsburg, PA: Stackpole, 2002.

Chase, William C. *Story of Stonewall Jackson.* Atlanta: D. E. Luther, 1901.

Chesnut, Mary Boykin. *A Diary from Dixie.* New York: Peter Smith, 1929.

Church, Beverly Reese. *Weddings Southern Style.* New York: Abbeville, 1993.

Clinton, Catherine. *Public Women and the Confederacy.* Milwaukee: Marquette University Press, 1999.

Coffin, Charles Carleton. *Stories of Our Soldiers: War Reminiscences.* Boston: Journal Newspaper, 1893.

Commire, Anne. *Historic World Leaders.* Detroit: Gale Research, 1994.

Cooling, Benjamin Franklin III, and Walton H. Owen. *Mr. Lincoln's Forts: A Guide to the Civil War Defenses of Washington.* Lanham, MD: Scarecrow, 2010.

Cooper, William J. *Jefferson Davis, American.* New York: Alfred A. Knopf, 2000.

Corsan, W. C. *Two Months in the Confederate States: An Englishman's Travels Through the South.* Baton Rouge: Louisiana State University Press, 1996.

Coulter, Merton E. *The Confederate States of America, 1861–1865.* Baton Rouge: Louisiana State University Press, 1950.

Cullen, Joseph P. *The Peninsula Campaign, 1862.* Harrisburg, PA: Stackpole, 1973.

Dabney, Virginius. *Richmond: The Story of a City.* Garden City, NY: Doubleday, 1976.

Dahlgren, John A. *Memoir of Ulric Dahlgren.* Edited by Mary V. Dahlgren. Philadelphia: J. B. Lippincott, 1872.

Dammann, Gordon. *Images of Civil War Medicine: A Photographic History.* New York: Demons, 2008.

Dannett, Sylvia. *She Rode with the Generals.* New York: T. Nelson, 1960.

Davis, Burke. *The Civil War: Strange and Fascinating Facts.* New York: Fairfax Press, 1960.

Davis, Curtis Carroll. Introduction to *Belle Boyd in Camp and Prison.* South Brunswick, NJ: T. Yoseloff, 1968.

————. "The 'Old Capitol' and Its Keeper: How William P. Wood Ran a Civil War Prison." *Records of the Columbia Historical Society of Washington, D.C.* 52 (1989): 206–34.

————. Papers. Martinsburg Public Library, Martinsburg, WV.

Davis, Jefferson. *The Papers of Jefferson Davis.* Vol. 8, *1862.* Edited by Lynda Lasswell Crist, Mary S. Dix, and Kenneth H. Williams. Baton Rouge: Louisiana State University Press, 1995.

———. *The Papers of Jefferson Davis*. Vol. 10, *October 1863–August 1864*. Edited by Lynda Lasswell Crist, Mary S. Dix, and Kenneth H. Williams. Baton Rouge: Louisiana State University Press, 1999.

Davis, Varina. *Jefferson Davis: Ex-President of the Confederate States of America*. New York: Belford, 1890.

Davis, William C. *Battle at Bull Run: A History of the First Major Campaign of the Civil War*. Garden City, NY: Doubleday, 1977.

———. *Jefferson Davis: The Man and His Hour*. New York: HarperCollins, 1991.

———. *Virginia at War, 1862*. Lexington: University Press of Kentucky, 2007.

Davis, William C., and Russ A. Pritchard. *The Battlefields of the Civil War*. New York: Smithmark, 1991.

———. *The Fighting Men of the Civil War*. New York: Gallery Books, 1989.

Dempsey, Jack. *Michigan and the Civil War: A Great and Bloody Sacrifice*. Charleston, SC: History Press, 2011.

Detzer, David. *Donnybrook: The Battle of Bull Run, 1861*. Orlando, FL: Harcourt, 2004.

de Vooght, Daniëlle. *Royal Taste: Food, Power, and Status at the European Courts After 1789*. Burlington, VT: Ashgate, 2011.

Dew, J. Harvie. *The Yankee and Rebel Yells*. New York: Century, 1892.

Dix, John Adams. *Memoirs of John Adams Dix*. Edited by Morgan Dix. New York: Harper & Brothers, 1883.

Doster, William Emile. *Lincoln and Episodes of the Civil War*. New York: G.P. Putnam's Sons, 1915.

Douglas, Henry Kyd. *I Rode with Stonewall*. Chapel Hill: University of North Carolina Press, 1940.

Du Bois, W. E. B. *Black Reconstruction in America*. New York: Atheneum, 1962.

Duffee, Dr. Washington. Dr. Washington Duffee to Edwin Stanton, July 30, 1862, Turner-Baker Papers (microfilm), Record Group 94, Adjutant General's Office, file 3751, National Archives and Records Administration.

Duncan, Richard R. *Beleaguered Winchester*. Baton Rouge: Louisiana State University Press, 2007.

Dunkelman, Mark H. *War's Relentless Hand*. Baton Rouge: Louisiana State University Press, 2006.

Duvall, Alice Cary Ball, *Recollections of the War by Grandmama*. Eleanor S. Brockenbrough Library, American Civil War Museum, Richmond, VA.

Edge, Frederick Milnes. *Major McClellan and the Campaign on the Yorktown Peninsula*. New York: Loyal Publication Society, 1865.

Edmonds, Emma. *Nurse and Spy in the Union Army*. Hartford, CT: W. S. Williams, 1865.

———. Pension File. National Archives, Washington, DC.

Eggleston, Larry G. *Women in the Civil War*. Jefferson, NC: McFarland, 2003.

Eicher, David J. *The Longest Night: A Military History of the Civil War*. New York: Simon & Schuster, 2002.

Ely, Alfred. *Journal of Alfred Ely: A Prisoner of War in Richmond.* New York: Appleton, 1862.

Evans, Eli N. *Judah P. Benjamin, The Jewish Confederate.* New York: Free Press, 1988.

Farwell, Byron. *Stonewall: A Biography of General Thomas J. Jackson.* New York: W.W. Norton, 1992.

Faust, Drew Gilpin. *This Republic of Suffering: Death and the American Civil War.* New York: Alfred A. Knopf, 2008.

Fay, William, Papers. Library of Virginia, Richmond.

Feis, William B. *Grant's Secret Service.* Lincoln: University of Nebraska Press, 2002.

Fishel, Edwin C. *The Secret War for the Union.* Boston: Houghton Mifflin, 1996.

Fisher, John C., and Carol Fisher. *Food in the American Military: A History.* Jefferson, NC: McFarland, 2011.

Fitzmaurice, Edmond. *The Life of Granville George Leveson Gower.* London: Longmans, Green, 1905.

Flannery, Michael A. *Civil War Pharmacy.* New York: Pharmaceutical Products Press, 2004.

Fonvielle, Chris E. *The Wilmington Campaign.* Campbell, CA: Savas, 1997.

Foote, Shelby. *The Civil War: A Narrative.* Vol. 1, *Fort Sumter to Perryville.* New York: Random House, 1958.

———. *The Civil War: A Narrative.* Vol. 2, *Fredericksburg to Meridian.* New York: Random House, 1963.

———. *The Civil War: A Narrative.* Vol. 3, *Red River to Appomattox.* New York: Random House, 1974.

Foreman, Amanda. *A World on Fire: Britain's Crucial Role in the American Civil War.* New York: Random House, 2011.

Fowler, John D., and David B. Parker. *Breaking the Heartland: The Civil War in Georgia.* Macon, GA: Mercer University Press, 2011.

Fowler, Orson Squire. *Fowler's Practical Phrenology.* New York: O. S. and L. N. Fowler, 1844.

Frémont, Jessie Benton. *The Letters of Jessie Benton Freemont.* Edited by Pamela Lee Herr and Mary Lee Spence. Champaign: University of Illinois Press, 1992.

Froude, James Anthony. *Thomas Carlyle: A History of His Life in London, 1834–1881.* New York: Charles Scribner's Sons, 1895.

Furgurson, Ernest B. *Ashes of Glory: Richmond at War.* New York: Alfred A. Knopf, 1996.

———. *Freedom Rising: Washington in the Civil War.* New York: Alfred A. Knopf, 2004.

Gaddy, David W. "Lee's Use of Intelligence." Lecture presented at the Museum of Natural History on November 21, 1994. Eleanor S. Brockenbrough Library, Museum of the Confederacy, Richmond, VA.

Gansler, Laura Leedy. *The Mysterious Private Thompson.* New York: Free Press, 2005.

Garrison, Webb B. *Civil War Curiosities.* Nashville: Rutledge Hill, 1994.

Gaskell, George Arthur. *Gaskell's Compendium of Forms.* St. Louis: R.S. Peale & Co., 1882.

Gay, Sydney Howard, Papers. Special Collections. Columbia University, New York.

Geary, John White. *A Politician Goes to War: The Civil War Letters of John White Geary*. Edited by William Alan Blair and Bell Irvin Wiley. University Park: Pennsylvania State University Press, 1995.

Gill, Miranda. *Eccentricity and the Cultural Imagination in Nineteenth Century Paris*. New York: Oxford University Press, 2009.

Gilmor, Harry. *Four Years in the Saddle*. New York: Harper, 1866.

Glazier, Willard W. *The Capture, the Prison Pen, and the Escape*. Hartford, CT: H. E. Goodwin, 1868.

Gold, David L. *Studies in Etymology and Etiology*. Alicante, Spain: Universidad de Alicante, 2009.

Goodheart, Adam. *1861: The Civil War Awakening*. New York: Alfred A. Knopf, 2011.

Gragg, Rod. *Confederate Goliath: The Battle of Fort Fisher*. New York: HarperCollins, 1991.

Graham, Eric J., and Richard D. Torrance. "Running the Blockade During the American Civil War: The Reminiscences of William Allan of Dundee." *History Scotland Magazine* 9, no. 6 (2009): 38.

Grant, Dorothy Lewis. "Lady of Refinement." *Torch*, Spring 1997.

Grant, Ulysses S. *The Papers of Ulysses S. Grant*. Edited by John Y. Simon. Carbondale, IL: Southern Illinois University Press, 1967.

Greene, A. Wilson. *Whatever You Resolve to Be: Essays on Stonewall Jackson*. Knoxville: University of Tennessee Press, 2005.

Greenhow, Rose O'Neal. "European Diary." North Carolina State Archives, Raleigh, NC. Her descendants also present the diary in full online: http://www.onealwebsite.com/RebelRose/.

———. *My Imprisonment and the First Year of Abolition Rule*. London: R. Bentley, 1863.

———. Papers. North Carolina State Archives, Raleigh, NC.

———. Papers. Special Collections Library, Duke University, Durham, NC.

———. Seized Correspondence, RG 59. National Archives, College Park, Maryland.

Guernsey, Alfred H., and Henry Mills Aiden. *Harper's Pictorial History of the Civil War*. 1866. Reprint, New York: Fairfax, 1977.

Hager, Thomas. *The Demon Under the Microscope*. New York: Harmony, 2006.

Hale, Laura. Virginia Archives. Warren Heritage Society, Front Royal, VA.

Hall, Richard. *Patriots in Disguise: Women Warriors of the Civil War*. New York: Paragon House, 1993.

Hambucken, Denis, and Matthew Payson. *Confederate Soldier of the American Civil War*. Woodstock, VT: Countryman, 2012.

Hansen, Harry. *The Civil War: A History*. New York: Penguin, 1991.

Harper, Judith E. *Women During the Civil War: An Encyclopedia*. New York: Routledge, 2004.

Hartje, Robert George. *Van Dorn: The Life and Times of a Confederate General*. Nashville: Vanderbilt University Press, 1994.

Hartzler, Dabiel D. *Marylanders in the Confederacy*. Silver Spring, MD: Family Line, 1986.

Harvey Baker, Jean. *Mary Todd Lincoln: A Biography*. New York: W. W. Norton, 2008.

Hassler, William Woods. *Colonel John Pelham: Lee's Boy Artillerist*. Chapel Hill: University of North Carolina Press, 1995.

Hastedt, Glenn P. *Spies, Wiretaps, and Secret Operations: An Encyclopedia of American Espionage*. Santa Barbara, CA: ABC-CLIO, 2011.

Hawkins, Rush C. "Early Coast Operations in North Carolina." In *Battles and Leaders in the Civil War*, edited by R. U. Johnson and C. C. Buel, 1:652–54. New York: Century, 1887.

Hawthone, Nathaniel. *The Complete Works of Nathaniel Hawthorne*. Boston: Houghton Mifflin, 1882–1899.

Haydon, Charles B. *For Country, Cause and Leader: The Civil War Journal of Charles B. Haydon*. New York: Ticknor & Fields, 1993.

Heidler, David Stephen, Jeanne T. Heidler, and David J. Coles. *Encyclopedia of the American Civil War*. Santa Barbara, CA: ABC-CLIO, 2000.

Hill, Roland. *Lord Acton*. New Haven, CT: Yale University Press, 2000.

Holsworth, Jerry W. *Civil War Winchester*. Charleston, SC: History Press, 2011.

Holmes, Emma. *The Diary of Miss Emma Holmes, 1861–1866*. Edited by John F. Marszalek. Baton Rouge: Louisiana State University Press, 1979.

Holzer, Harold. *Lincoln President-Elect: Abraham Lincoln and the Great Secession Winter 1860–1861*. New York: Simon & Schuster, 2007.

Horan, James David. *Desperate Women*. New York: G. P. Putnam's Sons, 1952.

Horner, Dave. *The Blockade-Runners*. New York: Dodd, Mead, 1968.

Horwitz, Tony. *Midnight Rising: John Brown and the Raid That Sparked the Civil War*. New York: Henry Holt, 2011.

Howard, H. S. Deposition. Secret Service Accounts, RG 110. National Archives, Washington, DC.

Jones, John B. *A Rebel War Clerk's Diary*. 1866. Reprint, New York: Sagamore, 1958.

Kaemmerlen, Cathy. *General Sherman and the Georgia Belles*. Charleston, SC: History Press, 2006.

Kagan, Neil, and Stephen G. Hyslop. *Eyewitness to the Civil War: The Complete History from Secession to Reconstruction*. Washington, DC: National Geographic, 2006.

Kennedy-Nolle, Sharon. Introduction to *Belle Boyd in Camp and Prison*. Baton Rouge: Louisiana State University Press, 1998.

Keyes, Erasmus D. *Fifty Years' Observation of Men and Events, Civil and Military*. New York: C. Scribner's Sons, 1884.

Keyes, John S. Unpublished autobiography. Curtis Carroll Davis Papers, Martinsburg Public Library, WV.

Kinchen, Oscar Arvle. *Women Who Spied for the Blue and the Gray*. Philadelphia: Dorrance, 1972.

Klein, Frederic S. "The Civil War Was a Pitchman's Paradise." *Civil War Times Illustrated*, January 1963.

Klingaman, William K. *Abraham Lincoln and the Road to Emancipation, 1861–1865*. New York: Viking, 2001.

Lamb, William, Papers. Special Collections Research Center. Swem Library, College of William & Mary.

Lebergott, Stanley. "Wage Trends, 1800–1900." In *Trends in the American Economy in the 19th Century*. Princeton, NJ: Princeton University Press, 1960.

Leech, Margaret. *Reveille in Washington, 1860–1865*. New York: Harper & Brothers, 1941.

Leonard, Elizabeth D. *All the Daring of the Soldier: Women of the Civil War Armies*. New York: W. W. Norton, 1999.

Lesser, W. Hunter. *Rebels at the Gate*. Naperville, IL: Sourcebooks, 2004.

Linden, Glenn M. *Voices from the Gathering Storm: The Coming of the American Civil War*. Wilmington, DE: Scholarly Resources, 2001.

Long, E. B., and Barbara Long. *Civil War Day by Day: An Almanac*. Garden City, NY: Doubleday, 1971.

Lossing, Benson John. *A History of the Civil War, 1861–1865*. New York: War Memorial Association, 1912.

Loughborough, Mary Ann Webster. *My Cave Life in Vicksburg: With Letters of Trial and Travel*. New York: Appleton and Company, 1864.

Lowry, Thomas P. *The Story the Soldiers Wouldn't Tell*. Mechanicsburg, PA: Stackpole, 1994.

Ludy, Robert B. *Historic Hotels of the World*. Philadelphia: David McKay, 1927.

Mahony, D. A. *The Prisoner of State*. New York: Carleton, 1863.

Markle, Donald E. *Spies and Spymasters of the Civil War*. New York: Hippocrene, 1994.

Marten, James Alan. *Civil War America: Voices from the Home Front*. Santa Barbara, CA: ABC-CLIO, 2003.

Marvin, Edwin E. *The Fifth Regiment, Connecticut Volunteers*. Hartford, CT: Wiley, Waterman & Eaton, 1889.

McArthur, Judith N., Orville Vernon Burton, and James B. Griffin. *A Gentleman and an Officer: A Military and Social History of James B. Griffin's Civil War*. New York: Oxford University Press, 1996.

McClellan, George Brinton. *The Civil War Papers of George B. McClellan: Selected Correspondence, 1860–1865*. Edited by Stephen W. Sears. New York: Ticknor & Fields, 1989.

———. *McClellan's Own Story*. New York: C. L. Webster, 1887.

———. *Report on the Organization and Campaigns of the Army of the Potomac*. New York: Sheldon, 1864.

McCreery, William B. *My Experience as a Prisoner of War, and Escape from Libby Prison*. 1893. Reprint, Bethesda, MD: University Publications of America, 1993.

McCurry, Stephanie. *Confederate Reckoning: Power and Politics in the Civil War South*. Cambridge, MA: Harvard University Press, 2010.

McGilchrist, John. *Lord Palmerston: A Biography*. London: George Routledge, 1865.

McKay, Ernest A. *Henry Wilson: Practical Radical*. Port Washington, NY: Kennikat, 1971.

McNeil, Jim. *Masters of the Shoals*. Cambridge, MA: Da Capo, 2003.

McNiven, Thomas. "Recollections." Library of Virginia, Richmond.

McPherson, James. *Battle Cry of Freedom: The Civil War Era*. New York: Oxford University Press, 1988.

———. *Crossroads of Freedom: Antietam*. New York: Oxford University Press, 2002.

———. *The Illustrated Battle Cry of Freedom: the Civil War Era*. New York: Oxford University Press, 2003.

———. *Tried by War: Abraham Lincoln as Commander in Chief*. New York: Penguin, 2008.

Miller, Neil. *Kartcher Caverns*. Tucson: University of Arizona Press, 2008.

Mitchell, Charles W., ed. *Maryland Voices of the Civil War*. Baltimore: Johns Hopkins University Press, 2007.

Moran, Frank E. "Colonel Rose's Tunnel at Libby Prison," *Century Magazine* 13 (March 1888): 770.

Morrow, Honoré. *Mary Todd Lincoln: An Appreciation of the Wife of Abraham Lincoln*. New York: William Morrow, 1928.

Mortimer, Gavin. *Double Death: The True Story of Pryce Lewis, the Civil War's Most Daring Spy*. New York: Walker, 2010.

Nevins, Allan. *Hamilton Fish: The Inner History of the Grant Administration*. New York: F. Ungar, 1957.

Northrop, John Worrell. *Chronicles from the Diary of a War Prisoner in Andersonville*. 1904. Bethesda, MD: University Publications of America, 1992.

O'Reilly, Francis Augustin. *The Fredericksburg Campaign*. Baton Rouge: Louisiana State University Press, 2003.

Parker, David B. *A Chatauqua Boy in '61 and Afterward*. Boston: Small, Maynard, 1912.

Pavlovsky, Arnold. "Riding in Circles: J.E.B. Stuart and the Confederate Cavalry, 1861–1862." Seattle: Amazon Digital Services, 2010.

Phillips, Edward Hamilton. "The Transfer of Jefferson and Berkeley Counties from Virginia to West Virginia." PhD diss., University of North Carolina at Chapel Hill, 1949.

Pickett, La Salle Corbell. *Pickett and His Men*. Atlanta: Foote & Davies, 1899.

Pinkerton, Allan. *The Spy of the Rebellion*. Chicago: A. G. Nettleton, 1884.

Pollard, Edward Alfred. *Observations in the North: Eight Months in Prison and On Parole*. Richmond: E. W. Ayres, 1865.

Proceedings of the Commission Relating to State Prisoners, Records of the Department of State, RG 59, E962, National Archives.

Pryor, Sara Agnes Rice. *Reminiscences of Peace and War*. New York: Macmillan, 1904.

Rable, George C. *God's Almost Chosen Peoples: A Religious History of the American Civil War*. Chapel Hill: University of North Carolina Press, 2010.

Ransom, John L. *Andersonville Diary*. Great Neck, NY: Great Neck Pub., 1883.

Reader, Francis Smith. *History of the 5th West Virginia Cavalry*. 1892. Reprint, Bethesda, MD: University Publications of America, 1992.

Reid Hanger, Michael. *Diary, 1861*. Special Collections. Washington and Lee University, Lexington, Virginia.

Reynolds, Davis S. *John Brown, Abolitionist*. New York: Alfred A. Knopf, 2005.

Reynolds, John P., Jr. "Biographical Sketch of Elizabeth Van Lew." Elizabeth Van Lew Papers, New York Public Library.

Robbins, Jerome J., Papers. Bentley Historical Library, University of Michigan, Ann Arbor. (Includes journal.)

Roberts, Nancy. *Civil War Ghost Stories and Legends*. Columbia: University of South Carolina Press, 1992.

Robertson, James I. *Civil War Virginia: Battleground for a Nation*. Charlottesville: University of Virginia Press, 1991.

———. *Stonewall Jackson: The Man, the Soldier, the Legend*. New York: Macmillan, 1997.

Robertson, Jonathan. *Michigan in the War*. Lansing: W. S. George Co. State Printers, 1882.

Rosen, Robert N. *The Jewish Confederates*. Columbia: University of South Carolina Press, 2000.

Ross, Ishbel. *First Lady of the South: The Life of Mrs. Jefferson Davis*. New York: Harper, 1958.

———. *Proud Kate: Portrait of an Ambitious Woman*. New York: Harper, 1953.

———. *Rebel Rose: Life of Rose O'Neal Greenhow*. New York: Harper, 1954.

Royce, Betsey. *A Genteel Spy*. Rockvale, TN: Two Peas, 2010.

Samuels, Shirley. *Facing America: Iconography and the Civil War*. New York: Oxford University Press, 2004.

Sandburg, Carl. *Abraham Lincoln: The Prairie Years and the War Years*. New York: Harcourt, Brace, 1954.

———. *Abraham Lincoln: The War Years*. 4 vols. New York: Harcourt, Brace, 1939.

Scarborough, Ruth. *Siren of the South*. Macon, GA: Mercer University Press, 1983.

Schroeder-Lein, Glenna R. *The Encyclopedia of Civil War Medicine*. Armonk, NY: M. E. Sharpe, 2008.

Sears, Stephen W. *To the Gates of Richmond: The Peninsula Campaign*. New York: Ticknor & Fields, 1992.

———. *The Young Napoleon*. New York: Ticknor & Fields, 1988.

Selby, John Millin. *Stonewall Jackson as Military Commander*. Princeton, NJ: Van Nostrand, 1968.

Selin, Helaine, and Pamela Kendall Stone. *Childbirth Across Cultures*. Dordrecht and New York: Springer, 2009.

Settles, Thomas Michael. *John Bankhead Magruder: A Military Reappraisal*. Baton Rouge: Louisiana State University Press, 2009.

Sigaud, Louis, *Belle Boyd, Confederate Spy*. Richmond, VA: Dietz, 1944.

———. "More About Belle Boyd." *Lincoln Herald* 64 (Winter 1962): 174–81.

———. Papers. Blennerhassett Museum of Regional History, Parkersburg, WV.

———. "When Belle Boyd Wrote Lincoln." *Lincoln Herald* 50 (February 1948): 15–22.

———. "William Boyd Compton, Belle Boyd's Cousin." *Lincoln Herald* 67 (Spring 1963): 22–23.

Silber, Nina. *Landmarks of the Civil War*. New York: Oxford University Press, 2003.

Smith, Frank. "The Polite War." *Coronet*, December 1937.

Speer, Lonnie R. *Portals to Hell: Military Prisons of the Civil War*. Mechanicsburg, PA: Stackpole, 1997.

Sperry, Kate, Papers. Stewart Bell Jr. Archives, #1578. Handley Regional Library, Winchester, VA.

Starr, Louis Morris. *Bohemian Brigade: Civil War Newsmen in Action*. New York: Knopf, 1954.

Stegman, Carolyn B., and Suzanne Nida Seibert. *Women of Achievement in Maryland History*. University Park, MD: Women of Achievement in Maryland History, 2002.

Stevenson, James H. *Boots and Saddles*. Harrisburg, PA: Patriot, 1879.

Stevenson, Thomas M. *History of the 78th Regiment Ohio Veteran Volunteer Infantry*. 1865. Reprint, Bethesda, MD: University Publications of America, 1993.

Stillwell, Leander. *The Story of a Common Soldier of Army Life in the Civil War, 1861–1865*. Alexandria, VA: Time-Life, 1983.

Strother, David Hunter. *A Virginia Yankee in the Civil War: The Diaries of David Hunter Strother*. Edited by Cecil D. Eby. Chapel Hill: University of North Carolina Press, 1998.

Stuart, Meriwether. "Colonel Ulric Dahlgren and Richmond's Union Underground." *Virginia Magazine of History and Biography* 72, no. 2 (1964): 152–204.

———. "Of Spies and Borrowed Names: The Identity of Union Operatives in Richmond Known as 'The Phillipses' Discovered." *Virginia Magazine of History and Biography* 89, no. 3 (1981): 308–27.

———. "Samuel Ruth and General R. E. Lee: Disloyalty and the Line of Supply to Fredericksburg." *Virginia Magazine of History and Biography* 71, no. 1 (1963): 35–109.

Sulick, Michael J. *Spying in America*. Washington, DC: Georgetown University Press, 2012.

Taft Bayne, Julia. *Tad Lincoln's Father*. Lincoln: University of Nebraska Press, 2001.

Tagg, Larry. *The Unpopular Mr. Lincoln*. El Dorado Hills, CA: Savas Beatie, 2009.

Tanner, Robert G. *Stonewall in the Valley*. Garden City, NY: Doubleday, 1976.

Taylor, Charles E. *The Signal and Secret Service of the Confederate States*. Harmans, MD: Toomey, 1986.

Taylor, Richard. *Destruction and Reconstruction*. New York: Bantam, 1992 (1879).

Taylor, Thomas. *Running the Blockade*. Freeport, NY: Books for Libraries, 1971.

Tendrich Frank, Lisa, ed. *An Encyclopedia of Women at War: From the Home Front to Battlefields*. Santa Barbara, CA: ABC-CLIO, 2013.

Thomas, Emory T. *The Confederate Nation, 1861–1865*. New York: Harper & Row, 1979.

———. *Robert E. Lee: A Biography*. New York: W. W. Norton, 1995.

Tidwell, William A. *April '65: Confederate Covert Action in the American Civil War*. Kent, OH: Kent State University Press, 1995.

Tidwell, William A., James O. Hall, and David W. Gaddy. *Come Retribution: The Confederate Secret Service and the Assassination of Lincoln*. Jackson: University Press of Mississippi, 1988.

Troiani, Don. *Don Troiani's Regiments and Uniforms of the Civil War*. Mechanicsburg, PA: Stackpole, 2002.

Turner-Baker Papers, RG 94, National Archives and Record Administration.

US Congress. *Report of the Joint Committee on the Conduct of the War*. Washington, DC: Government Printing Office, 1863.

US War Department. *The War of the Rebellion: A Compilation of the Official Records of the Union and Confederate Armies*. Washington, DC: Government Printing Office, 1880–1901.

Van Doren Stern, Philip. *Secret Missions of the Civil War*. Chicago: Rand McNally, 1959.

Van Lew, Elizabeth. Album. Virginia Historical Society, Richmond.

———. Papers. Manuscripts and Archives Division, New York Public Library. (Includes "Occasional Journal.")

———. Papers. Special Collections Research Center, Swem Library, College of William & Mary.

Varhola, Michael. *Life in Civil War America*. Cincinnati, OH: Family Tree, 2011.

Varon, Elizabeth R. *Southern Lady, Yankee Spy*. New York: Oxford University Press, 2003.

Wakeman, Sarah Rosetta. *An Uncommon Soldier: The Civil War Letters of Sarah Rosetta Wakeman*. Edited by Lauren Cook Burgess. New York: Oxford University Press, 1995.

Wallenstein, Peter, and Bertram Wyatt-Brown. *Virginia's Civil War*. Charlottesville: University of Virginia Press, 2005.

Ward, Andrew. *The Slaves' War: The Civil War in the Words of Former Slaves*. Boston: Houghton Mifflin, 2008.

Ward, Geoffrey C., Ken Burns, and Ric Burns. *The Civil War: An Illustrated History*. New York: Knopf, 1990.

Waugh, Charles G., and Martin Harry Greenberg. *The Women's War in the South*. Nashville: Cumberland House, 1999.

Waugh, John C. *Lincoln and McClellan*. New York: Palgrave Macmillan, 2010.

Wayne, Tiffany K. *Women's Roles in Nineteenth Century America*. Westport, CT: Greenwood, 2007.

Weitz, Mark A. *A Higher Duty: Desertion Among Georgia Troops During the Civil War*. Lincoln: University of Nebraska Press, 2000.

Welsh, Jack D. *Medical Histories of Confederate Generals*. Kent, OH: Kent State University Press, 1999.

Wheelan, Joseph. *Libby Prison Breakout*. New York: Public Affairs, 2010.

Wheeler, Richard. *Voices of the Civil War*. New York: Crowell, 1976.

White, Ronald C., Jr. *A. Lincoln: A Biography*. New York: Random House, 2009.

Whitman, Walt. *Walt Whitman's Civil War*. Edited by Walter Lowenfels. New York: Knopf, 1960.

Wiley, Bell Irvin. *The Life of Billy Yank: The Common Soldier of the Union*. Indianapolis: Bobbs-Merrill, 1952.

———. *The Life of Johnny Reb: The Common Soldier of the Confederacy*. Indianapolis: Bobbs-Merrill, 1943.

Wilkie, Curtis. *Dixie: A Personal Odyssey Through Events That Shaped the Modern South.* New York: Scribner, 2001.

Williams, Thomas Harry. *Lincoln and His Generals.* New York: Knopf, 1952.

———. *P. G. T. Beauregard: Napoleon in Gray.* Baton Rouge: Louisiana State University Press, 1954.

Williamson, James Joseph. *Prison Life in the Old Capitol and Reminiscences of the Civil War.* West Orange, NJ: [n.p.], 1911.

Wilson, Thomas K. *Sufferings Endured for a Free Government; or, A History of the Cruelties and Atrocities of the Rebellion.* Philadelphia: Smith & Peters, 1864.

Wineapple, Brenda. *Ecstatic Nation: Confidence, Crisis, and Compromise, 1848–1877.* New York: HarperCollins, 2013.

Wise, Stephen R. *Lifeline of the Confederacy.* Columbia: University of South Carolina Press, 1988.

Wixson, Neal E. *From Civility to Survival: Richmond Ladies During the Civil War.* Bloomington, IN: iUniverse, 2012.

Wood, Don C. *Documented History of Martinsburg and Berkeley County.* Berkeley County, WV: Berkeley County Historical Society, 2004.

Woodworth, Steven E. *Davis and Lee at War.* Lawrence: University Press of Kansas, 1995.

Wright, John. *The Language of the Civil War.* Westport, CT: Oryx, 2001.

Wright, Mike. *City Under Siege: Richmond in the Civil War.* Lanham, MD: Cooper Square, 2002.

Wright Morris, Martha Elizabeth, Papers. Manuscripts Division. Library of Congress, Washington, DC.

Wuttge, Frank, Jr. Research Files. Schomburg Center for Research in Black Culture, New York Public Library.

INDEX

About the Author

___

Karen Abbott is the *New York Times* bestselling author of *Sin in the Second City* and *American Rose*. She is a regular contributor to Smithsonian.com, and also writes for *Disunion*, the *New York Times* series about the Civil War. A native of Philadelphia, where she worked as a journalist, she now lives with her husband and two African Grey parrots in New York City.

Visit her online at karenabbott.net.